# NUCLEAR-CYTOPLASMIC
## INTERACTIONS
## IN THE CELL CYCLE

This is a volume in
CELL BIOLOGY
A series of monographs
Editors: *D. E. Buetow, I. L. Cameron, and G. M. Padilla*

A complete list of the books in this series appears at the end of the volume.

# NUCLEAR-CYTOPLASMIC INTERACTIONS IN THE CELL CYCLE

Edited by

## Gary L. Whitson

*Department of Zoology*
*The University of Tennessee*
*Knoxville, Tennessee*

1980

ACADEMIC PRESS

*A Subsidiary of Harcourt Brace Jovanovich, Publishers*

New York   London   Toronto   Sydney   San Francisco

ACADEMIC PRESS, INC.
111 Fifth Avenue, New York, New York 10003

*United Kingdom Edition published by*
ACADEMIC PRESS, INC. (LONDON) LTD.
24/28 Oval Road, London NW1 7DX

**Library of Congress Cataloging in Publication Data**
Main entry under title:

Nuclear–cytoplasmic interactions in the cell cycle.

(Cell biology)
Includes index.
1. Cell cycle. 2. Cell nuclei. 3. Cytoplasm.
I. Whitson, Gary L.
QH605.N8      574.87'62      80–763
ISBN 0–12–747750–0

PRINTED IN THE UNITED STATES OF AMERICA

80 81 82 83    9 8 7 6 5 4 3 2 1

*To my parents Robert and Margaret*
*To my wife Mary Ellen*
*To my sons Randall Scott, Steven Kent, and Derek Alan*

# Contents

12 **Calcium and Cyclic Nucleotide Interactions during the Cell Cycle**

**Paul A. Charp and Gary L. Whitson**

# List of Contributors

*Numbers in parentheses indicate the pages on which the authors' contributions begin.*

**Paula T. Beall** (223), Departments of Physiology and Pediatrics, Baylor College of Medicine, Houston, Texas 77030

**R. C. Bird** (203), Department of Zoology, University of Toronto, Toronto, Ontario M5S 1A1, Canada

**Robert C. Briggs** (181), Department of Biochemistry, Vanderbilt University School of Medicine, Nashville, Tennessee 37232

**D. E. Buetow** (9), Department of Physiology and Biophysics, University of Illinois, Urbana, Illinois 61801

**I. L. Cameron** (249), Department of Anatomy, The University of Texas Health Science Center at San Antonio, San Antonio, Texas 78284

**Paul A. Charp** (309), Department of Zoology, The University of Tennessee, Knoxville, Tennessee 37916

**T. Gallagher** (9), C. K. Byrnes Research Center, St. Laurences Hospital, Dublin 7, Ireland

**W. J. George** (293), Departments of Pharmacology and Anatomy, Tulane University School of Medicine, New Orleans, Louisiana 70112

**B. G. Grubbs** (249), Department of Anatomy, The University of Texas Health Science Center at San Antonio, San Antonio, Texas 78284

**Roger Hand** (167), McGill Cancer Center, and Department of Medicine, McGill University, Montreal, Quebec H3G 1Y6, Canada

**S. Henry** (57), Department of Cell and Molecular Biology, Medical College of Georgia, Augusta, Georgia 30912

**Lubomir S. Hnilica** (181), Department of Biochemistry, Vanderbilt University School of Medicine, Nashville, Tennessee 37232

**L. D. Hodge** (57), Department of Cell and Molecular Biology, Medical College of Georgia, Augusta, Georgia 30912

**J. R. Jeter, Jr.** (249, 293), Department of Anatomy, Tulane University Medical School, New Orleans, Louisiana 70112

**Wanda M. Krajewska** * (181), Department of Biochemistry, Vanderbilt University School of Medicine, Nashville, Tennessee 37232

**Gloria Lincoln** (181), Department of Biochemistry and Molecular Biology, University of Florida School of Medicine, Gainesville, Florida 32610

**George M. Padilla** (1), Department of Physiology, Duke University Medical Center, Durham, North Carolina 27710

**T. B. Pool** (249), Department of Anatomy, The University of Texas Health Science Center at San Antonio, San Antonio, Texas 78284

**Rose Sheinin** (105), Department of Microbiology and Parasitology, University of Toronto, Toronto, Ontario M5S 1A1, Canada

**T. Simmons** (57), Department of Life Sciences, Polytechnic Institute of New York, Brooklyn, New York 11201

**Jesse E. Sisken** (271), Department of Pharmacology, College of Medicine, University of Kentucky, Lexington, Kentucky 40536

**N. K. R. Smith** (249), Department of Anatomy, The University of Texas Health Science Center at San Antonio, San Antonio, Texas 78284

**R. L. Sparks** (249), Department of Anatomy, The University of Texas Health Science Center at San Antonio, San Antonio, Texas 78284

**Gary Stein** (181), Department of Biochemistry and Molecular Biology, University of Florida School of Medicine, Gainesville, Florida 32610

**Janet Stein** (181), Department of Immunology and Medical Microbiology, University of Florida School of Medicine, Gainesville, Florida 32610

**L. A. White** (293), Departments of Pharmacology and Anatomy, Tulane University School of Medicine, New Orleans, Louisiana 70112

**Gary L. Whitson** (1, 309), Department of Zoology, The University of Tennessee, Knoxville, Tennessee 37916

**E. A. Wurtz** † (9), Department of Physiology and Biophysics, University of Illinois, Urbana, Illinois 61801

**A. M. Zimmerman** (203), Department of Zoology, University of Toronto, Toronto, Ontario M5S 1A1, Canada

**S. Zimmerman** (203), Glendon College, Division of Natural Science, York University, Toronto, Ontario, Canada

*Present Address: Department of Biochemistry, University of Lodz, Lodz, Poland

†Present Address: Laboratory of Radiation Ecology, College of Fisheries, University of Washington, Seattle, Washington 98195

# Preface

As in previous works on this subject, this book reflects the existing diversity in cell cycle studies. The general theme of this work concerns the growth of our knowledge of the underlying macromolecular basis for the regulation of cellular activities responsible for the traverse of cells through the cell cycle. It emphasizes certain complex nucleocytoplasmic interactions which, in relation to environmental changes, are extremely important to phase-specific activities, cell cycle check points, and phase transitions in the cell cycle.

Interest in cell cycle studies is gaining momentum, so much so that in May 1980, during the Eighth International Cell Cycle Conference held at Duke University, "The International Cell Cycle Society" was formalized. It is hoped that through this organization and through the studies presented in this book research on cell cycle-related phenomena will be stimulated and that a new group of scientists will be attracted to this growing field of investigation.

This book should meet the needs of the graduate students and teacher as well as researcher. Since the introductory chapter summarizes most of the salient features of the topics covered there is no need to reiterate them.

Cell cycle studies have been personally rewarding to me in innumerable ways, for example, in forming lasting friendships with individuals who have improved the quality of my life and with whom I have shared new ideas. The death of two of these individuals is indeed regretted: Shuhei Yuyama, shortly after the cell cycle meeting in 1978, and Erik Zeuthen, early this year. The high quality of their work recorded in their many publications will continue to influence cell cycle researchers for a long time to come.

Gary L. Whitson

# Nuclear–Cytoplasmic Interactions in the Cell Cycle: An Overview

## GEORGE M. PADILLA and GARY L. WHITSON

The cell cycle is an assembly of variable events involving complex nuclear–cytoplasmic and environmental interactions. In eukaryotic cells, the cell cycle may thus be considered an ordered sequence of interrelated processes distinguishable by physiological and cytological transitions into recognizable phases of activity (e.g., DNA synthesis, mitosis, cytokinesis). In order to unravel the cause–effect relationships and assign a determinant role, if any, to any given process, two general approaches have been employed in the past: (1) the cell cycle, of cells whose transit is synchronized, is interrupted at specific points with a variety of experimental interventions and the consequences of the interruptions are examined, or (2) cells that are quiescent (noncycling) are stimulated to reenter the cell cycle and the nature of the stimulus, together with a delineation of the reentry point, are also scrutinized. Both approaches operate under the assumption that the cell cycle is, in fact, composed of discrete processes linked, not necessarily in a strict cause–effect relationship, but for all intent and purposes, in a mutually dependent sequence. Variations of this model may recognize the presence of processes which operate in a parallel fashion to the events subject to an interruptive thrust.

A third approach considers the inherent variability in the duration of individual cell cycles as an expression of underlying controlling mechanisms. As noted recently by Pardee *et al.* (1979), most of the variability in transit times is found within the $G_1$ phase, the period of the cell cycle where it is thought most growth-controlling mechanisms are operative. It is during this phase that cells are most responsive to interference

**1**

NUCLEAR–CYTOPLASMIC INTERACTIONS
IN THE CELL CYCLE

from external sources. In fact the $G_1$ phase (as well as the other phases of the cell cycle) may be thought of as an assembly of "transition" points (Pardee *et al.*, 1979); Mitchison (1971) prefers to consider the concept of cycle "markers" which are broadly defined as "discrete events, chemical structure and physiological . . . ." Also included would be the stages of mitosis as well as the transition and restriction points; the latter may represent a locus for control (Pardee, 1974). In terms of the present discussion, we wish to consider to what extent present investigations on the control of the cell cycle address themselves to a description of the events which comprise the cell cycle, and whether we can formulate a model that explains the inherent properties of cells as they pass through the various phases of the cell cycle and the variability in their times.

Pardee *et al.* (1979) have developed the thesis that a particular $G_1$ event, which is synonymous to a growth control point, may itself contribute to the variability in the $G_1$ phases of a population of cells and that establishing a model that describes how these points are crossed by individual cells one may obtain a "quantitative kinetic description" for the individual cell cycle variabilities and lay the foundation for future experimentation on the control of cell growth. These investigators describe and compare the properties of two kinetic models. The first is a "deterministic" one in which the variability between cells is attributed to the totality of many cellular properties which are different and have a cumulative effect which is expressed in terms of the cell cycle variability. That is to say, the cell cycle is really made up of a large number of individual segments or steps and that the rate at which the cells pass through each step is variable and normally distributed as a function of differences between cells. The second, a "transition probability" model proposed by Smith and Martin (1973) has as its central idea the notion that a single random process which is independent of other events is responsible for the variability with which cells pass from one phase of the cell cycle into the next. In other words, the cells entering the cell cycle are inherently identical but enter it with constant but individually distinct probabilities. Pardee *et al.* (1979) discuss and compare the properties of these two models and note that kinetic data alone are not sufficient to allow a choice between the two models. They conclude that we need to determine the properties that render cells different and to what extent these properties have a metabolic basis; hence, the need for more investigations on the cell cycle in order to comprehend the nature of cell cycle regulation.

In this volume we have assembled a series of investigations which explore some of the important nuclear–cytoplasmic interactions that take place during the cell cycle. Emphasis is placed on new approaches to cell cycle regulation, particularly the control of cell reproduction. We readily admit that a great deal of work remains to be done to characterize the molecular basis for control mechanisms which involve such things as reproduction and function of organelles during the cell cycle, but this is a beginning.

It is generally accepted that chloroplasts and mitochondria are semiautonomous cell organelles, containing DNA, RNA, and their own class of ribosomes which are separate and distinct in nature from that found in the cells in which they reside. It is apparent that the cycles of duplication of these organelles do not always coincide with the nuclear replication cycle (S phase) of the cell that houses them. Therefore, when it comes to reproduction of these organelles, there appears to be cycles within cycles. Patterns of DNA synthesis in mitochondria vary among different cell types (Prescott, 1976) and in a line of synchronized liver cells occurs late in the cell cycle (Koch and Stokstad, 1967). DNA replication in mitochondria within the same cell does not always occur in synchrony either (Hanson *et al.*, 1970). There is much to be gained concerning mitochondriobiogenesis during the cell cycle. For instance, Brunk (1979) recently found that inhibitors of mitochondrial DNA transcription inhibited rat myoblast fusion. Fusion of myoblasts does not occur in cycling cells, so there may be a definite link here between mitochondrial transcription and cell cycle regulation. This fertile area for study, however, is not covered in this volume. There is some new and exciting information on chloroplast biogenesis that is covered in this volume and is a good place to start.

Buetow *et al.* (Chapter 2) present a very thorough analysis on chloroplast biogenesis in *Euglena, Chlamydomonas,* and *Chlorella.* These photosynthetic unicellular flagellates have been the subject of many diverse investigations because of their relative ease of growth as well as their ability to be synchronized by repetitive regimens of alternating light–dark cycles. As with mitochondria, nuclear DNA replication and chloroplast DNA replication are not in synchrony with one another. For instance, in *Euglena,* DNA replication occurs during the latter half of the light cycle, whereas chloroplast DNA synthesis occurs throughout the light and dark periods (Cook, 1966). Synchronized cell division is confined to the dark period in *Euglena* and chloroplast duplication also occurs "at or near the time of cell division" (Buetow *et al.*, this volume, Chapter 2). Buetow and co-workers also discuss similar patterns that exist for DNA replication and chloroplast duplication in *Chlamydomonas.* They present a thorough review of the current literature concerning experimentation on transcription and translation of both cytoplasmic and chloroplast proteins during the cell cycle. Particular emphasis is given to those proteins and cofactors associated with the light and dark reactions of photosynthesis. In addition, they include current information concerning the biogenesis of chloroplast membranes. They conclude that very little is known yet concerning the biochemical or metabolic events which trigger chloroplast replication and further suggest that cytogenetic studies "are necessary to define more accurately a chloroplast cycle" which should also help provide the necessary information as to whether the nucleus plays any role in this event. The remainder of the book deals with nuclear–cytoplasmic interaction in animal cells.

Chapter 3 is concerned with events that occur during the transition from mitosis to $G_1$ in mammalian cells. Although much is known concerning biochemical and structural changes that occur in $G_1$, little is known about the metabolic phenomena involved in the transition to this phase. In this chapter, Simmons *et al.* deal with this subject in a thorough and systematic approach. They discuss, in particular, the implications of regulatory or inducing proteins responsible for events associated with cell division with some discussion on the possible role of microtubules and microfilaments and the interactions of calcium and cyclic nucleotides at this time. They also present an approach to the study of nuclear membrane reformation by the use of lipid precursors as probes. Moreover, they discuss marker polypeptides that survive mitosis and the possible reutilization of mRNP particles that provide the protein synthetic machinery for $G_1$ cells. Apparently some portion of premitotic mRNA survives mitosis in certain mammalian cells and is used by early $G_1$ cells in the next cycle to synthesize proteins. These investigators propose a unique approach to the isolation and study of various mRNAs in order to establish this fact and further describe future directions for this important area of research.

As stated by Prescott (1976), "the genetic analysis of the cell cycle is a relatively new area of research that is rapidly becoming a major part of the study of cell reproduction." Using a genetic approach in Chapter 4 Sheinen presents a long list of temperature-sensitive mutants of mammalian cells which provides a unique model for the analysis of cell cycle progression which she effectively argues will reveal "the nature and extent of interweaving and overlapping control mechanisms between different stages of the cell cycle." The advantage , she explains, of using temperature-sensitive mutants is that they generally carry missense mutations which are masked at a permissive temperature but expressed at a nonpermissive temperature. Growth is restricted at the nonpermissive temperature so biochemical patterns can be readily compared between cycling and noncycling cells at various checkpoints with different subcultures of the same cells. Sheinen goes on to show that temperature-sensitive mutants provide temporal maps for cell cycle progression, particularly progression through $G_1$ and S. In a very comprehensive way, she has provided us with a new basis to assess nuclear–cytoplasmic interactions and a challenge to identify biochemical gene products via this probe.

Hand, in Chapter 5, presents a model for the regulation of DNA synthesis in mammalian cells. He presents evidence for proteins and low molecular weight heat stable substances synthesized in $G_1$ that are believed to play a role in the initiation of replication. Hand suggests, "that a combination of nucleotide sequences and DNA–protein interactions serve to define an initiation point." He raises the important question of whether there is evidence for "functional subgenomic units" that regulate the patterns of DNA replication. With his model, he purports that replication factors (triphosphates and enzymes) exert positive control, whereas conformational changes after replication act as a negative control. He further contends that the use of x-irradiation and uv-irradiation would be useful probes for the analysis of this model.

In Chapter 6, Briggs *et al.* use an immunological approach in the elucidation of tissue- and species-specific nuclear antigens. They used HeLa cells synchronized by a double thymidine block in their studies. With this approach they have been able to detect and isolate specific nuclear antigens during cell cycle progression. Using immune sera, they specifically detected "an uncharacterized antigen common to human $G_0$ cells." They propose that this antigen may regulate cell cycle transition. This discovery was made while they were characterizing heterogeneous non-histone proteins. Their interest in this area of research comes from the implication that such proteins may be important regulators of gene transcription. So far they have not discovered with immune sera any qualitative changes in nuclear antigens during the cell cycle, but they admit that the discovery of the uncharacterized antigen is an indication that this probe may be useful in the assessment or detection of other "nuclear components involved in cell cycle related nuclear–cytoplasmic interactions."

Bird *et al.* (Chapter 7) report on tubulin synthesis during the cell cycle. In this chapter, they review the natural synchrony system associated with early sea urchin development, flagellar development in the polymorphic cycle of *Naegleria,* flagellar regeneration in *Chlamydomonas,* but mainly cilia regeneration in *Tetrahymena* and tubulin changes involved in the synchronized cell cycle of *Tetrahymena.* In both *Chlamydomonas* and *Naegleria* they report that induction of tubulin synthesis occurs without depletion of soluble pools, whereas in *Tetrahymena* pool depletion always preceeds the induction of tubulin. In this study, these investigators use a calcium-shock treatment to cause shedding of cilia which is followed by regeneration in normal media. They observed that both actinomycin D (an RNA synthesis inhibitor) and cycloheximide (a protein synthesis inhibitor) inhibit cilia regeneration. Analysis of ribosomal profiles by these investigators showed the active assembly of new polysomes is required for new tubulin synthesis. In normal synchronized *Tetrahymena,* they observed that the greatest amount of tubulin synthesis occurs during $G_2$, which, they explain, coincides with oral primordium development that requires the synthesis of large numbers of microtubules. It would be interesting to ascertain whether they are distinct and separate pools of tubulin, e.g., one which may be necessary for ciliary structures and one which may be necessary for cell division. These investigators propose the future use of cDNA probes to identify mRNA sequences for tubulin. Perhaps such probes would reveal subtle differences in tubulin properties.

A new and perhaps promising area for research on the cell cycle is presented in Chapter 8 by Beall on the measurements and assessment of the role of water in cell structure and function. There are only a few reports on this topic, probably because of the nature of this kind of experimentation. She explains, however, that nuclear magnetic resonance spectroscopy (nmr) can be used successfully to measure water–macromolecular interactions. Some general and specific observations have already been made. For instance, she has observed that water moves very freely around condensed chromatin, but less freely about diffuse chromatin in mammalian cells. In studies on the cell cycle, Beall has observed that the water content of cells is lowest during S and $G_2$ but highest in dividing cells. In another interesting observation, Beall also noticed a difference in the mobility of water in certain selected human breast cancer cell lines. She observed that fast-dividing cells demonstrated high $T_1$ values (nmr relaxation times), whereas slow dividing cells had low $T_1$ values. High values are related to greater mobility of water and low values with less mobility of water. In respect to actin-like proteins, Beall observed increased water accompanying depolymerization.

It is perhaps fitting to explain that we follow next (Chapter 9 by Cameron *et al.*) with the role of selected elements in the regulation of cell reproduction. These investigators explore the new field of electron probe analysis for quantification of various selected elements in cells. This apparatus permits measurements in cells within the micron–submicron range. Cameron and co-workers discuss various theories and the application of this probe on the role of distribution of such ions as Na, Mg, K, Ca, and Cl in the regulation of cell reproduction. Although in its infancy as a probe, the following general and perhaps important observations have been made. Tumor cells and cells with short generation times have higher Na and Cl than cells with low mitotic activity. Elevated K and Mg are associated with high mitotic activity in nontumor cell populations but not for tumor cells. The physiological significance of these differences remains to be elucidated.

Chapters 10–12 of this volume are concerned with the possible role of calcium and cyclic nucleotides in the control of mitosis and cell division. Calcium in particular is gaining renewed attention because of its known role in the interactions of proteins involved in regulation of cell motility. Since mitosis and cell division are both expressions of cell motility functions, there is good reason to suspect where common mechanisms are involved there should be shared properties.

Sisken (Chapter 10) discusses the role of calcium in the regulation of mitosis. This is a very exciting area for research and Sisken presents some new insights concerning the intracellular localization and movement of calcium ions during the various phases of mitosis. He presents data that supports the idea that calcium pools are closely associated with the cell membrane (plasma membrane) and that membrane changes associated with cell division involve the release of calcium from such sights and that activation of a calcium-dependent ATPase involves the transport of calcium to the region of the mitotic apparatus and also perhaps to microfilament complexes associated with the division furrow. He concerns himself primarily with the transport of calcium to the mitotic apparatus and its possible role along with calcium-dependent regulatory proteins involved in the movement of chromosomes during mitosis. He brings into the discussion the relevant literature concerning calmodulin (CDR) and cyclic nucleotides and how these substances could interact in the assembly–disassembly of microtubular proteins involved in spindle dynamics.

Chapter 11 by White *et al.* presents a discussion of the action of erythropoietin and changes in cyclic nucleotide levels in relation to the control of erythroid cell proliferation in mouse fetal liver cells. They noted about a twofold decrease in cAMP during cell division in unstimulated cells but an increase in cAMP in erythropoietin (EP) stimulated cells. Cyclic AMP was highest in $G_1$ in both EP-treated and untreated cells. Cyclic GMP was highest during division in untreated cells but showed about a fivefold decrease in EP-treated cells. However, they noted an elevation of cGMP by EP following a lag period of 4 hours.

Chapter 12 by Charp and Whitson discusses some of the important interactions of calcium and cyclic nucleotides in relation to cell division. The data they present are consistent with the idea that calcium is an important regulator of cyclic nucleotide production. A calcium influx is associated with a twofold decrease in cAMP and a severalfold increase in cGMP during the initiation of cell division. Cyclic GMP is highest during cell division and is accompanied by a large efflux in calcium. Calcium, calmodulin, and cyclic nucleotides are the focus of much recent attention concerning cell motility functions. It is hopeful that current interest in these important modifiers of cell function will become of interest to future investigators of the cell cycle.

As we stated at the onset, this volume of selected topics is clearly only at the threshold of new beginnings on the study of nucleocytoplasmic and environmental interactions of the cell cycle.

## REFERENCES

Brunk, C. F. (1979). *J. Cell Biol.* **83,** 364a.

Cook, J. R. (1966). *J. Cell Biol.* **29,** 369.

Hanson, K. P., Ivanova, L. V., Nikitina, Z. S., Shutko, A. N., and Komar, V. E. (1970). *Biokhimiya* **35,** 635.

Koch, J., and Stokstad, E. L. R. (1967). *Eur. J. Biochem.* **3,** 1.

Mitchison, J. M. (1971). "The Biology of The Cell Cycle." Cambridge Univ. Press, New York and London.

Pardee, A. B. (1974). *Proc. Natl. Acad. Sci. U.S.A.* **71,** 1286.

Pardee, A. B., Shilo, B.-Z., and Koch, A. L. (1979). *Hormones Cell Culture Cold Spring Harbor Conf. Cell Proliferation,* **6,** p. 373.

Prescott, D. M. (1976). "Reproduction of Eukaryotic Cells", p. 130. Academic Press, New York.

Smith, J. A., and Martin, L. (1973). *Proc. Natl. Acad. Sci.* **70,** 1263.

# 2

# Chloroplast Biogenesis during the Cell Cycle

D. E. BUETOW, E. A. WURTZ, and T. GALLAGHER

## I. INTRODUCTION

Chloroplasts are specific plant cell organelles which trap light energy and, via the process of photosynthesis, make it available to living systems. Light microscopists discovered early that plastids proliferate by growth and division within the cells that contain them, without the visible participation or intervention of other cell structures. Further, early genetic studies clearly showed that certain plastid defects are inherited in

**9**

such a way that the mutated genes appear to reside in the organelle rather than in the nucleus. Such cytological and genetic observations (e.g., Kirk and Tilney-Bassett, 1978) led to the view that plastids are autonomous, self-sufficient entities.

In the 1940s and 1950s a series of discoveries established that genetic information is stored in living organisms in the form of DNA (and in certain viruses, RNA). The search for DNA in plastids was spurred on by these discoveries and, in the 1960s, it was finally proved that these organelles do indeed also store genetic information in the form of DNA (e.g., Tewari, 1971). Also, it has been shown that this plastid genetic information is transcribed into RNA and translated into protein, apparently all within the plastid itself (although the idea that some plastid DNA transcripts are translated in the cytoplasm outside the organelle has not been totally ruled out yet). Further, it is known that many of the chemical components of the plastids are synthesized within the organelles themselves (Kirk and Tilney-Bassett, 1978).

Genetic studies, especially over the past 2 to 3 decades, have shown that some chloroplast proteins are inherited in a non-Mendelian fashion, i.e., coded by the organelle, while the majority are inherited in a Mendelian fashion, i.e., coded by the nucleus (e.g., Sager, 1972). Therefore, plastids are not autonomous bodies. Rather they are semiautonomous structures which require transcripts from both their own genome and that of the nucleus in order to develop and function.

An extensive elucidation of nuclear–cytoplasmic interrelationships will be required to understand the cellular controls regulating the formation, function, replication, and division of chloroplasts. Light, of course, is an ultimate external factor that influences the operation of these cellular controls. Most studies concerned with the structural and functional development of chloroplasts have used algae or plant seedlings first maintained in the dark and then exposed to the light (e.g., Kirk and Tilney-Bassett, 1978; Akoyonoglou and Argyroudi-Akoyonoglou, 1978). In the dark, plastids exist as rudimentary structures in the organisms used in these greening studies. In the light, the rudimentary plastids develop into mature chloroplasts over a period of time. Also, such studies are usually done under conditions wherein cell division does not take place. In contrast, the biogenesis of chloroplasts in dividing cells is considered in the present chapter. Specifically, chloroplast structural and biosynthetic events are examined in terms of their occurrence at various times during the cell cycle. Ultimately such studies should bring about an understanding of the events leading to and controlling the replication and division of *mature* chloroplasts. Replication and division are aspects of chloroplast biogenesis which are largely unstudied thus far.

## II. EUGLENA

### A. Synchronization of Cell Division

#### 1. Techniques

The cell cycle of *Euglena gracilis* has been studied with mass cultures synchronized for cell division through the use of alternating light and dark cycles (Cook and James, 1960; Edmunds, 1964, 1965a; Petropulos, 1964; Schantz *et al.*, 1972; Boasson and Gibbs, 1973; Chotkowska and Konopa, 1973; Brandt, 1975; Falchuk *et al.*, 1975b; Valencia and Bertaux, 1975), by heat shocks (Pogo and Arce, 1964), by temperature cycles consisting of a low temperature which inhibits cell division followed by a temperature which is optimal for cell division (Neal *et al.*, 1968), and by growth on a medium with lactate as an organic carbon source for growth (Calvayrac, 1972; Calvayrac and Ledoight, 1975).

The heat shock technique is useful for the synchronization of cell division in *Euglena* cultures but is not useful for chloroplast studies since significant numbers of the cells are bleached during the heat shocks at 34°C or higher (Pogo and Arce, 1964). It is well known from studies with nonsynchronized *Euglena* that, at the high temperature, these cells lose their chloroplasts and become permanently bleached (e.g., Brawerman and Chargaff, 1960; Mego and Buetow, 1967). The temperature-cycling technique of Neal *et al.* (1968) uses an alternating cycle of 17.5 hours at 14.5°C (no cell division) and 6.5 hours at 28.5°C (synchronized cell division), but so far has been done only in the dark. In continuous darkness, the *Euglena* chloroplast regresses to a proplastid (Schiff and Epstein, 1968).

Conditions for synchronizing cell division of *Euglena* cultures in a medium containing 33 m$M$ lactate as carbon source have been given by Calvayrac (1972) and Calvayrac and Ledoight (1975). The synchrony obtained is dependent upon the present of lactate and occurs either in complete darkness or in alternating cycles of 6 hours of light and 6 hours of darkness. Such synchrony has not been observed when other carbon sources are used. Exponential growth of *Euglena* with lactate also is unusual in that it occurs in the dark in the absence of added vitamin $B_{12}$ (Bre and Diamond, 1975; Bre *et al.*, 1975). Usually this vitamin is an essential growth factor for *E. gracilis* (Hutner *et al.*, 1956).

Most studies on *Euglena* synchronized for cell divsion have been done with alternating light and dark cycles at a constant temperature. Details of the development of this technique have been given by Padilla and Cook (1964). Usually, the autotrophic medium of Cramer and Myers

(1952) is used. Cell number per milliliter remains constant in the light period but doubles or nearly doubles in the dark period (Fig. 1). Autotrophic cultures can be synchronized repetitively with cell division confined to the dark phase of each light:dark cycle up to a cell density of about $10^5$ per milliliter. Falchuk *et al.* (1975b) have synchronized *Euglena* on a light:dark cycle at densities of up to $2 \times 10^6$ cells per milliliter with a medium containing sucrose, glutamate, and malate as carbon sources (Falchuk *et al.*, 1975a). No chloroplast studies have been done using this latter technique, however.

The first strain of *Euglena* to be synchronized was cultured at 20°C with a repetitive cycle of 16 hours of light followed by 8 hours of darkness (Cook and James, 1960; Cook, 1961a,b). The most commonly used strain of *Euglena*, i.e., strain Z, is usually synchronized with 14 hours of light followed by 10 hours of darkness (Edmunds, 1964, 1965a; Cook, 1966a,b, 1968; Schantz *et al.*, 1972; Boasson and Gibbs, 1973; Brandt,

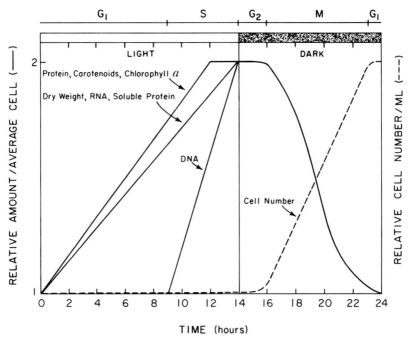

**Fig. 1.** Summary of biosynthetic patterns in *Euglena gracilis,* strain Z, grown synchronously on autotrophic medium with a light:dark cycle of 14:10 hours. Adapted from Cook (1961b, 1966b), Edmunds (1964, 1965b), Chotkowska and Konopa (1973), Falchuk *et al.* (1975b), and LaFarge-Frayssinet *et al.* (1978). Cell number per milliliter remains constant through the light period and doubles in the dark period.

1975; Bertaux *et al.,* 1978; LaFarge-Frayssinet *et al.,* 1978). Light intensities and temperature used vary with the laboratory involved and range from 3,500 to 13,000 lx and from 20° to 25°C, respectively. In order to maintain a doubling of cell number in the dark period of each cycle, the length of the light and dark periods can be varied only within fairly narrow limits (Edmunds, 1966; Edmunds and Funch, 1969a,b). A light intensity of 13,000 lx is reported as saturating for photosynthesis in strain Z (Cook, 1963, 1966c), but a lower intensity appears to be optimal for *E. gracilis* var. *bacillaris* (Stern *et al.,* 1964). Cell division is only poorly synchronized in dim light, i.e., 800 lx (Edmunds, 1966). Either sterile air or an air–$CO_2$ mixture (95%–5%) is bubbled through the cultures which are kept uniformly suspended by a spinning magnetic bar placed at the bottom of the culture vessel. The generation time for *Euglena* is more rapid in the air–$CO_2$ mixture than it is in air alone, especially at pH 7.4 (Jones and Cook, 1978). At pH 6.8, high external levels of inorganic phosphate are necessary for synchrony (Cook, 1971).

### 2. General Biochemistry of the Synchronized Cell

When *Euglena* are synchronized on a light:dark cycle of 14:10 hours, the average cell's total dry weight and content of RNA, DNA, protein, and soluble protein double in a linear fashion during the light period (Fig. 1; Table I) and then decline in the dark as cell division occurs (Fig. 1). Some variation is observed in the division times of individual cells since it usually takes 6 to 8 hours for all the cells to divide in the dark. This variation results in an essentially exponential increase in the number of cells per milliliter in the dark period. The degree of variation

Table I    Metabolic Parameters of *E. gracilis*, Strain Z, during the Light Period of One Light:Dark Synchronization Cycle of 14:10 Hours[a]

| Parameter | pg per cell | |
|---|---|---|
| | 0 Hour | 14 Hours |
| Dry weight | 500 | 1040 |
| Total RNA | 21 | 42 |
| Total protein | 220 | 410 |
| Total soluble protein | 33 | 65 |
| Total DNA | 2.25 | 4.50 |
| Chlorophyll *a* | 15 | 30 |
| Carotenoids | 5.7 | 11.6 |

[a] Adapted from Edmunds (1965b).

in division times of individual cells during the burst of cell division in the dark period is equivalent to that found in individual generation times of *Euglena* in exponential cultures (Cook and Cook, 1962). In sum, the doubling of the individual cell's dry weight and content of RNA, etc., during the light period and the pattern of cell divisions in the dark period closely mimic the normal situation for the individual cell in an exponentially growing culture. Therefore, each 24-hour light:dark synchronization cycle (Fig. 1) is at least a close approximation of one cell cycle (Cook, 1966b).

## B.  Chloroplast Biogenesis during the Cell Cycle

This discussion of the biogenesis of chloroplasts in *Euglena* during the cell cycle is restricted to those published studies in which cell division was synchronized by means of a repetitive light:dark cycle such as described in Sections II,A,1 and 2 and shown in Fig. 1. Considerable research also has been done on the biogenesis of *Euglena* chloroplasts in cells grown exponentially in the dark on a nutrient medium containing organic carbon and then transferred to the light in a nonnutrient ("resting") medium. In the "resting" medium, cell division is inhibited but chloroplast development still takes place. Such studies have been reviewed frequently (Schiff, 1973; Schmidt and Lyman, 1976; Nigon and Heizmann, 1978) and will not be covered here.

### 1.  Number and Morphology of Chloroplasts

Cook (1966a) reported that the number of chloroplasts per synchronized *Euglena* cell remained at six throughout the 14-hour light period and then increased in the dark period to eight per cell at the time of cell division. The number then declined to six by the end of the dark period when cell division was completed (Table II). In contrast, Boasson and Gibbs (1973) reported about eight chloroplasts per cell in both the light and the dark periods (Table II). The *range* of numbers of chloroplasts per cell, however, reportedly increased in the dark in both studies (Cook, 1966a; Boasson and Gibbs, 1973), but the latter study showed a range about twice that of the former study. The differences in the two studies could be due to the difficulty in making direct counts of *Euglena* chloroplasts *in situ*. The difficulty arises because numerous paramylum granules obscure the chloroplasts which are connected to each other by ribbonlike structures (Epstein and Schiff, 1961; Orcival-Lafont *et al.*, 1972; Calvayrac and Lefort-Tran, 1976).

Cook (1966a,b) suggested that in cultures of *E. gracilis*, strain Z, synchronized by a repetitive light:dark cycle (Fig. 1), the chloroplasts divide just prior to cytokinesis. Orcival-Lafont *et al.* (1972) also provided data consistent with the view that the chloroplasts divide just prior to cell division in *Euglena* cultures synchronized by lactate. Boasson and Gibbs (1973), however, reported that the chloroplasts divide synchronously at the time of cytokinesis in *Euglena* synchronized by the same light–dark cycles as Cook (1966a,b) used. Further, the apparent tight coupling between chloroplast replication and cell division could not be disturbed by shifting the phase of the light:dark cycles nor by inhibiting chloroplast division with streptomycin (Boasson and Gibbs, 1973). In any case, it is clear that chloroplast division occurs in the dark in synchronized cells at or near the time of cell division.

Cyclic changes occur in chloroplast structure in *E. gracilis* synchronized by light:dark cycles (Könitz, 1965; Cook *et al.*, 1976). Chloroplasts are relatively compact with closely appressed lamellae during most of the light period but become distended during the dark (division) period (Cook *et al.*, 1976). This change in ultrastructure of the chloroplasts persists for at least one cycle even when the cells are left in continuous light, a result suggesting that the periodicity is more closely related to the age of the cell rather than to a direct effect of light. The pyrenoid exists only transiently, being present in the cells only during the first half of the light period (Könitz, 1965; Cook *et al.*, 1976).

Development of the presumptive chloroplast from the proplastid in *Euglena* occurs both by lateral growth of lamellae and by multiplication of the number of lamellae (Ben-Shaul *et al.*, 1964). In *mature* chloroplasts, however, such as found in synchronized cells, the lamellae appear only to grow laterally (Cook *et al.*, 1976). Young cells early in the light period have chloroplasts with about six to eight lamellae each. Similarly, older cells (late in the light period), which have completed chlorophyll synthesis and are about ready for cell and chloroplast division, still contain chloroplasts with only six to eight lamellae each.

**Table II  Numbers of Chloroplasts per *Euglena* Cell during One Light:Dark Synchronization Cycle of 14:10 Hours**

| Period | Time (hr) | Number per cell | Range per cell | References |
|--------|-----------|-----------------|----------------|------------|
| Light | 0 | 5.9 | 2–8 | Cook (1966a,b) |
| Light | 14 | 6 | 1–8 | |
| Dark | 18 | 7.6 | 4–11 | |
| Dark | 24 | 6 | — | |
| Light | 7 | 8.2 | 4–17 | Boasson and Gibbs (1973) |
| Dark | 18 | 8.4 | 2–20 | |

## 2. Nucleic Acid Synthesis

*Euglena* nuclear DNA doubles in the latter half of the light period of a light:dark cycle of 14:10 hours (Fig. 1). In contrast, chloroplast DNA synthesis occurs *throughout* the light and dark periods as determined by autoradiographic analysis of the incorporation of [$^3$H]adenine into this DNA (Cook, 1966a,b). The continuous synthesis of *Euglena* chloroplast DNA correlates with the finding that a significant portion of this DNA is "labile" (Manning and Richards, 1972; Richards and Manning, 1975; Walfield and Hershberger, 1978; Lyman and Srinivas, 1978). The DNA synthesis associated with this lability is in excess of one duplication of the chloroplast DNA and represents replacement of a significant fraction of the organelle DNA that is being turned over and/or repaired during each cell generation. The existence of such labile organelle DNA was earlier suggested by Iwamura (1966) for *Chlorella* and by Sampson *et al.* (1963) and Hotta *et al.* (1965) for higher plants.

In addition to the continuous synthesis of chloroplast DNA in *Euglena*, there are two periods of enhanced synthesis of this DNA in the light cycle (Cook, 1966a,b; Brandt, 1975). The first peak occurs early in the light period (hour 2) and the second later in the light, i.e., at hours 13–14 according to Cook (1966a,b) or at hour 8 according to Brandt (1975). Cook (1966a,b) originally interpreted each peak as representing the replication of one of two postulated different species of chloroplast DNA in *Euglena*. More recent data eliminate this interpretation, however, since renaturation kinetics of chloroplast DNA (Stutz *et al.*, 1975) and the patterns of chloroplast DNA fragments generated by restriction enzymes both show only one species of DNA (Gray and Hallick, 1977; Kopecka *et al.*, 1977; Mielenz *et al.*, 1977). Another interpretation is that the two peaks reflect an S period for chloroplast DNA which is interrupted. Such an S period has no precedent, however. Further, *Euglena* is bleached most easily by ultraviolet light at an age which corresponds to the first peak of enhanced chloroplast DNA synthesis and is most resistant to bleaching during the second peak (Cook and Hunt, 1965; Cook, 1966b). These latter results suggest then that the first peak represents DNA replication and that the second peak may represent increased turnover of the "labile" portion of chloroplast DNA. They also suggest that the second peak is not involved in chloroplast development.

No studies have been done on the synthesis of chloroplast RNA in *Euglena* synchronized with repetitive light:dark cycles. It is known, however, that light stimulates RNA synthesis in nonsynchronized *Euglena* (Verdier *et al.*, 1973; Cohen and Schiff, 1976; Heizmann, 1976; Freyssinet, 1977). Chloroplast rRNA synthesis is most effectively stimulated by blue light and red light whereas cytoplasmic rRNA synthesis is stimulated most effectively by blue light (Cohen and Schiff, 1976). Only a short light period is needed to obtain maximal synthesis of *Euglena* chloroplast rRNA (Cohen and Schiff, 1976; Heizmann, 1976). In light-stimulated *Euglena,* as much as 50% of the double-stranded chloroplast DNA is transcribed (Chelm and Hallick, 1976; Rawson and Boerma, 1976).

### 3. Total Protein and Photosynthetic Pigment Syntheses

The synthesis and turnover of *Euglena* chloroplast proteins continue throughout the light period (Table III) as indicated by the isolation of radioactively labeled chloroplasts from cells incubated at 27°C with [$^{14}$C]leucine for 1 hour at various times through the light period (Brandt, 1976). Inhibitor studies show that synthesis of chloroplast proteins occurs on both chloroplast and cytoplasmic ribosomes throughout the light period. This synthesis is sensitive to chloramphenicol, an inhibitor of protein synthesis on organelle ribosomes, and to cycloheximide, an inhibitor of protein synthesis on cytoplasmic ribosomes, at all points in the light period (Table III). No studies have been done on the synthesis of chloroplast proteins of *Euglena* during the dark period.

The amounts of chlorophyll (Cook, 1966c; Walther and Edmunds, 1973; Laval-Martin *et al.,* 1979) and total carotenoids (Cook, 1966c) *per*

Table III  Chloroplast Protein Synthesis during the Light Period of *E. gracilis*, Strain Z, Synchronized at 27°C on Light:Dark Cycles of 14:10 Hours[a,b]

| Light period (hours) | Control (cpm/mg chlorophyll) | Treated with | |
|---|---|---|---|
| | | Chloramphenicol (% of control) | Cycloheximide (% of control) |
| 4 | 859 | 41 | 59 |
| 5 | 588 | 55 | 45 |
| 6 | 629 | 60 | 40 |
| 7 | 793 | 35 | 65 |
| 8 | 755 | 52 | 48 |
| 9 | 667 | 48 | 52 |
| 10 | 720 | 37 | 63 |
| 11 | 443 | 54 | 46 |
| 12 | 454 | 43 | 57 |
| 13 | 835 | 52 | 48 |
| 14 | 615 | 41 | 59 |

[a] From Brandt (1976).
[b] Cells were exposed to [$^{14}$C]leucine for 1 hour at various times. Chloroplasts were then isolated and leucine incorporation was determined with an isotope counter.

*cell* double during the light period and then decline in half during the dark period as cell number per milliliter doubles. The increase in amounts of pigments per cell is essentially linear throughout the light period (Cook, 1966c) or for at least 12 of the 14 hours of light (Edmunds, 1965b). When *Euglena* are synchronized on a repetitive light:dark cycle of 16:8 hours, chlorophyll *a* and total carotenoids per cell double in amount in a linear fashion for 14 of the 16 hours of light (Cook, 1961b).

## 4. *Photosynthetic Activity*

Photosynthetic activity in a single *Euglena* cell as measured by a Cartesian diver technique (Lövlie and Farfaglio, 1965) increases in a sigmoidal fashion throughout the interdivision period. In *Euglena* cultures synchronized on repetitive light:dark cycles, photosynthetic activity per milliliter of culture (or per cell) increases throughout the light period (Cook, 1966c; Codd and Merrett, 1971a; Laval-Martin *et al.*, 1979) or at least through 80% of the light period (Walther and Edmunds, 1973). The reported amount of increase in photosynthetic activity per cell during the light period varies, however, 132% according to Codd and Merrett (1971a) and 148% according to Walther and Edmunds (1973), though the latter laboratory later reported a 92% increase (Laval-Martin *et al.*, 1979). In any case, these three reports indicate that the photosynthetic activity per *Euglena* cell doubles or somewhat more than doubles in the light period. In contrast, Cook (1966c) reported only a 61% increase in photosynthetic activity per cell in the light period. Possible reasons for the difference between Cook's study (1966c) and the other studies will be discussed in Section II,B,6. In all the reports, on a *per cell* basis, photosynthetic activity declines in the dark period (the cell division period) and reaches a minimal level (about one-half of the maximal) before or by the end of the dark period when cell number per milliliter has doubled.

There is no correspondence between photosynthetic capacity and chlorophyll content. Edmunds' laboratory reported that the capacity for photosynthesis increases at a faster rate than chlorophyll synthesis during the first 3 hours of light and then falls behind chlorophyll synthesis (Walther and Edmunds, 1973; Laval-Martin *et al.*, 1979). Cook (1966c) also showed no linear correlation between chlorophyll content and photosynthesis in the light period in synchronized *Euglena,* but found that photosynthetic capacity always lagged behind the chlorophyll increase even during the early part of the light period. In sum, during the light period (or at least beyond 3 hours of light), synchronized *Euglena* possess a chlorophyll content capable of absorbing more light than can be utilized by the cells. The ability to capture energy is not as limiting as the ability to utilize the energy captured (Cook, 1966c).

The content of paramylum *per cell* increases sevenfold in the light period (Cook, 1966c). It is known that paramylum increases in exponentially growing cultures of *Euglena* only when energy (as light) is present in excess of immediate needs for synthesis (Cook, 1963). The large increase in paramylum in the light in synchronized *Euglena* shows, therefore, that the amount of light (11,000 lx) used was not limiting to photosynthesis in this study. During the dark period, the paramylum is essentially all used up, apparently being freely utilized as an energy source for general metabolism and cell division.

In general, the rate of increase in photosynthetic activity in the light correlates with the rate of increase in total protein and RNA per cell (Cook, 1961b, 1966c; Edmunds, 1965b; Codd and Merrett, 1971a; Walther and Edmunds, 1973; Laval-Martin *et al.*, 1979). More specifically, however, one can ask if the rate of increase in photosynthetic activity is limited by the rates of increase in the light or the dark reactions of photosynthesis. As noted above, photosynthetic activity in a synchronized *Euglena* cell is not limited by the cell's content of chlorophyll at least beyond the first 3 hours of the light period. Further, the light reactions of photosynthesis also do not seem to control the photosynthetic activity (Walther and Edmunds, 1973). The activity of photosystem II (dichlorophenol–indole–phenol reduction) increased only by a small amount during the light period, a time when photosynthetic activity more than doubled in this study. Similarly, the activity of photosystem I (methyl red reduction) hardly changed during the light period.

Since the light reactions apparently do not control the increase in photosynthetic activity in the light period, control of this activity by dark reactions is indicated. Walther and Edmunds (1973) showed that the activity of the Calvin cycle enzyme, ribulose-1,5-bisphosphate carboxylase (RuBPCase), increased by 138% in a linear fashion from hour 2 to hour 10 of the light period (Fig. 2). In other words, RuBPCase activity continued to increase at a time (last 2 hours of the light period) when photosynthetic activity was already declining (Fig. 2). Also, the rate of photosynthetic activity increased faster than the rate of RuBPCase activity during the first 80% of the light cycle. In contrast, Codd and Merrett (1971a) found that RuBPCase activity paralleled changes in photosynthetic rate. Even so, the activity of the enzyme was not great enough to satisfy the rates of $CO_2$ fixation at all times investigated (Codd and Merrett, 1971a). In sum, the activity of RuBPCase does not directly control photosynthetic activity in synchronized *Euglena*.

In contrast to RuBPCase activity, both the phase and pattern of the activity (Fig. 2) of another Calvin cycle enzyme, NADP-dependent glyceraldelyde-3-phosphate dehydrogenase (GPD), closely mimics the

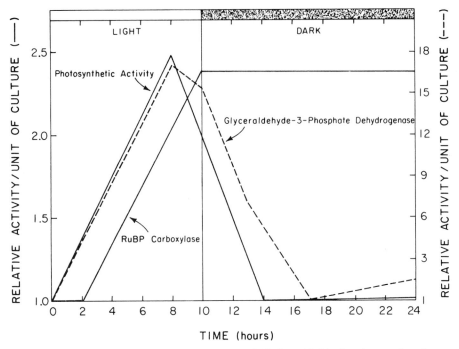

**Fig. 2.** Photosynthetic activity and activities of ribulose-1,5,-bisphosphate carboxylase and NADP-dependent glyceraldehyde-3-phosphate dehydrogenase over the cell cycle in a culture of *E. gracilis,* strain Z, grown synchronously on autotrophic medium with air and a light:dark cycle of 10:14 hours. Adapted from Walther and Edmunds (1973).

photosynthetic activity of *Euglena* in the light period (Walther and Edmunds, 1973). Therefore, GPD activity functions as a possible control of photosynthetic activity.

The specific activity of phosphopyruvate carboxylase (PC) is greater than that of RuBPCase at all stages of the cell cycle in *Euglena* synchronized on light:dark cycles (Codd and Merrett, 1971c). In $C_4$ plants, PC is associated with the outer membrane of mesophyll chloroplasts (Slack *et al.,* 1969). However, in *Euglena,* a $C_3$ plant (Chollet and Ogren, 1975), PC is located in the cytosol (Codd and Merrett, 1971c). This location restricts the role of PC in *Euglena* to providing an anaplerotic sequence to the tricarboxylic cycle.

### 5.  *Products of Photosynthesis*

Codd and Merrett (1971a,b) determined the products of $^{14}CO_2$ fixation in division synchronized *Euglena* bubbled with air at hours 3, 12, 17, and 23 of a light:dark cycle of 14:10 hours. In this study, the rate of $^{14}CO_2$ fixation doubled *per cell* over the light period and declined about

50% by the end of the dark period when the number of cells had doubled. Little difference is seen between the percentage of the total photosynthetically fixed $^{14}C$ in a given fraction from cells harvested at different stages of the light:dark cycle (Table IV). An exception occurs with 12-hour cells, however. At this stage, the total incorporation into the polysaccharide–nucleic acid fraction is increased and that in the water-soluble fraction is decreased.

In contrast to the gross cell fractions shown in Table IV, individual products of $^{14}CO_2$ fixation vary considerably over the light:dark cycle (Table V). For example, labeled sucrose was not detected in 12-hour cells. However, incorporation into sucrose peaked at 17 hours and then was considerably reduced at 23 hours. Labeled maltose was not detected in 17-hour cells, but was found in 3-, 12-, and 23-hour cells. The percentage of label found in phosphate esters, amino acids, or organic acids also differed at different stages of the cycle. With 12-hour (late light period) and 17-hour cells (early dark period, cells dividing), there was a rapid initial $^{14}C$ incorporation, i.e., by 1 minute of incubation, into phosphate esters. However, this initial incorporation was reduced in 23-hour (late dark period, cell division completed) and 3-hour cells (early light period, no cell division).

The labeling of glycolate and glycerate differed during the cycle (Table V). At one time, algae were thought to be incapable of metabolizing glycolate (Hess and Tolbert, 1967). However, *Euglena* contain all of the enzymes of the glycolate cycle (Codd and Merrett, 1971b; Murray *et al.*, 1971) and thus can readily metabolize photosynthetically formed glycolate. The relationships between the glycolate and the glycerate pathways and the dark reactions of photosynthesis (Calvin cycle) are shown in Fig.

**Table IV** Photosynthetic $^{14}CO_2$ Incorporation into Cell Fractions of *Euglena* Bubbled with Air during One Light:Dark Synchronization Cycle of 14:10 Hours[a,b]

| Cell sample (hours) | Total $^{14}C$ incorporation in cell fractions[c] (%) | | | | |
|---|---|---|---|---|---|
| | Medium | Water soluble | Ethanol soluble | Polysaccharides–nucleic acids | Cell residue |
| 3 | 23.3 | 65.6 | 7.1 | 4.0 | 0.4 |
| 12 | 13.2 | 49.7 | 6.2 | 24.6 | 6.3 |
| 17 | 22.1 | 69.0 | 3.8 | 5.0 | 0.2 |
| 23 | 23.5 | 62.6 | 9.3 | 4.4 | 0.2 |

[a] Adapted from Codd and Merrett (1971a).
[b] At each sampling time, a volume of culture was taken and cells collected by centrifugation, washed, and resuspended in growth medium. $NaH^{14}CO_3$ was added. Cells were incubated and fractionated.
[c] After 20-min incubation with $NaH^{14}CO_3$.

**Table V  Photosynthetic Incorporation of ¹⁴C-Labeled Bicarbonate into Ethanol-Soluble Compounds by *Euglena* during One Light:Dark Synchronization Cycle of 14:10 Hours[a,b]**

| | Total ¹⁴C incorporation into ethanol-soluble compounds (%) | | | | | | | |
| | 3-Hour cells | | 12-Hour cells | | 17-Hour cells | | 23-Hour cells | |
| Compound | 1 minute[c] | 20 minute[c] | 1 minute | 20 minute | 1 minute | 20 minute | 1 minute | 20 minute |
|---|---|---|---|---|---|---|---|---|
| 3-Phosphoglycerate | 7.7 | 1.2 | 8.8 | 6.0 | 5.2 | 3.1 | 4.0 | 3.5 |
| Other phosphates | 59.1 | 35.4 | 70.3 | 41.7 | 73.2 | 22.7 | 58.5 | 28.1 |
| Glycolate | 4.4 | 1.0 | ND | 0.2 | 0.8 | 0.4 | 0.6 | 4.0 |
| Glycerate | ND[d] | 0.4 | 2.6 | 1.0 | 0.6 | ND | 0.4 | 0.4 |
| Other organic acids | 3.4 | 11.8 | 1.8 | 9.1 | 2.7 | 1.8 | 13.6 | 7.8 |
| Glycine + serine | 12.4 | 10.0 | 7.7 | 8.6 | 5.4 | 3.3 | 12.2 | 14.3 |
| Other amino acids | 9.6 | 16.2 | 6.5 | 18.6 | 5.1 | 7.7 | 6.1 | 18.9 |
| Sucrose | ND | 3.4 | ND | ND | 5.4 | 21.3 | 0.7 | 10.3 |
| Maltose | ND | 1.1 | 1.4 | 1.2 | ND | ND | 2.9 | 0.5 |
| Unidentified and origin | 3.4 | 19.4 | 0.8 | 13.6 | 1.6 | 39.6 | 1.0 | 12.1 |

[a] Adapted from Codd and Merrett (1971a).

[b] Cells were sampled at 3, 12, 17 and 23 hours (see Table IV) and incubated with NaH¹⁴CO₃. After incubation with isotope, cells were extracted with absolute ethanol at −40°C. ¹⁴C-Labeled, ethanol-soluble compounds were separated by chromatography and the radioactivity of each was determined.

[c] Time of incubation with NaH¹⁴CO₃.

[d] Not detected.

3. The incorporation of $^{14}C$ into glycolate (Table V) is highest in 23-hour cells (late dark period, cell division completed) and in 3-hour cells (early light period). Similarly, in synchronized *Scenedesmus*, glycolate production is maximal at the time of cell division or immediately thereafter (Nelson *et al.*, 1969). The labeling of glycine and serine (Table V) is inversely related to the labeling of glycolate at 12 and 17 hours of the cycle, times when the labeling of glycolate is relatively low (Codd and Merrett, 1971a). Unlike glycolate, labeling of glycerate is heaviest in 12-hour cells. Indeed, by 7 seconds of incubation, [$^{14}C$]glycerate accounted for 33% of the radioactivity found in products of photosynthesis (Codd and Merrett, 1971a).

In a study of glycolate metabolism during the cell cycle of *Euglena* synchronized on light:dark cycles of 14:10 hours, Codd and Merrett (1971b) showed that enzyme activity regulated the flow of carbon via the glycolate pathway. Phosphoglycolate phosphatase activity *per cell* (Fig. 3) remained nearly constant throughout the light and dark cycles. In contrast, 3-phosphoglycerate phosphatase activity *per cell* (Fig. 3) decreased by half in the light period and then doubled in the dark period. Glycolate dehydrogenase activity *per cell* (Fig. 3) increased in the light and decreased in the dark.

With cells at the end of the light period (12 hours) and early dark period (17 hours), high levels of glycolate dehydrogenase are accompanied by an early and rapid labeling of glycerate (see also Table V), by the synthesis of uniformly labeled [$^{14}C$]glycolate and glycerate, and by minimal excretion of glycolate by the cells. Therefore, in the late light and early dark periods, a time when the cells are preparing for division, the glycolate pathway is used preferentially to form glycerate. However, the flow of carbon via this pathway is limited from at least the last hour of the dark period (23 hours) through the early part of the next light period (3 hours), a 4-hour period of time immediately following cell division. During this time, glycolate dehydrogenase activity per cell is minimal and 3-phosphoglycerate phosphatase activity per cell is maximal. Also, at this time, though glycolate is still uniformly labeled with $^{14}C$, both phosphoglycerate and glycerate are predominantly carboxyl-labeled. Therefore, following cell division, the glycerate pathway is used preferentially to form glycerate (Fig. 3).

As a whole, the marked differences in the labeling of individual photosynthetic products show that the flow of photosynthetically fixed carbon is controlled along various pathways to different extents over the cell cycle of synchronized *Euglena* (Codd and Merrett, 1971a,b). This finding agrees in general with those of Stange *et al.* (1960) on synchronously growing *Chlorella*.

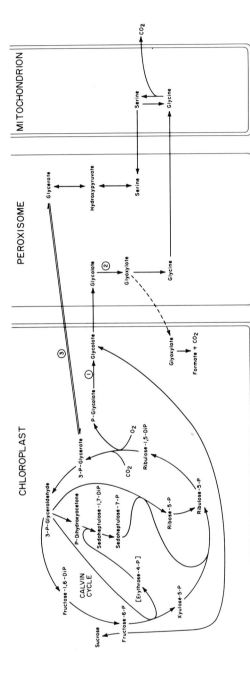

**Fig. 3.** Relationships between the Calvin cycle (dark reactions of photosynthesis) and the glycolate and glycerate pathways. At high $CO_2$ concentrations, e.g., 5%, the synthesis of 3-phosphoglycerate is favored and the synthesis of glycolate is inhibited. At low $CO_2$ concentrations, i.e., < 1%, glycolate is produced in large amounts. The glycolate → serine pathway (photorespiration) is irreversible. The glycerate → serine pathway is reversible. Enzymes: (1) phosphoglycolate phosphatase; (2) glycolate dehydrogenase; (3) 3-phosphoglycerate phosphatase. Adapted from Bassham and Calvin (1957), Chollet and Ogren (1975), and Zelitch (1972).

## 6. $C_1$ Metabolism

Lor and Cossins (1973, 1978a,b) have shown a relationship between glycolate metabolism and folate metabolism in autotrophic *Euglena* synchronized by light–dark (14:10 hour) cycles. Further, the degree of this relationship depends on whether the cultures are supplemented with air alone or with air plus 5% $CO_2$. In cultures with air alone, glycolate is a precursor of formyl folates and, consequently, operation of the glycolate pathway is accompanied by a drain of $C_1$ units (Lor and Cossins, 1978a,b). In cultures with air plus 5% $CO_2$, glycolate formation (Andrews *et al.*, 1973) and oxidation (Codd and Merrett, 1971b; Lor and Cossins, 1978a) are repressed. Therefore, in 5% $CO_2$-supplemented cultures, folate derivatives must be formed by other, not fully defined, pathways (Lor and Cossins, 1978a,b).

Folate metabolism is critical to the development and functioning of a cell. Folate derivatives are involved in the metabolism of lipids, in the synthesis of purine and pyrimidine nucleotides, and in the metabolism of amino acids including the biosynthesis of formylmethionine which is used in the initiation of protein synthesis on chloroplast 70 S ribosomes (Blakely, 1969). It is not surprising, therefore, that the rate of increase in photosynthetic activity correlates with the rate of increase in total RNA and protein content per cell (Section II,B,4).

As already mentioned, in *Euglena* cultures supplemented with air plus 5% $CO_2$, the glycolate pathway leading to folate derivatives is repressed. It is possible that the other pathways used in this case for the production of folate derivatives are less efficient than the glycolate pathway. If so, this would explain the lower rate of increase (Section II,B,4) in photosynthetic activity and in total RNA and protein per cell in synchronized *Euglena* cultures supplemented with air plus 5% $CO_2$ (Cook, 1966c) compared to cultures supplemented with air alone (Cook, 1961b; Edmunds, 1965b; Codd and Merrett, 1971a; Walther and Edmunds, 1973; Laval-Martin *et al.*, 1979).

## C. Summary

Cell division in cultures of *E. gracilis* can be synchronized by a variety of techniques. When synchronized on alternating light:dark cycles, cell division is confined to the dark period. One light:dark cycle is, at the least, a close approximation of one cell cycle. Chloroplasts replicate by lateral growth of the lamellae in the light period and divide in the dark period at or near the time of cell division (Fig. 4). Chloroplast morphology changes during the cell cycle and the pyrenoid is present only during the first half of the light period.

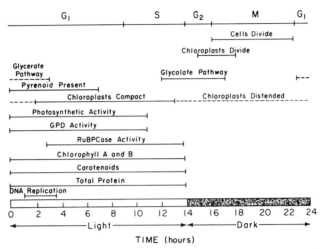

**Fig. 4.** Summary of the times during the cell cycle of synchronized *E. gracilis*, strain Z, when chloroplast DNA and total protein are synthesized, when carotenoids and chlorophylls increase in amount per cell, when photosynthetic activity and chloroplast enzyme activities increase, when morphological changes occur in chloroplasts, when glycerate is formed predominantly by the glycerate pathway and the glycolate pathway, and when chloroplasts divide. It should be noted that chloroplast total protein synthesis was measured in the light period only (Section II,B,3). All other parameters were measured throughout the cell cycle.

Chloroplast DNA replicates early in the light period (Fig. 4) but some chloroplast DNA synthesis occurs throughout the cell cycle. The nature of the latter synthesis is not defined, but may represent repair synthesis or synthesis associated with the "turn over" of a significant fraction of the organelle DNA during each cell cycle. Chloroplast proteins and photosynthetic pigments are synthesized throughout the light period. When cultures are bubbled with air, photosynthetic activity doubles over the first 80% of the light cycle (Fig. 4) as the chloroplasts replicate. However, if the cultures are bubbled with "high" $CO_2$, i.e., 5% $CO_2$:95% air, the rate of increase in photosynthetic activity is slowed and continues in the dark period. The difference in photosynthetic activity under the two environmental conditions appears related to the availability of folate derivatives. Photosynthetic activity is controlled by dark reactions and not by light reactions or by the cellular levels of chlorophyll. The products of $CO_2$ fixation vary considerably over the cell cycle. Apparently, the flow of photosynthetically fixed carbon is controlled along various pathways to different extents over the cell cycle. For example, the glycolate pathway is maximally used in the hours preceding chloroplast division and cell division whereas the glycerate pathway is used in the hours following these divisions (Fig. 4).

## III. *CHLAMYDOMONAS*

### A. Synchronization of Cell Division

As in the case of *Euglena* (Section II,A,2), one light:dark synchronization cycle is considered the equivalent of one cell cycle. Phototrophically grown cultures of *Chlamydomonas* can be synchronized to a high degree by a single dark to light shift every 24 hours. Under a light:dark regimen of 12 hours:12 hours, the cells divide in the dark period (Bernstein, 1960). The mother cell may divide into four, eight, and even higher numbers of daughter cells or autospores depending upon the cell density of the culture and the light intensity. The daughter cells are retained in the parental cell wall until late in the dark period and then liberated synchronously (Kates and Jones, 1967). Recently, a cell wall-less mutant of *C. reinhardtii* was synchronized under mixotrophic growth conditions by alternating light and dark periods of 12 hours each (Grant *et al.*, 1978). Because the mutant lacks a cell wall, the daughter cells separate immediately following cytokinesis. Therefore, the degree of synchrony of the mixotrophically grown cell wall-less *Chlamydomonas* is not directly comparable to that of phototrophically grown wild-type cells.

### B. Effects of Inhibitors during the Cell Cycle

Howell *et al.* (1975) used a unique technique for determining transition points for a number of different inhibitors in asynchronous cultures of *C. reinhardtii*. Transition points were determined by measuring the amount of division which takes place in an exponential asynchronous culture after the addition of an inhibitor. By using the concept of the age distribution of an exponentially growing culture (Blum and Buetow, 1963; Cook and James, 1964), the authors calculated the transition point in the cell cycle after which the inhibitor would no longer prevent cell division. The result was that the transition points were grouped into two broad periods of the cell cycle. The zero point of the cell cycle is defined as the time in the cell cycle which occurs immediately after cytokinesis. Those inhibitors that block chloroplast protein synthesis, chloramphenicol and spectinomycin, have transition points in the second quarter of the cell cycle as do the organellar RNA and DNA inhibitors, rifampin and ethidium bromide. On the other hand, actinomycin D, acridine orange, and hydroxyurea, all of which may affect both nuclear and organellar RNA and DNA synthesis, have a transition point in the fourth quarter of the cell cycle. The antibiotic cycloheximide, which inhibits cytoplasmic protein synthesis, also has a late fourth quarter transition point. This latter finding indicates that it is necessary for some protein synthesis to occur in the cytoplasm late in the cycle for cell division to take place. The electron transport inhibitors, cyanide and sodium azide, and the uncoupler 2,4-dinitrophenol have fourth quarter transition points. Therefore, energy for cell division appears to be required late in the cell cycle for division to occur. Second quarter transition points exist for placing photoautotrophically growing cultures in the dark or for adding the Hill reaction inhibitor, dichlorophenyldimethylurea. These studies are of interest in that they allow some measure, however crude, of the cell cycle without the induction of synchrony and any artifacts which may be introduced by the synchrony-inducing growth regimen. This seems to be an especially important point with regard to algae where the method used for induction of synchrony is usually a light:dark cycle. A number of the events taking place in the cell cycle may be energy dependent, i.e., requiring light and being cell cycle-independent (c.f., RuBPCase, Section III,D,3).

The authors are careful to point out that caution must be used when interpreting the effects of inhibitors on specific cell cycle events. First of all, the permeability of the cells to different inhibitors may vary during the cell cycle. Second, due to the pleiotrophic effects of inhibitors, the cell cycle event being measured may not be directly affected by the inhibitor. Instead, the inhibitor may be blocking the cell from progressing through the cell cycle to the event which is being measured.

## C.  Cell Cycle Mutants

Howell and Naliboff (1973) isolated several temperature-sensitive conditional cell cycle-blocked mutants of *C. reinhardtii.* The temperature of an asynchronous exponentially growing culture was raised from 21°C to the nonpermissive temperature of 33°C and then the residual amount of cell division in the culture was determined. From the age distribution of the culture and from the residual amount of cell division, "block points" in the cell cycle were determined. In some cases these block points were verified by temperature-shift experiments with synchronous cultures of the mutants. For the most part, the mutants could not resume cell division after exposure to the nonpermissive temperatures and, therefore, could not be used for temperature shift-down experiments. Mutants were found which had block points throughout the cell cycle in contrast to the results of inhibitor studies which showed that the transition points for the inhibitors were confined to the second and fourth quarters of the cell cycle (Section III,B).

## D. Chloroplast Biogenesis during the Cell Cycle

### 1. DNA Synthesis

An AT-rich satellite is resolved in CsCl buoyant-density gradients of *Chlamydomonas* whole-cell DNA (Chun *et al.*, 1963). Sager and Ishida (1963) showed that this satellite is enriched in chloroplast preparations. The satellite has a buoyant density in CsCl gradients of 1.695 gm/cm$^3$ compared to a density of 1.723 gm/cm$^3$ for nuclear DNA. This fortuitous difference in buoyant densities has made it possible to study the replication of the satellite, i.e., the chloroplast DNA, without the need for the isolation of pure chloroplasts free of contamination with nuclear DNA.

Chiang and Sueoka (1967) used an $^{15}$N to $^{14}$N isotope-transfer experiment with synchronous cultures of *Chlamydomonas* to study the replication of chloroplast DNA. They showed that this DNA replicates in a semiconservative manner similar to the nuclear DNA. However, the chloroplast DNA replicates at a time, i.e., early to midlight period, dis-

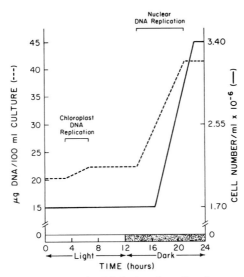

**Fig. 5.** DNA synthesis in *C. reinhardtii* during the cell cycle. ——, cell number per milliliter; - - - - -, micrograms DNA per 100 ml of culture. Adapted from Chiang and Sueoka (1967).

tinct from that of the nuclear DNA, i.e., midpart of the dark period shortly before cytokinesis (Fig. 5).

Recently, Grant *et al.* (1978) studied DNA synthesis in synchronous *Chlamydomonas* cultures using radioactively labeled DNA precursors ($^{32}PO_4^{2-}$, [$^3$H]thymidine, [$^{14}$C]adenine). Their experiments measured all types of DNA syntheses (replication, repair, and synthesis associated with recombination). Chloroplast DNA synthesis was not confined to the light part of a light:dark cycle of 12:12 hours, but also took place during the dark period. The nature of this additional DNA synthesis was not determined, i.e., whether it represented repair, replication, or recombination. It remains unknown whether this "unscheduled" DNA synthesis is related to the continuous chloroplast DNA synthesis observed in *Euglena* (Section II,B,2).

## 2. Transcription

Howell and Walker (1977) studied RNA transcription through the cell cycle by means of RNA:DNA hybridization. Total cell RNA was hybridized in solution to radioactively labeled chloroplast DNA. The amount of DNA hybridizing to saturating amounts of RNA was used as the measure of the complexity of the RNA transcripts. Total cell RNA from asynchronous exponentially growing cultures of *Chlamydomonas* hybridized 60% of the chloroplast DNA at saturation. Since 50% of the chloroplast DNA would hybridize if one entire strand was transcribed this result suggests that (1) one complete strand is transcribed during the cell cycle, and (2) that the second strand also is at least partly transcribed.

When total cell RNA from synchronously grown cells was hybridized to chloroplast DNA, the lowest complexity of RNA transcripts (39%) was found 2 hours into the light period. The complexity steadily increased to 53% at 8 hours in the light period, reached a maximum of 62% at 2 hours into the dark period, and then dropped to 47% at 8 hours into the dark period. These results (Howell and Walker, 1977) indicate that the chloroplast genome is transcribed to a great extent *throughout* the cell cycle. It should be noted, however, that these hybridization experiments were done with saturating amounts of RNA and that a slight asynchrony of the cells in the culture would lead to an overestimate of the complexity of the RNA transcripts hybridizing to chloroplast DNA at any given time during the cell cycle.

Cattolico *et al.* (1973) determined the total ribosomal RNA content of synchronously grown cells. The RNA extracted at several different points of the cell cycle was resolved by polyacrylamide gel electrophoresis (Fig. 6). The amount of chloroplast ribosomal RNA *per cell*

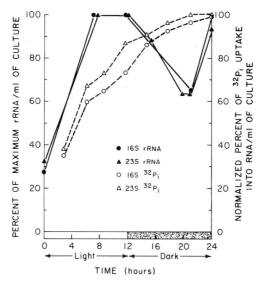

**Fig. 6.** Relative amounts of chloroplast ribosomal RNA per milliliter of culture. Adapted from the results of Cattolico *et al.* (1973) based upon absorbance measurements (●————●, 16 S; ▲————▲, 23 S) and the results of Wilson and Chiang (1977) based upon $^{32}P_i$ uptake into ribosomal RNA (○- - -○, 16 S; △- - -△, 23 S).

was lowest immediately after cell division, then increased in a nearly linear fashion until the midlight period, remained constant through the rest of the light period, and declined during the dark period reaching a minimum again at the time of cell division. Wilson and Chiang (1977) verified the results of Cattolico *et al.* (1973) by showing that ribosomal RNA synthesis *per cell* increases during the light period of the cycle and then decreases *per cell* during the dark period (Fig. 6). These findings imply that the chloroplast is being programmed for protein synthesis immediately after cell division and that its capacity for protein synthesis increases throughout the light period of the cell cycle at least.

### 3. Protein Synthesis

Intact chloroplasts cannot be isolated from *Chlamydomonas* with present cell fractionation techniques. Therefore, the amount of chloroplast protein being synthesized at any time must be estimated indirectly. Baumgartel and Howell (1977) studied changes in polypeptide initiation and elongation rates during the cell cycle of *C. reinhardtii*. Synchronously grown cultures were pulse-labeled with [³H]arginine and total cell polyribosomes were fractionated on a sucrose gradient. Samples of the large polyribosomes ($N > 6$, where $N$ is the number of constituent ribosomes), the less polymerized polyribosomes ($2 \leq N \leq 6$), and the monosomes were pooled separately. Ribosomal RNA was extracted with phenol from the samples and fractionated into individual ribosomal RNA species by polyacrylamide gel electrophoresis. This procedure allows a determination of the fraction of the total cell polyribosomes and monosomes that are cytoplasmic (25 S and 18 S ribosomal RNA) and the fraction which are of chloroplast origin (23 S and 16 S ribosomal RNA). Their results show that the fraction of chloroplast ribosomes in the total cell polyribosomes is low and that the sizes of the chloroplast polyribosomes are small. If chloroplast-specific ribonuclease activity can be ruled out, the latter result suggests that the chloroplast polyribosomes have a low packing density (number of ribosomes per message) and that, correspondingly, chloroplast protein synthesis is being initiated at a low rate.

Baumgartel and Howell (1977) also found that the proportion of chloroplast ribosomes in total cell polyribosomes increases at the beginning of the light period. Also at this time, the largest chloroplast polyribosomes are recovered. This result indicates that the initiation rate in chloroplasts also is highest at the beginning of the light period. The fraction of chloroplast ribosomes in polyribosomes declines during the rest of the cell cycle as does the size of the chloroplast polyribosomes. These latter results indicate that the rate of initiation of protein synthesis falls behind the rate of elongation during the cell cycle. The authors could not measure the elongation rate of the chloroplast polyribosomes apart from that of the cytoplasmic polyribosomes. Therefore, the amount of chloroplast protein synthesis could not be quantitated.

Several groups have investigated the production of specific chloroplast proteins during the cell cycle in *Chlamydomonas*. Armstrong *et al.* (1971) attempted to identify the site of synthesis of several components of the photosynthetic electron transport chain as well as two enzymes of the Calvin cycle. The amount of cytochrome 553 (c-553) increases linearly from about 1 to 10 hours of the light period and then levels off (Fig. 7). Spectinomycin and chloramphenicol, inhibitors of protein synthesis on organelle 70 S ribosomes, prevented this increase during the light period as did the chloroplast transcriptional inhibitor, rifampicin. Cytochrome 563 shows a different pattern of synthesis increasing in amount only slightly up to hour 6 of the light period, then increasing rapidly to hour 10, and then leveling off (Fig. 7). Spectinomycin, chloramphenicol, and rifampicin inhibit the increase in c-563 as does cycloheximide (an inhibitor of protein synthesis on cytoplasmic 80 S ribosomes). Ferredoxin increases in amount in an approximately linear fashion from hour 3 to about hour 10 of the light period (Fig. 7). This increase is inhibited only by cycloheximide. The activity of ferredoxin

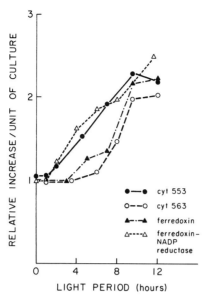

**Fig. 7.** Relative increases in amounts of components of the photosynthetic electron transport chain per unit of culture. Adapted from Armstrong *et al.* (1971).

NADP–reductase increases linearly from hour 1 to the end of the light period (Fig. 7). This increase in enzyme activity is significantly inhibited only by cycloheximide.

The Calvin cycle enzymes, ribulose-1,5-bisphosphate carboxylase (RuBPCase) and phosphoribulokinase, both show a nearly linear increase (Armstrong *et al.*, 1971) in activity throughout the light period. Cycloheximide and spectinomycin, but not rifampicin, inhibit the increase in RuBPCase activity. Only cycloheximide inhibits the increase of phosphoribulokinase activity.

Iwanij *et al.* (1975) examined the synthesis of RuBPCase (as opposed to its activity) in synchronous cultures of *Chlamydomonas* pulse-labeled with [³H]arginine. The rate of synthesis of RuBPCase increased throughout the light period and then dropped to a low but measurable amount in the dark period (Fig. 8).

Howell *et al.* (1977) found that synchronous cultures of *Chlamydomonas,* taken from the middle part of the dark period and placed in the light, immediately synthesized the large and small subunits of RuBPCase. This result indicates that the synthesis of RuBPCase may not be cell cycle-dependent, but rather dependent upon conditions of illumination.

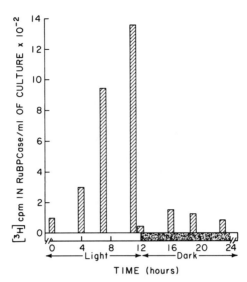

**Fig. 8.**   The incorporation of [³H]arginine into ribulose bisphosphate carboxylase during a 30-minute labeling period at various times in the cell cycle. Adapted from Iwanij *et al.* (1975).

### 4.  *Lipid Synthesis*

The thylakoid membranes are the site of the light reactions of photosynthesis and form a major component of chloroplasts. Therefore, lipid synthesis during the cell cycle should be linked to chloroplast membrane biogenesis specifically and to photosynthetic activity in general.

Armstrong *et al.* (1971), in studies of synchronous cultures of *Chlamydomonas,* found that chlorophyll synthesis occurs approximately from the third to the tenth hour of the light period. Spectinomycin and rifampicin did not affect chlorophyll synthesis while cycloheximide was substantially inhibitory.

Beck and Levine (1977) studied the synthesis of several lipid components of the chloroplast membranes in synchronous cultures of *Chlamydomonas.* Sulfolipid synthesis was measured by the uptake of $^{35}SO_4^{2-}$, phospholipid synthesis by the uptake of $^{32}PO_4^{2-}$, and galactolipid synthesis by the uptake of $H^{14}CO_3^-$ into lipid fractions of isolated chloroplast membranes. The rate of incorporation of $^{35}S$ into sulfolipids is low at the beginning of the light period, markedly increases from hour 3 to hour 7, and then decreases, reaching a minimum by 2–3 hours into the dark period (Fig. 9). Sulfolipid synthesis precedes chlorophyll syn-

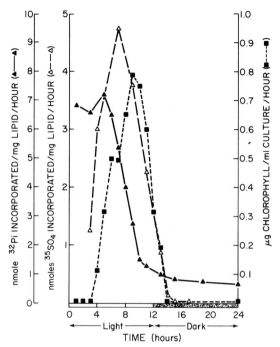

**Fig. 9.** Synthesis of chloroplast membrane lipid components, i.e., [$^{35}$S]sulfolipids, $^{32}$P-labeled phospholipids, and chlorophyll, during the cell cycle. Samples were labeled with $^{35}$SO$_4^{2-}$ or $^{32}$P$_1$ for 1 hour at various times during the cell cycle. Chlorophyll was extracted and then quantitated by absorbance measurements. Adapted from Beck and Levine (1977).

thesis by approximately 2 hours. In contrast, the rate of phospholipid (phosphatidylglycerol) synthesis is nearly maximal at the beginning of the light period, begins decreasing after hour 3 in the light and becomes minimal during the dark period. For galactolipid synthesis, the incorporation of H$^{14}$CO$_3^-$ into both mono-(MGD) and digalactosyldiacylglycerol (DGD) was measured. Both MGD and DGD show two peaks of synthesis in the light period (Fig. 10). Both are synthesized most rapidly at the beginning of the light period and then to less than half this maximum rate by hour 3 or 4. Then both are synthesized at an increasing rate again to a second maximum at hour 7 of the light period and then again at a declining rate. This bimodal synthesis is not due to changes in the cells' incorporation of the isotope during the cell cycle because the specific activity of $^{14}$C-incorporation into chlorophyll remains constant during the light period. As a whole, these findings (Figs. 9 and 10) imply

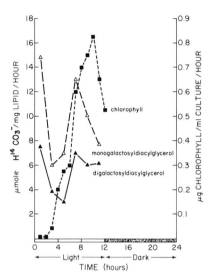

**Fig. 10.** Synthesis of chloroplast membrane galactolipids and chlorophyll during the cell cycle. Samples were labeled with $H^{14}CO_3^-$ for 1 hour at various times during the cell cycle. Chlorophyll was extracted and then quantitated by absorbance measurements. Adapted from Beck and Levine (1977).

that the composition of the chloroplast membranes changes dynamically during the cell cycle with components being made and incorporated into the membrane in a sequential manner.

## 5. Biogenesis of Photosynthetic Membranes

Beck and Levine (1974) examined polypeptide synthesis in chloroplast membranes during synchronous growth of *Chlamydomonas* using [3]H- and [14]C-labeled acetate and one-dimensional sodium dodecyl sulfate gel electrophoresis. They found that the rate of labeling of a group of chloroplast membrane polypeptides (molecular weight, 50,000–55,000 daltons) is highest during the early light period and then decreases after the sixth hour of light. Another polypeptide with molecular weight of 40,000 daltons shows nearly constant labeling during the light. Two other polypeptides with molecular weights of 31,000 and 27,000, respectively, show little synthesis before hour 3 of the light period and then an increasing synthesis until hour 10 of the light period. The authors did not determine the isotopic labeling of individual polypeptides in the dark, but they did show that the majority of the membrane protein complement was labeled with radioactive acetate during the light period.

In contrast to the results of Beck and Levine (1974), Bourguignon and

Palade (1976) found that the major chloroplast membrane polypeptides were labeled to a greater extent in the middle of the dark period than at other times in the cell cycle. Although Bourguignon and Palade used an experimental approach similar to that of Beck and Levine, they used a different radioactive precursor, i.e., arginine. It is quite probable that the differences in the labeling patterns between the two studies are due to the different precursors used. The permeability of the cell and the sizes of its pools may not only vary during the cell cycle but also be different for different precursors (Iwanij *et al.*, 1975).

The single large chloroplast of *Chlamydomonas* occupies a large portion of the cell's volume. Both cell size and chloroplast volume increase substantially during the light period. Therefore, this period would appear to be the most likely time for the synthesis of chloroplast membranes to occur. If so, the question then is whether or not the membrane proteins are synthesized in the dark and inserted into the membranes in the light when lipid synthesis appears to be taking place (Beck and Levine, 1977). If the answer to this question is no, then it would appear that membrane proteins are synthesized and inserted into the membrane concurrently with the lipids. Chua *et al.* (1973) provide evidence suggesting that concurrent synthesis does indeed take place. Some membrane proteins appear to be synthesized within the chloroplast on 70 S ribosomes (Hoober *et al.*, 1969; Hoober, 1970; Eytan and Ohad, 1970; Armstrong *et al.*, 1971). Chua *et al.* (1973) found polyribosomes bound to chloroplast membranes of *Chlamydomonas* at hour 4 of the light period, a time when lipids are synthesized and inserted into these membranes. They suggested that these membrane-bound polyribosomes synthesize thylakoid membrane proteins and that the nascent polypeptides are inserted directly into the membranes. Thus, the membrane proteins would be synthesized and inserted into the membranes concurrently with membrane lipids.

### 6. *Photosynthetic Activity*

If chloroplast membranes are being synthesized during the light period of the cell cycle, then it is of interest to determine whether or not photosynthetic activity is correlated with the insertion of specific components into the membrane. Schor *et al.* (1970) measured variations in the activity of photosystems I (PS I) and II (PS II) in chloroplast thylakoid membranes of *C. reinhardtii* as well as the cellular content of c-559 + c-553, c-563, and chlorophyll during the light:dark cycle. The ratios of c-563 and of c-559 + c-553 to chlorophyll and of PS I and PS II activity to chlorophyll increase during the first 4 to 6 hours of the light period (Fig. 11). The chlorophyll content of the cells remains constant for at

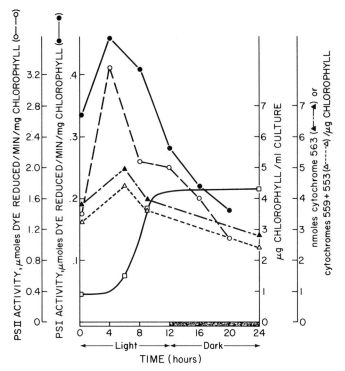

**Fig. 11.**  Specific activity of PS I and PS II and the amounts of cytochromes per milligram chlorophyll during the cell cycle. PS I activity was measured by the photoreduction of methyl red and PS II activity by the photoreduction of dichloroindophenol. Cytochrome 563 was measured by the dithionite-reduced minus ascorbate-oxidized difference spectrum and C-559 + C-553 by the ascorbate-reduced minus ferricyanide-oxidized difference spectrum. Adapted from Schor *et al.* (1970).

least the first 3 hours of the light period. Therefore, at least during the early portion of the light period, the change in the ratio of these cytochromes to chlorophyll must be interpreted as an insertion of additional cytochrome into the chloroplast membranes. There is a decrease in this ratio during the dark period when chlorophyll content in the culture again is nearly constant. In the dark then, there is an apparent loss of c-563 and of c-559 + c-553 from the chloroplast membranes and this loss is associated with a decrease in PS I and PS II activities. From these measurements it appears that there is a correlation between photosynthetic activity and the insertion of cytochromes into the chloroplast membrane. However, from a number of determinations throughout the cell cycle, Schor *et al.* (1970) concluded that PS II activity correlates more closely with the amount of c-559 + c-553 present than with the amount of c-563 present.

Armstrong *et al.* (1971) also measured the rate of oxygen evolution per unit culture of synchronized *Chlamydomonas*. This rate remained constant in the early part of the light period to hours 5 or 6, increased in a linear fashion to hour 11 of the light period, and then remained constant throughout the dark period. These findings indicate that photosynthetic activity reaches a maximum once the membranes have attained a full complement of components (Figs. 9 and 10).

### E. Chloroplast Division during the Cell Cycle

Division of the single, large, cup-shaped chloroplast of *Chlamydomonas* has been studied in asynchronous cultures using electron microscopy (Goodenough, 1970). Chloroplast division takes place immediately after nuclear division has occurred and cytokinesis has been initiated by a cleavage furrow. The chloroplast constricts in the plane of the cleavage furrow eventually forming two daughter chloroplasts. The daughter chloroplasts are subsequently separated from one another as the cleavage furrow passes between them. This process can occur two or more times as the mother cell divides into four or more daughter cells.

### F. Summary

Cell division, in cultures of *Chlamydomonas* synchronized with light: dark cycles, is confined to the dark period and is immediately preceded by the replication of nuclear DNA (Fig. 12). As is the case with *Euglena* (Section II,A,2), one light:dark cycle is considered the equivalent of one cell cycle for *Chlamydomonas*. A variety of temperature-sensitive conditional cell cycle mutants have been identified which are blocked at various points throughout the cell cycle. Chloroplast DNA replicates in the early to midportion of the light period, a time distinct from that when nuclear DNA replicates (Fig. 12). However, as is the case with *Euglena*, some chloroplast DNA synthesis occurs throughout the cell cycle. The nature of this additional organelle DNA synthesis is unknown. The single chloroplast divides immediately after the nucleus divides.

Over the cell cycle, the entire chloroplast genome (one strand of DNA) appears to be transcribed, and at certain times during the cycle, the second strand of DNA also appears to be partially transcribed. The complexity of chloroplast-coded RNA transcripts is lowest early in the light period, increases through the light period, and is maximal early in the dark period just before cell division occurs. Chloroplast ribosomal

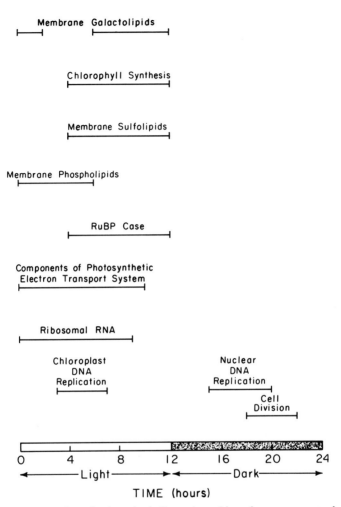

**Fig. 12.** Periods of maximal synthesis for various chloroplast components during the cell cycle of *C. reinhardtii.* Periods of synthesis are approximated from the results cited in Section III.

RNA is mainly synthesized during the first two-thirds of the light period (Fig. 12). Therefore, it appears that the chloroplast is programmed for protein synthesis immediately after cell division and that its capacity for protein synthesis increases through the light period and into the early dark period.

During replication of the chloroplast, the formation of membranes appears to occur by the sequential addition of constituent components during the light period of the cell cycle. Phospholipids and galactolipids are synthesized first (Fig. 12). Concommittantly with the early synthesis of these lipids, components of the electron transport system are synthesized. Next, sulfolipids, more galactolipids, and chlorophyll are synthesized. It is of interest to note here that a major function of galactolipids and sulfolipids is thought to be the stabilization of lipoprotein membranes (Weier and Benson, 1966). Some synthesis of the stromal enzyme, ribulose bisphosphate carboxylase occurs throughout the cell cycle. However, the rate of synthesis of this enzyme greatly increases (Fig. 8) after membrane formation has started (Fig. 12). Through the cell cycle, PS II activity correlates with the amount of c-559 + c-553 present in the chloroplast membranes. Overall photosynthetic activity correlates with the insertion of cytochromes into the chloroplast membranes and is maximal when the membranes have been fully formed.

## IV. *CHLORELLA* AND *SCENEDESMUS*

### A. Synchronization of Cell Division

*Chlorella elipsoidea,* grown autotrophically, may be synchronized by the use of light–dark cycles. Tamiya *et al.* (1953) obtained synchronous cultures by first growing cells to a stage at which they were mainly small in size. The cells were then harvested, subjected to a cycle of 35 hours of light and 17 hours of dark, and subsequently to cycles of 17 hours light and 9 hours dark. Cell division was synchronized and was confined to the dark period.

Sorokin (1957) synchronized cultures of a thermophilic strain of *C. pyrenoidosa*. A random culture was grown under conditions of low light intensity at 25°C for 3 to 4 days. The temperature was then raised to 38.5°C, the light intensity was increased, and the culture was subjected to a light–dark cycle of 9:9 hours. After three to four such cycles, almost complete synchrony was obtained. Hare and Schmidt (1968) modified this technique by using continuous dilution to provide cells with a nearly constant environment during synchronous growth. Pirson and Lorenzen (1958) used alternating light:dark cycles of 14:10 hours and obtained synchronous cultures of *C. pyrenoidosa* after two to three such cycles. In the dark, *Chlorella* divides into four or eight autospores depending upon the intensity of light used during the light cycle.

Alternating light:dark cycles also have been used to synchronize cultures of *Scenedesmus obtusiusculus* (Das, 1973) with cell division occurring in the dark period.

As with *Euglena* (Section II,A,2) and *Chlamydomonas* (Section III,A), one light:dark cycle is considered the equivalent of one cell cycle for *Chlorella* or *Scenedesmus*.

## B.  Chloroplast Biogenesis during the Cell Cycle

### 1.  RNA Synthesis

Galling (1975) has studied the synthesis of ribosomal RNA in synchronized *C. pyrenoidosa*. RNA was isolated from synchronized cells at different stages during the cell cycle. Total RNA synthesis in synchronized *C. pyrenoidosa* increases in an almost linear manner per unit of culture throughout the cell cycle (Galling, 1975). Similar results also were obtained by Senger and Bishop (1966). Compared to cytoplasmic 17 S ribosomal RNA, relatively more chloroplast 16 S ribosomal RNA per unit of culture is synthesized during the first half of the light period (see Fig. 13). After this, the relative amount of chloroplast 16 S ribosomal RNA declines reaching a minimum after 6 hours in the dark period.

Hirai *et al.* (1979) also have studied the synthesis of chloroplast and cytoplasmic ribosomal RNAs during the cell cycle of synchronized *C. ellipsoida*. Net incorporation of labeled uridine or uracil into and specific activity of chloroplast 23 S + 16 S and cytoplasmic 25 S + 18 S ribosomal RNAs were measured. Net incorporation into chloroplast ribosomal RNA was greatest early in the light phase and then declined throughout the rest of the cell cycle. These results are similar to the measurements of Senger and Bishop (1966) and Galling (1975) already described. However, some chloroplast ribosomal RNA synthesis does occur throughout the cell cycle (Galling, 1973; Hirai *et al.*, 1979). Net incorporation of isotopic precursors into cytoplasmic ribosomal RNA increases throughout the cell cycle.

### 2.  DNA Synthesis

Stange *et al.* (1962) found that *C. pyrenoidosa*, synchronized on a light:dark cycle of 18:18 hours, synthesized total DNA around the onset of the dark phase. Senger and Bishop (1966) found essentially the same result with *C. pyrenoidosa* synchronized on a light:dark cycle of 16:12 hours. In the same species synchronized on a light:dark cycle of 16:10 hours, nuclear DNA is synthesized between 10 and 18 hours after the beginning of the light period (Wanka *et al.*, 1970).

DNA synthesis during the cell cycle of strain 211/8b of *Chlorella* is shown in Fig. 14. Replication of chloroplast DNA begins early in the cell

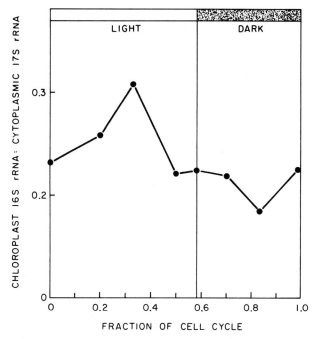

**Fig. 13.** Ratio of chloroplast 16 S rRNA to cytoplasmic 17 S rRNA per unit of culture during the cell cycle of *C. pyrenoidosa*. One light:dark cycle equals 24 hours. Adapted from Galling (1975).

cycle and is completed in the light period (Dalmon *et al.*, 1975). However, a low but persistent synthesis of chloroplast DNA occurs even in the dark period of the cell cycle. Nuclear DNA synthesis commences around hour 10 of the cycle and continues for 4 to 7 hours depending on the strain used.

### 3. Photosynthetic Activity

The photosynthetic activity of synchronized *C. ellipsoidea* is maximal in the middle of the light period and minimal in the middle of the dark period (Tamiya, 1966). Similar results have been obtained for other species of *Chlorella* (Pirson and Lorenzen, 1966) and for *Scenedesmus* (Senger and Bishop, 1969).

PS I activity of *S. obliquus* remains constant during the cell cycle, photosynthetic activity parallels the changes in PS II activity (Senger and Bishop, 1967), and the Emerson enhancement effect (Fig. 15) changes in a manner similar to both PS II activity and photosynthetic activity (Senger and Bishop, 1969).

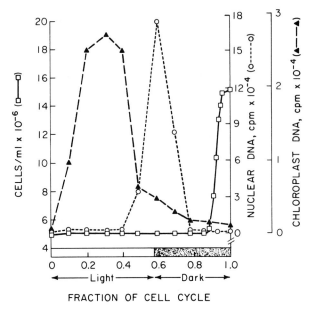

**Fig. 14.** The synthesis of chloroplast and nuclear DNA and the increase in number of cells per milliliter of *C. pyrenoidosa* strain 211/8b during one light:dark cycle. Adapted from Dalmon *et al.* (1975).

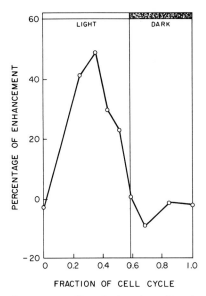

**Fig. 15.** Emerson enhancement effect during the cell cycle of *S. obliquus*. Adapted from Senger and Bishop (1969).

During the cell cycle of synchronized *C. pyrenoidosa,* the ratio of PS I:PS II activities, as measured in preparations from broken cells, is inversely proportional (Fig. 16) to the ratio of the corresponding low temperature pigment emission signals from these systems in intact cells (Nelle *et al.,* 1975). These results were interpreted as indicating that only partially active chlorophyll was present at certain stages of the cell cycle. Also, the activities of both PS I and PS II varied throughout the cell cycle in contrast to *Scenedesmus* as discussed above. Both activities were maximal in the middle of the light phase (Fig. 16).

The relative quantum efficiency of photosynthesis changes during the cell cycle of *C. ellipsoidea* (Nihei *et al.,* 1954), *C. pyrenoidosa* (Sorokin and Krauss, 1961), and *S. obliquus* (Senger and Bishop, 1967). The highest relative quantum efficiency occurs in the middle of the light period and the lowest in the middle of the dark period.

Molloy and Schmidt (1970) measured RuBPCase activity during the cell cycle of synchronized *C. pyrenoidosa.* The cells were grown at a low light intensity (5,400 lx) and at a high light intensity (11,800 lx). Cells grown at the low light intensity divide into four cells. In this case, RuBP-Case activity per milliliter of culture increases during the light period and remains constant during the dark period. Cells grown at the high

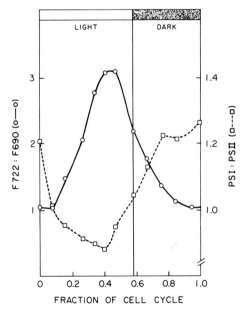

**Fig. 16.** The ratio of PS I:PS II activities of chloroplast particles and the ratio of PS I/PS II pigment fluorescence (F722:F690) during the cell cycle of *C. pyrenoidosa.* Adapted from Nelle *et al.* (1975).

light intensity divide into eight cells. In this case, RuBPCase activity per milliliter of culture increases during the light period and continues to increase during the dark period.

Stange *et al.* (1960) measured the incorporation of photosynthetically fixed $^{14}CO_2$ into products during the cell cycle of synchronized *Chlorella*. The only major difference noted during the cell cycle occurred in the incorporation of $^{14}C$ into sucrose and alanine. Alanine was most heavily labeled with $^{14}C$ during the early part of the cell cycle whereas sucrose was most heavily labeled during the stage of cell division in the dark period.

### 4. Protein Synthesis

The rates of total protein synthesis and of dark fixation of $^{14}CO_2$ (Galloway *et al.*, 1974) correlate with one another throughout most of the cell cycle of synchronized *C. fusca* with the exception of the latter part of the dark period (Fig. 17).

The levels of several different biosynthetic enzymes involved in the ribonucleotide biosynthetic pathway(s) and the deoxythymidine triphosphate pathway of the cell have been measured during the cell cycle of synchronized *C. pyrenoidosa* (Cole and Schmidt, 1964; Johnson and Schmidt, 1966; Shen and Schmidt, 1966; Schmidt, 1969). However, these studies did not measure biosynthetic events in the chloroplast specifically.

Adler (1976) studied the synthesis of chloroplast polypeptides during

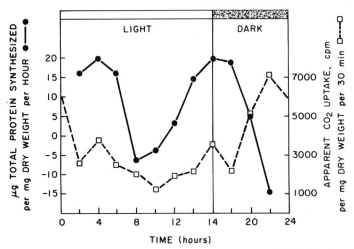

**Fig. 17.** The rates of total cellular protein synthesis and of dark fixation of $^{14}CO_2$ during the cell cycle of synchronized *C. fusca*. Adapted from Galloway *et al.* (1974).

the first 5 hours of the cell cycle of synchronized *C. pyrenoidosa.* Cells were labeled for 15 minutes with [$^3$H]leucine at different times after the start of the light period. Chloroplast membranes were isolated and their constituent polypeptides were resolved on polyacrylamide gels. There was an increased rate of leucine incorporation into chloroplast membranes between 4 and 5 hours into the light period. Three groups of membrane polypeptides were distinguished on the basis of their rates of synthesis: (1) those whose rates of synthesis showed only minor fluctuations during the first 5 hours of the cell cycle; (2) those whose rates of synthesis increased; and (3) those whose rates of synthesis declined. In particular, peptides with molecular weights corresponding to the peptides found in chlorophyll–protein complex I (CP I) and in chlorophyll–protein complex II (CP II) showed increased rates of synthesis after 4 hours of illumination. The labeling of the peptide corresponding to CP I then declined at 5 hours whereas the peptide corresponding to CP II continued to increase.

## C. Chloroplast Division during the Cell Cycle

The single chloroplast of *C. fusca* divides during cytokinesis with chloroplast division being mediated by the developing septum after the nucleus has divided (Atkinson *et al.,* 1974). In *S. obtusiusculus,* the chloroplast divides similarly being constricted by the developing cell plate (Das, 1975). Nilshammar and Walles (1974), however, reported that in some cells of synchronized *S. obtusiusculus* chloroplast division occurs prior to nuclear division, whereas in other cells it occurs at a later stage.

## D. Starch Formation and the Pyrenoid

In *Scenedesmus,* the pyrenoid does not divide but disappears between the first and second nuclear divisions (Bisalputra and Weier, 1964). In *Chlorella,* the single chloroplast of each autospore contains a small protein body which develops into the pyrenoid during the subsequent light period (Wanka, 1975). Each autospore also contains some small starch grains in its chloroplast. During the last 6 hours of the next dark period, these starch grains disappear at the same time as starch layers are formed around the pyrenoid.

During the cell cycle of synchronized *Chlorella,* the main accumulation of starch in the pyrenoid occurs between hours 4 and 12 of the light period. Subsequently, this starch is rapidly degraded. Starch grains, separately from the starch in the pyrenoid, begin to form at 4 hours into the light period. The main accumulation of these grains in the cell occurs between hours 8 and 16. These latter grains then disappear during the dark phase.

### E.   Summary

When *Chlorella* or *Scenedesmus* are synchronized on light:dark cycles, cell division occurs in the dark period. In each case, one light:dark cycle is the approximate equivalent of one cell cycle as already noted for *Euglena* and *Chlamydomonas* (Sections II,A,2 and III,A). Maximal chloroplast DNA synthesis in *Chlorella* occurs in the light period well before cell division occurs (Fig. 18). However, as is the case with *Euglena* (Sections II,B,2 and II,C) and *Chlamydomonas* (Sections III,D,1 and III,E), chloroplast DNA synthesis continues throughout the cell cycle. The period of maximal synthesis likely represents the period of chloroplast DNA replication (Fig. 18), but the nature of the additional organelle DNA synthesis remains obscure.

In *Chlorella*, chloroplast ribosomal RNA is synthesized to the largest

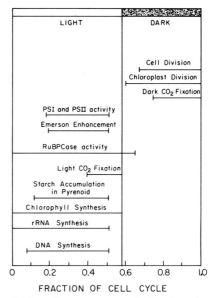

**Fig. 18.**   Formation of chloroplast components during the cell cycle of *Chlorella*. Shown are the times of maximal chloroplast DNA and rRNA synthesis, maximal light and dark $CO_2$ fixations, maximal Emerson enhancement, and maximal PS I and PS II activities, the period of chlorophyll synthesis, the period of starch accumulation in the pyrenoid, the time of increasing RuBPCase activity, and the times of chloroplast and cell divisions.

extent from the early to the late light period (Fig. 18). The rate of synthesis of chloroplast membrane proteins is high in the early portion of the light period (Section IV,B,4). After this time, a period of maximal activity of PS I and PS II begins. Concomitant with the time of maximal activity of the two photosystems, the Emerson enhancement effect is maximal. Toward the end of the light period, light fixation of $CO_2$ is optimal. RuBPCase activity increases through the light period and into the dark period, but dark fixation of $CO_2$ is greatest in the latter half of the dark period when chloroplast and cell division are well underway. As a whole the data suggest that the single chloroplast of *Chlorella* becomes programmed for replication shortly after cell division is completed. The continuous synthesis of chlorophyll throughout the light period further suggests that chloroplast replication is completed only shortly before the organelle divides.

The pattern of photosynthetic activity of *Scenedesmus* during its cell cycle is similar to that of *Chlorella* (Section III,B,3). However, photosynthetic activity in *Scenedesmus* appears more closely related to the activity of PS II than to PS I, whereas in *Chlorella* it appears related to the activities of both PS I and PS II.

Starch, which accumulates in the pyrenoid during the light period (Fig. 18), is rapidly degraded thereafter. The single chloroplast of *Chlorella* and *Scenedesmus* divides after the nucleus divides. The septum or cell plate which develops during cytokinesis mediates the division of the chloroplast.

## V. CONCLUSIONS

A cycle of replication and division of chloroplasts, proposed on the basis of the studies considered in this chapter, is given in Figs. 19 and 20. Cell cycles, generalized from the same studies, are also given in these figures. The lengths of the individual phases of the chloroplast cycles of *Euglena, Chlorella,* and *Chlamydomonas* differ from each other. The period of chloroplast DNA synthesis in *Euglena* or *Chlamydomonas* is shorter than in *Chlorella*. Chloroplast division in *Euglena* occurs in a shorter period of time than it does in either *Chlamydomonas* or *Chlorella*. This difference in chloroplast division times probably occurs because a synchronized *Euglena* cell divides only once into two cells, whereas a synchronized *Chlamydomonas* or *Chlorella* cell divides more than once into four or more autospores. Since cell division occurs repeatedly in synchronized *Chlorella* or *Chlamydomonas,* so does chloroplast division. However, in all three cell types, chloroplast DNA synthesis occurs well

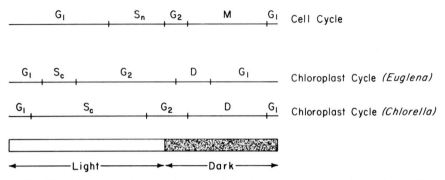

**Fig. 19.** Relationship between the cell cycle and a proposed chloroplast cycle in cells synchronized for cell division on light:dark cycles of 14:10 hours. One light:dark cycle represents one complete cell cycle. D, time of division of chloroplasts; M, time of mitosis; $S_n$, time of synthesis of nuclear DNA; $S_c$, time of synthesis of chloroplast DNA; $G_1$, $G_2$ = times before and after, respectively, of DNA synthesis.

before nuclear DNA synthesis. Thus, the syntheses of these two genomes are not closely coordinated in time.

Chloroplast DNA synthesis also occurs well before chloroplast division occurs (Figs. 19 and 20). Indeed, the period of chloroplast replication, in contrast to chloroplast division, encompasses the entire light period and the early portion of the dark period. In *Euglena,* chloroplast replication even begins about halfway through the dark period before mitosis in a synchronized culture is completed. However, since the time of division of individual *Euglena* cells varies in the dark period, the time of mitosis probably also varies. Thus, in the *individual Euglena* cell, chloroplast replication most likely begins as soon as mitosis is completed.

Photosynthetic activity is controlled by dark reactions and not by light

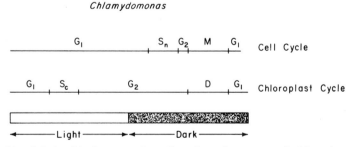

**Fig. 20.** Relationship between the cell cycle and a proposed chloroplast cycle in *Chlamydomonas* synchronized for cell division on light:dark cycles of 12:12 hours. One light:dark cycle represents one complete cell cycle. Symbols as in Fig. 19.

reactions, at least in *Euglena*. In *Euglena, Chlorella,* and *Chlamydomonas,* photosynthetic activity does not correlate with chlorophyll synthesis. However, photosynthetic activity does correlate with PS II activity in *Chlamydomonas* and in *Scenedesmus* and with both PS I and PS II activities in *Chlorella*. In any case, the highest levels of photosynthetic activity in any of the synchronized cells studied occurs when the chloroplast membranes have attained a full complement of their constituent components.

As a whole, the data reviewed in this chapter indicate that chloroplasts are programmed for protein synthesis immediately after mitosis (cell division) and their capacity for protein synthesis increases in the light period. During replication, membranes appear to be formed by the sequential addition of constituent components. Chloroplast replication apparently is completed just before the organelle divides. The precise metabolic events which trigger chloroplast division remain unknown, however. Further, there are no published cytogenetic studies which have investigated the segregation and distribution of organelle genomes at the time of organelle division. Such studies are necessary to define more accurately a chloroplast cycle such as depicted in Figs. 19 and 20.

As stated in the Introduction, chloroplasts are semiautonomous structures which, in order to develop (replicate), rely on transcripts from both the nuclear genome and their own genome. The molecular mechanism(s) underlying this shared responsibility for chloroplast replication has not been defined. Further, whether or not the nucleus plays any role in the events comprising chloroplast division is completely unknown. Much research remains to be done.

## REFERENCES

Adler, K. (1976). *Plant Sci. Lett.* **6,** 261–266.
Akoyonoglou, G., and Argyroudi-Akoyonoglou, J. H., eds. (1978). "Chloroplast Development." Elsevier/North-Holland, Amsterdam.
Andrews, T. J., Lorimer, G. H., and Tolbert, N. E. (1973). *Biochemistry* **12,** 11–18.
Armstrong, J. J., Surzycki, S. J., Moll, B., and Levine, R. P. (1971). *Biochemistry* **10,** 692–701.
Atkinson, A. N., John, P. L. C., and Gunning, B. E. S. (1974). *Protoplasma* **81,** 77–109.
Bassham, J. A., and Calvin, M. (1957). "The Path of Carbon in Photosynthesis." Prentice-Hall, Englewood Cliffs, New Jersey.
Baumgartel, D. M., and Howell, S. H. (1977). *Biochemistry* **16,** 3182–3189.
Beck, D. P., and Levine, R. P. (1974). *J. Cell Biol.* **63,** 759–772.
Beck, J. C., and Levine, R. P. (1977). *Biochim. Biophys. Acta* **489,** 360–369.
Ben-Shaul, Y., Schiff, J. A., and Epstein, H. T. (1964). *Plant Physiol.* **39,** 231–240.
Bernstein, E. (1960). *Science* **131,** 1528–1530.
Bertaux, O., Moyne, G., LaFarge-Frayssinet, C., and Valencia, R. (1978). *J. Ultrastruct. Res.* **62,** 251–269.

Bisalputra, T., and Weier, T. E. (1964). *Am. J. Bot.* **51**, 881–892.

Blakely, R. L. (1969). "The Biochemistry of Folic Acid and Related Pteridines." Elsevier, New York.

Blum, J. J., and Buetow, D. E. (1963). *Exp. Cell Res.* **29**, 407–421.

Boasson, R., and Gibbs, S. P. (1973). *Planta* **115**, 125–134.

Bourguignon, L. Y. W., and Palade, G. E. (1976). *J. Cell Biol.* **69**, 327–344.

Brandt, P. (1975). *Planta* **124**, 105–107.

Brandt, P. (1976). *Planta* **130**, 81–83.

Brawerman, G., and Chargaff, E. (1960). *Biochim. Biophys. Acta* **37**, 221–229.

Bre, H.-H., and Diamond, J. (1975). *In* "Les Cycles Cellulaires et Leur Blocage chez Plusieurs Protistes" (M. Lefort-Tran and R. Valencia, eds.), Colloq. Int. C.N.R.S. No. 240, pp. 345–348. Éditions du C.N.R.S., Paris.

Bre, M.-H., Diamond, J., and Jacques, R. (1975). *J. Protozool.* **22**, 432–434.

Calvayrac, R. (1972). Thesis, C.N.R.S. No. A.O. 7183, University of Paris 6, Orsay, France.

Calvayrac, R., and Ledoight, G. (1975). *In* "Les Cycles Cellulaires et Leur Blocage chez Plusieurs Protistes" (M. Lefort-Tran and R. Valencia, eds.), Colloq. Int. C.N.R.S. No. 240, pp. 47–51. Éditions du C.N.R.S., Paris.

Calvayrac, R., and Lefort-Tran, M. (1976). *Protoplasma* **89**, 353–358.

Cattolico, R. A., Senner, J. W., and Jones, R. F. (1973). *Arch. Biochem. Biophys.* **156**, 58–68.

Chelm, B. K., and Hallick, R. B. (1976). *Biochemistry* **15**, 593–599.

Chiang, K.-S., and Sueoka, N. (1967). *Proc. Natl. Acad. Sci. U.S.A.* **57**, 1506–1513.

Chollet, R., and Ogren, W. L. (1975). *Bot. Rev.* **41**, 137–179.

Chotkowska, E., and Konopa, J. (1973). *Arch. Immunol. Ther. Exp. Engl. Transl.* **21**, 767–774.

Chua, N.-H., Blobel, G., Siekevitz, P., and Palade, G. E. (1973). *Proc. Natl. Acad. Sci. U.S.A.* **70**, 1554–1558.

Chun, E. H. L., Vaughan, M. H., and Rich, A. (1963). *J. Mol. Biol.* **7**, 130–141.

Codd, G. A., and Merrett, M. J. (1971a). *Plant Physiol.* **47**, 635–639.

Codd, G. A., and Merrett, M. G. (1971b). *Plant Physiol.* **47**, 640–643.

Codd, G. A., and Merrett, M. J. (1971c). *Planta* **100**, 124–130.

Cohen, D., and Schiff, J. A. (1976). *Arch. Biochem. Biophys.* **177**, 201–216.

Cole, F. E., and Schmidt, R. R. (1964). *Biochim. Biophys. Acta* **90**, 616–618.

Cook, J. R. (1961a). *Plant Cell Physiol.* **2**, 199–202.

Cook, J. R. (1961b). *Biol. Bull.* **121**, 277–289.

Cook, J. R. (1963). *J. Protozool.* **10**, 436–444.

Cook, J. R. (1966a). *J. Cell Biol.* **29**, 369–373.

Cook, J. R. (1966b). *In* "Cell Synchrony: Studies in Biosynthetic Regulation" (I. L. Cameron and G. M. Padilla, eds.), pp. 153–168. Academic Press, New York.

Cook, J. R. (1966c). *Plant Physiol.* **41**, 821–825.

Cook J. R. (1971). *Exp. Cell Res.* **69**, 207–211.

Cook, J. R., and Cook B. (1962). *Exp. Cell Res.* **28**, 524–530.

Cook, J. R., and Hunt, W. (1965). *Photochem. Photobiol.* **4**, 877–880.

Cook, J. R., and James, T. W. (1960). *Exp. Cell Res.* **21**, 583–589.

Cook, J. R., and James, T. W. (1964). *In* "Synchrony of Cell Division and Growth" (E. Zeuthen, ed.), pp. 485–495. Wiley, New York.

Cook, J. R., Haggard, S. S., and Harris, P. (1976). *J. Protozool.* **23**, 368–373.

Cramer, M., and Myers, J. (1952). *Arch. Mikrobiol.* **17**, 384–402.

Dalmon, J., Bayen, M., and Gilet, R. (1975). *In* "Les Cycles Cellulaires et Leur Blocage chez Plusieurs Protistes" (M. Lefort-Tran and R. Valencia, eds.), Colloq. Int. C.N.R.S. No. 240, pp. 179–183. Éditions du C.N.R.S., Paris.

Das, G. (1973). *Can. J. Bot.* **51**, 113–120.

Das, G. (1975). *Protoplasma* **84,** 175–180.

Edmunds, L. N., Jr. (1964). *Science* **145,** 266–268.

Edmunds, L. N., Jr. (1965a). *J. Cell. Comp. Physiol.* **66,** 147–158.

Edmunds, L. N., Jr. (1965b). *J. Cell. Comp. Physiol.* **66,** 159–182.

Edmunds, L. N., Jr. (1966). *J. Cell. Physiol.* **67,** 35–44.

Edmunds, L. N., Jr., and Funch, R. R. (1969a). *Science* **165,** 500–503.

Edmunds, L. N., Jr., and Funch, R. (1969b). *Planta* **87,** 134–163.

Epstein, H. T., and Schiff, J. A. (1961). *J. Protozool.* **8,** 427–432.

Eytan, G., and Ohad, I. (1970). *J. Biol. Chem.* **245,** 4297–4307.

Falchuk, K. H., Fawcett, D. W., and Vallee, B. L. (1975a). *J. Cell Sci.* **17,** 57–78.

Falchuk, K. H., Kirshnan, A., and Vallee, B. L. (1975b). *Biochemistry* **14,** 3439–3444.

Freyssinet, G. (1977). *Physiol. Veg.* **15,** 519–550.

Galling, G. (1973). *Biochem. Physiol. Pflanz.* **164,** 575–581.

Galling, G. (1975). *In* "Les Cycles Cellulaires et Leur Blocage chez Plusieurs Protistes" (M. Lefort-Tran and R. Valencia, eds.), Colloq. Int. C.N.R.S. No. 240, pp. 225–231. Éditions du C.N.R.S., Paris.

Galloway, R. A., Rolle, I., and Soeder, C. J. (1974). *Arch. Hydrobiol.* **73,** 1–20.

Goodenough, U. W. (1970). *J. Phycol.* **6,** 1–6.

Grant, D., Swinton, D. C., and Chiang, K.-S. (1978). *Planta* **141,** 259–267.

Gray, P. W., and Hallick, R. B. (1977). *Biochemistry* **16,** 1665–1671.

Hare, T. A., and Schmidt, R. R. (1968). *Appl. Microbiol.* **16,** 496–499.

Heizmann, R. (1976). *Ann. Biol.* **15,** 197–226.

Hess, J. L., and Tolbert, N. E. (1967). *Plant Physiol.* **42,** 371–379.

Hirai, A., Nishimura, T., and Iwamura, T. (1979). *Plant Cell Physiol.* **20,** 93–102.

Hoober, J. K. (1970). *J. Biol. Chem.* **245,** 4327–4334.

Hoober, J. K., Siekevitz, P., and Palade, G. E. (1969). *J. Biol. Chem.* **244,** 2621–2631.

Hotta, Y., Bassel, A., and Stern, H. (1965). *J. Cell Biol.* **27,** 451–457.

Howell, S. H., and Naliboff, J. A. (1973). *J. Cell Biol.* **57,** 760–772.

Howell, S. H., and Walker, L. L. (1977). *Dev. Biol.* **56,** 11–23.

Howell, S. H., Blaschko, W. J., and Drew, C. M. (1975). *J. Cell Biol.* **67,** 126–135.

Howell, S. H., Posakony, J. W., and Hill, K. R. (1977). *J. Cell Biol.* **72,** 223–241.

Hutner, S. H., Bach, M. K., and Ross, G. I. M. (1956). *J. Protozool.* **3,** 101–112.

Iwamura, T. (1966). *Prog. Nucleic Acid Res.* **5,** 133–156.

Iwanij, V., Chua, N.-H., and Siekevitz, P. (1975). *J. Cell Biol.* **64,** 572–585.

Johnson, R. A., and Schmidt, R. R. (1966). *Biochim. Biophys. Acta* **129,** 140–144.

Jones, C. R., and Cook, J. R. (1978). *J. Cell. Physiol.* **96,** 253–260.

Kates, J. R., and Jones, R. F. (1967). *Biochim. Biophys. Acta* **145,** 153–158.

Kirk, J. T. O., and Tilney-Bassett, R. A. E. (1978). "The Plastids." Elsevier/North-Holland, Amsterdam.

Könitz, W. (1965). *Planta* **66,** 345–373.

Kopecka, H., Crouse, E. J., and Stutz, E. (1977). *Eur. J. Biochem.* **72,** 525–535.

LaFarge-Frayssinet, C., Bertaux, O., Valencia, R., and Frayssinet, C. (1978). *Biochim. Biophys. Acta* **539,** 435–444.

Laval-Martin, D. L., Shuch, D. J., and Edmunds, L. N., Jr. (1979). *Plant Physiol.* **63,** 495–502.

Lor, K. L., and Cossins, E. A. (1973). *Phytochemistry* **12,** 9–14.

Lor, K. L., and Cossins, E. A. (1978a). *Phytochemistry* **17,** 659–665.

Lor, K. L., and Cossins, E. A. (1978b). *Phytochemistry* **17,** 1711–1715.

Lövlie, A., and Farfaglio, G. (1965). *Exp. Cell Res.* **39,** 418–434.

Lyman, H., and Srinivas, U. K. (1978). *In* "Chloroplast Development" (G. Akoyonoglou and J. H. Argyroudi-Akoyonoglou, eds.), pp. 593–607. Elsevier/North-Holland, Amsterdam.

Manning, J. E., and Richards, O. C. (1972). *Biochemistry* **11**, 2036–2043.

Mego, J. L., and Buetow, D. E. (1967). *In* "Le Chloroplaste: Croissance et Viellissement" (C. Sironval, ed.), pp. 274–290. Masson, Paris.

Mielenz, J. R., Milner, J. J., and Hershberger, C. L. (1977). *J. Bacteriol.* **130**, 860–868.

Molloy, G. R., and Schmidt, R. R. (1970). *Biochem. Biophys. Res. Commun* **40**, 1125–1133.

Murray, D. R., Giovanelli, J., and Smillie, R. M. (1971). *Aust. J. Biol. Sci.* **24**, 23–33.

Neal, W. K., II, Funkhouser, E. A., and Price, C. A. (1968). *J. Protozool.* **15**, 761–763.

Nelle, R., Tischner, R., Harnischfeger, C., and Lorenzen, H. (1975). *Biochem. Physiol. Pflanz.* **167**, 463–472.

Nelson, E. B., Tolbert, N. E., and Hess, J. L. (1969). *Plant Physiol.* **44**, 55–59.

Nigon, V., and Heizmann, P. (1978). *Int. Rev. Cytol.* **53**, 211–290.

Nihei, T., Sasa, T., Miyachi, S., Suzuki, K., and Tamiya, H. (1954). *Arch Mikrobiol.* **21**, 156–166.

Nilshammar, M., and Walles, B. (1974), *Protoplasma* **79**, 317–332.

Orcival-Lafont, A. M., Pineau, B., Ledoight, G., and Calvayrac, R. (1972). *Can. J. Bot.* **50**, 1503–1508.

Padilla, G. M., and Cook, J. R. (1964). *In* "Synchrony in Cell Division and Growth" (E. Zeuthen, ed.), pp. 521–535. Wiley, New York.

Petropulos, S. F. (1964). *Science* **145**, 268–270.

Pirson, A., and Lorenzen, H. (1958). *Z. Bot.* **46**, 53–67.

Pirson, A., and Lorenzen, H. (1966). *Annu. Rev. Plant Physiol.* **17**, 439–458.

Pogo, A. O., and Arce, A. (1964). *Exp. Cell Res.* **36**, 390–458.

Rawson, J. R. Y., and Boerma, C. L. (1976). *Biochemistry* **15**, 588–592.

Richards, O. C., and Manning, J. E. (1975). *In* "Les Cycles Cellulaires et Leur Blocage chez Plusieurs Protistes" (M. Lefort-Tran and R. Valencia, eds.), Colloq. Int. C.N.R.S. No. 240, pp. 213–221. Éditions du C.N.R.S., Paris.

Sager, R. (1972). "Cytoplasmic Genes and Organelles." Academic Press, New York.

Sager, R., and Ishida, M. R. (1963). *Proc. Natl. Acad. Sci. U.S.A.* **50**, 725–730.

Sampson, M., Katch, A., Hotta, Y., and Stern, H. (1963). *Proc. Natl. Acad. Sci. U.S.A.* **50**, 459–563.

Schantz, R., Salaün, J.-P., Schantz, M.-L., and Duranton, H. (1972). *Physiol. Veg.* **10**, 133–151.

Schiff, J. A. (1973). *Adv. Morphog.* **10**, 265–312.

Schiff, J. A., and Epstein, H. T. (1968). *In* "The Biology of *Euglena*" (D. E. Buetow, ed.), Vol. II, pp. 285–333. Academic Press, New York.

Schmidt, G., and Lyman, H. (1976). *In* "Genetics of Algae" (R. Lewin, ed.), pp. 257–299. Univ. of California Press, Berkeley, California.

Schmidt, R. R. (1969). *In* "The Cell Cycle" (G. M. Padilla, G. L. Whitson, and I. L. Cameron, eds.), pp. 159–177. Academic Press, New York.

Schor, S., Siekevitz, P., and Palade, G. E. (1970). *Proc. Natl. Acad. Sci. U.S.A.* **66**, 174–180.

Senger, H., and Bishop, N. I. (1966). *Plant Cell Physiol.* **7**, 441–455.

Senger, H., and Bishop, N. I. (1967). *Nature (London)* **214**, 140–142.

Senger, H., and Bishop, N. I. (1969). *Nature (London)* **221**, 975.

Shen, S. R.-C., and Schmidt, R. R. (1966). *Arch. Biochem. Biophys.* **115**, 13–20.

Slack, C. R., Hatch, M. D., and Goodchild, D. J. (1969). *Biochem. J.* **114**, 489–498.

Sorokin, C. J. (1957). *Physiol. Plant.* **10**, 659–666.

Sorokin, C., and Krauss, R. W. (1961). *Biochem. Biophys. Acta* **48**, 314–319.

Stange, L., Bennett, E. L., and Calvin, M. (1960). *Biochim. Biophys. Acta* **37**, 92–100.

Stange, L., Kirk, M., Bennett, E. L., and Calvin, M. (1962). *Biochim Biophys. Acta* **61,** 681–695.

Stern, A. I., Epstein, H. T., and Schiff, J. A. (1964). *Plant Physiol.* **39,** 226–231.

Stutz, E., Crouse, E. J., and Graf, L. (1975). *In* "Les Cycles Cellulaires et Leur Blocage chez Plusieurs Protistes" (M. Lefort-Tran and R. Valencia, eds.), Colloq. Int. C.N.R.S., No. 240, pp. 255–260. Éditions du C.N.R.S., Paris.

Tamiya, H., (1966). *Annu. Rev. Plant Physio.* **17,** 1–26.

Tamiya, H., Iwamura, T., Shibata, K., Hase, E., and Nihei, T. (1953). *Biochim. Biophys. Acta* **12,** 23–40.

Tewari, K. K. (1971). *Annu. Rev. Plant Plysiol.* **22,** 141–168.

Valencia, R., and Bertaux, O. (1975). *In* "Les Cycles Cellulaires et Leur Blocage chez Plusieurs Protistes" (M. Lefort-Tran and R. Valencia, eds.), Colloq. Int. C.N.R.S. No. 240, pp. 69–75. Éditions du C.N.R.S., Paris.

Verdier, G., Trabuchet, G., Heizmann, P., and Nigon, G. (1973). *Biochim. Biophys. Acta* **312,** 528–539.

Walfield, A. M., and Hershberger, C. L. (1978). *J. Bacteriol.* **133,** 1437–1443.

Walther, W. G., and Edmunds, L. N., Jr. (1973). *Plant Physiol.* **51,** 250–258.

Wanka, F. (1975). *In* "Les Cycles Cellulaires et Leur Blocage chez Plusieurs Protistes" (M. Lefort-Tran and R. Valencia, eds.), Colloq. Int. C.N.R.S. No. 240, pp. 131–136. Éditions du C.N.R.S., Paris.

Wanka, F., Joosten, H. F. P., de Grip, W. J. (1970). *Arch Mikrobiol.* **75,** 25–36.

Weier, T. E., and Benson, A. A. (1966). *In* "Biochemistry of Chloroplasts" (T. W. Goodwin, ed.), Vol. I, pp. 91–113. Academic Press, New York.

Wilson, R., and Chiang, K.-S. (1977). *J. Cell Biol.* **72,** 470–481.

Zelitch, I. (1972). *Plant Physiol.* **50,** 109–113.

# Events Associated with the Mitosis to G$_1$ Transition in Mammalian Cells

### T. SIMMONS, S. HENRY, and L. D. HODGE

## I. INTRODUCTION

Historically, the life cycle of the eukaryotic cell has been divided into periods based on the recognition of two major events: cell division and the replication of the genetic material. With these two events as landmarks, four phases of the cell cycle have been defined: a time from the completion of mitosis to the initiation of DNA replication called G$_1$; a time of DNA replication called S phase; a time following the completion of DNA replication to the initiation of cell division called G$_2$; and a time of cell division called mitosis. Although these phases of the cell cycle are thought to consist of an ordered series of biochemical and structural

**57**

NUCLEAR–CYTOPLASMIC INTERACTIONS
IN THE CELL CYCLE

events, our knowledge of the precise nature and number of the events, critical for the progression of cells through the cycle, is far from complete. In addition, there is a paucity of information concerning the detailed relationships between biochemical and structural rearrangements, as well as key regulatory phenomena that could influence the progression of cells through the mitotic cycle.

The most easily recognizable nuclear and cytoplasmic rearrangements that occur in cycling cells are the events at the time of cell division. Along with the nuclear dispersal into chromosomal units and division of the cytoplasm, there are significant effects on macromolecular synthesis. Mitosis itself can be thought of as being composed of two transition periods: $G_2$ to metaphase and metaphase to $G_1$. Of the two, the metaphase to $G_1$ transition, is the most readily investigated since large numbers of cells synchronous for metaphase can be prepared with several mammalian cell lines. With HeLa $S_3$ cells, we have routinely been able to obtain $5-10 \times 10^7$ cells of which 90% or more cells are in metaphase. In the following discussion, we shall review structural and biochemical events in the cell cycle from the perspective of the mitosis to $G_1$ transition and, in addition, we shall describe data representative of our current studies of cytoplasmic and nuclear events. The unifying point of view utilized throughout this discussion is the attempt to correlate visualized structural changes in dividing cells with possible biochemical mechanisms underlying these changes. It is with this bias in mind that we have organized, presented, and speculated about the available information concerning events at mitosis with regard to the metaphase–$G_1$ transition. This bias will also be evident in the experiments that will be described since a conscious effort has been made to relate, where possible, the biochemical data to structural alterations visualized in individual cells. In order to do this, we have relied on the technique of electron microscope autoradiography. It is hoped that by this dual approach significant insights will be obtained about the complex relationships that exist between structural and biochemical alterations. It is also conceivable that by studying individual events and structure–function relationships in a defined period of the cell cycle, new insights into regulatory phenomena involved in the control of continuously dividing cells may also be obtained.

It should be noted that other approaches to the study of the underlying basis(es) of the cell cycle may also prove to be productive. In fact, given the current status of our knowledge of the cell cycle and its control, a variety of approaches are both necessary and advisable. One approach, the isolation of conditional-lethal mutants, has reinforced the concept of unique regulatory events and has defined putative key transition points, particularly in the $G_1$ phase. To date, however, there seems to be a lack of useful regimens for the induction of a diversity of mutations with respect to specific times in the cell cycle or individual cell cycle processes. Thus, the majority of cell cycle mutants are defined as having lesions of unknown biochemical origins which lead to the arrest of cells somewhere in G$_1$. As long as this remains the case, it will not be possible to fully exploit this genetic approach in the analysis of the cell cycle and its control.

A second, currently popular approach to the study of the cell cycle, involves the use of quiescent cells stimulated to proliferate as a means of probing the so-called G$_0$/G$_1$ to S transition. Leaving aside arguments as to the reality of a defineable G$_0$ state, these studies have been useful in providing evidence for the possible existence of a cell cycle control point(s) in G$_1$, as well as insights into biochemical events in the progression of cells to and into S phase. One major drawback is the relatively poor degree of synchrony obtained following the induction of proliferation. This is evident by the broad variation in time over which cell division occurs. Thus, in this system a precise study of cell structure and function at mitosis would not be possible and, in fact, studies of the G$_0$/G$_1$ to S period may also be somewhat compromised by the degree of synchrony obtainable with quiescent cells stimulated to proliferate.

Although our primary interest in the mitotic cycle of mammalian cells is the mitosis to G$_1$ transition, data concerning other stages of the cell cycle or analogous processes in lower eukaryotes, where pertinent, will also be discussed. Furthermore, since many of the biochemical and structural changes correlated with mitosis apparently represent cyclic alterations, mechanisms suggested for the entrance of cells into mitosis, i.e., the G$_2$ to metaphase transition, will also be reviewed. In view of the many advances in our understanding of the molecular biology and biochemistry of randomly growing cells, it is also important to consider the extent to which this information can be extrapolated to the mitosis to G$_1$ transition. In all cases, our bias is to correlate cell structure and biochemistry in order to form an integrated view of the status of the cell at this point in the mammalian cell cycle. In this way, we should be able to generate plausible mechanisms underlying mitotic events that will then form the basis of future experimental approaches.

## II. STRUCTURAL CHANGES IN THE MITOSIS TO G$_1$ Transition

### A. Visualization of the Rearrangements and Reformation of Preformed Components

The mitosis to G$_1$ transition in mammalian cells can be described as a series of ordered structural events that results in the partition of chromosomes and cytoplasm into two progeny cells along with the reestablishment of an interphase morphology. Structural events of this period in HeLa cells, which are representative of mammalian cells generally, have been described in detail (Robbins and Gonatos, 1964; Ehrlandson and deHarven, 1971). Briefly, the major structural changes are: (1) chromosome movement; (2) breakdown of the mitotic apparatus; (3) reformation of the nuclear envelope; (4) chromosome decondensation; (5) reformation of nucleoli; (6) cytokinesis; (7) partitioning of cytoplasmic organelles such as mitochondria and lysosomes into progeny cells; (8) reassembly of cytoplasmic polyribosomes from single ribosomes, and, presumably (9) the reappearance of an interphase cytoplasmic microtubule system and cytoskeleton.

Many of these events of late mitosis can be considered to represent mirror images of the changes that occurred as cells traversed the $G_2$ to metaphase period. Therefore, an important feature of reformation of such structures as the nuclear envelope, nuclear pores, nucleoli, cytoplasmic polyribosomes, and cytoplasmic microtubule system is that their respective reformation makes use of preexisting components; some of which have been visualized by microscopy in dividing cells. Components of the nuclear envelope have been reported to persist in arrays or large sheets, or attached to mitotic chromosomes (Ehrlandson and deHarven, 1971; reviewed in Wunderlich et al., 1976), although the latter has not been rigorously proved. It is clear that the beginning of nuclear envelope reformation in anaphase of HeLa cells is marked by the appearance of recognizable double-membrane fragments on the periphery of the fused chromosome mass, distal to the equator of the dividing cell (Ehrlandson and deHarven, 1971). The nuclear pore complex has likewise been observed to persist in association with metaphase chromosomes (Maul, 1977). In addition, the reassembly of nuclear pore components into the reforming interphase nucleus has been demonstrated, with fluorescent antibodies raised against a lamina–pore complex fraction, to parallel the rearrangements observed for the nuclear envelope in late mitosis (Ely et al., 1978). This result does not necessarily indicate that these rearrangements are linked events since the nuclear envelope does not appear to be essential for either the attachment of the pore complex to the nucleus or the maintenance of its structural integrity (Aaronson and Blobel, 1974; Wunderlich et al., 1976).

Nucleolar remnants have been seen to be associated with chromosomes and specific chromosomes bearing these remnants have been identified in several mammalian cell lines by chromosome banding and molecular hybridization techniques (Ehrlandson and deHarven, 1971; Henderson et al., 1972; Phillip, 1972; deCapoa et al., 1976; Hughes et *al.,* 1979). During late telophase–early G$_1$, the nucleolar fragments become associated with the already reformed nuclear envelope (Ehrlandson and deHarven, 1971). Although the centriole has been reported to be associated, possibly via microfilaments, with the nuclear envelope (Bornens, 1977), the time of initiation of this association has not been studied. It may or may not occur in late mitosis concomitant with envelope reformation. Lastly, based on laser beam destruction of the centriole, it seems that this structure need not be present for the completion of mitosis (Berns and Richards, 1977).

## B. Visualization of Contractile Elements in Mitotic Structures

A major finding in the mitotic apparatus and cleavage furrow of non-muscle cells has been the visualization of actin, myosin, as well as microtubules. The localization of actin, myosin, and tubulin by light and/or electron microscopy in dividing cells has revealed differential distributions of these proteins at least on a quantitative basis. Tubulin has been localized with specific antibody in astral rays, in kinetochore-attached fibers running from chromosomes to the poles, in fibers extending into the interzone between separating chromosomes, and in spindle remnants running between progeny cells (Cande *et al.,* 1977). Although myosin was detected in the mitotic apparatus, it seemed to be concentrated in the cleavage furrow when purified anti-myosin antibodies were used (Fujiwara and Pollard, 1976). These studies require confirmation by other laboratories. In addition to myosin, α-actinin also has been reported to be concentrated in the cleavage furrow (Herman and Pollard, 1979).

The localization of actin in late mitotic cells has been studied by a number of laboratories with differing results (compare Schroeder, 1973, 1976; Sanger, 1975a,b; Cande *et al.,* 1977; Herman and Pollard, 1978, 1979). There is general agreement that actin is present in the mitotic apparatus and contractile ring regardless of the cell type studied or the techniques used. However, major discrepancies exist in reports of the specific localization of actin during late mitosis. While some of the differences were quantitative, contradictory reports also exist concerning its location at particular sites. Specifically, reports differ as to whether actin is concentrated in the cleavage furrow during cytokinesis (Schroeder, 1973; Sanger, 1975b; Sanger and Sanger, 1976); concentrated in the kinetochore (Sanger, 1975a,b; Sanger and Sanger, 1976); localized in the chromosome spindle fibers running from kinetochore to spindle poles during chromosome segregation (Sanger, 1975b; Cande *et al.,* 1977), or localized in the interzone during chromosome segregation (Herman and Pollard, 1978, 1979). As suggested by others (Sanger, 1975b; Cande *et al.,* 1977; Herman and Pollard, 1978, 1979), some of the discrepancies may reflect differences in method. In most of the studies, cells were extracted with glycerol or detergents which, in addition to leading to a loss of 10–40% of cellular actin, may produce artifacts in the distribution of retained actin as well. This in turn would affect the results in fluorescent antibody and heavy meromyosin binding studies. Furthermore, the thickness of cytoplasm at any one point can be a major determinant of fluorescence intensity and, thus, lack of uniformity of this parameter in cell preparations could affect quantitative results. The precision of fluorescent techniques also depends upon minimalization of background fluorescence and the nature of the fluorescent probes. For example, results obtained employing purified anti-actin antibodies or a IgG fraction may have differed simply because of the nature of the immunological probe (Fujiwara and Pollard, 1976).

As of this writing, only one report has appeared in which purified anti-actin antibodies and direct immunofluorescence have been used in combination with a cell fixation procedure that does not involve glycerol or detergents (Herman and Pollard, 1979). Of importance, cells so treated retain 90% of the cellular actin and demonstrate that actin occurs in the interzone spindle region during anaphase and does not concentrate in the cleavage furrow during cytokinesis in telophase. These results were confirmed by decoration with heavy meromyosin (Herman and Pollard, 1979) and supported an earlier electron microscope study of the distribution of actin (Schroeder, 1976).

Thus, cell fixation procedures are of critical importance in studies of actin–myosin localization in dividing cells. The distribution of these macromolecules in mitotic cells has direct bearing on mechanisms that could be hypothesized to be responsible for karyokinesis and cytokinesis.

## III. BIOCHEMICAL CHANGES IN THE MITOSIS TO $G_1$ TRANSITION

### A. Relationship to Structural Changes

Several biochemical events and molecules have been suggested as significant to the progression of cells through mitosis because they display activity or concentration maxima in metaphase cells or just prior to the entry of cells into mitosis, or because they are associated with mitotic cell structures. In general, most of these events or molecules are hypothesized to fall into one of three major categories: (1) inducers of mitotic events such as chromosome condensation; (2) mediators of chromosome movements; and (3) mediators of cytokinesis. Because of the lack of precise synchrony as populations of cells progress from late S phase through $G_2$ into metaphase, the assignment of events or molecules to times just prior to mitosis must be cautiously accepted. Nevertheless, in an attempt to emphasize the possible relationship of biochemical events and molecules to demonstrable structural changes at mitosis, phenomena are grouped for discussion in these categories. In addition, we comment briefly on potential interactions between the various putative mediators of mitosis induction, karyokinesis and cytokinesis, as well as on the underlying bases of three structural events in late mitosis, nuclear envelope reformation, chromosome decondensation, and nucleolar reassembly.

## 1. Inducers of Mitotic Events

It has been hypothesized that one or more labile moieties accumulate, presumably in the cytoplasm, prior to mitosis and when a threshold level is reached, cells enter prophase (Rao *et al.*, 1975). Both cell fusion experiments between mammalian cells at different phases of the cell cycle and nuclear transplantation experiments have implicated a role for a cytoplasmic component(s) in the induction of nuclear envelope breakdown and chromosome condensation (Sperling and Rao, 1974; Rao *et al.*, 1975, 1978; Al-Bader *et al.*, 1978). One external peptide signal, neuronal growth factor, has been reported to induce disorganization of the nucleolus and disruption of the nuclear membrane (Sabatini *et al.*, 1965; Levi-Montalcini *et al.*, 1969; Mobley *et al.*, 1977). In lower eukaryotes, a cytoplasmic protein that regulates nuclear envelope breakdown in cleavage of amphibian oocytes has been reported (Wasserman and Smith, 1978), as has an "unstable" inducer of nuclear division in *Physarum* (Tyson *et al.*, 1979). The latter is either a protein or requires protein synthesis for activation. It would be reasonable to suspect that extracts of mammalian $G_2$ cells should contain evidence for the accumulation of putative inducer molecules, and as judged by the criterion of two-dimensional gel electrophoresis of total cellular protein extracts, the presence of nine $G_2$-specific proteins in the molecular weight range of $4-5 \times 10^4$ daltons has been reported (Al-Bader *et al.*, 1978).

An obligatory characteristic of any putative inductive event or molecule is that it have a relatively short functional half-life, so that the cell would not be in perpetual mitosis. Thus, the loss or inactivation of an inducer molecule(s) or an inductive event(s) could lead to a completion of a cyclic mitotic process such as nuclear envelope breakdown and reformation and in this way contribute to the termination of mitosis. While this type of mechanism suggests that events in late mitosis can occur passively as a by-product of the withdrawal of active components, it seems more likely that critical processes such as karyokinesis, cytokinesis, and the restoration of nuclear and cytoplasmic functions would be under more direct control. Studies such as those that indicate the insertion of chromatin into the nuclear envelope with a defined configuration (Comings and Okado, 1970a,b; Zentgraf *et al.*, 1975), the association of the centriole with the reformed nuclear envelope (Bornens, 1977), and the reutilization of premitotic components to reestablish an interphase nucleus (see Section II,A), all suggest that the attainment of a $G_1$ state is a series of at least finely coordinated, if not controlled, events. Thus, the role of putative inducers of mitosis should also be considered in the context of late mitotic events.

Numerous specific proteins, biochemical reactions, or small molecules have been reported to be characteristic of metaphase or to be involved in the establishment of the metaphase state through events such as the induction of chromosome condensation or the regulation of microtubule polymerization. Several may also have relevance to either karyokinesis or cytokinesis and there exists, as will be indicated in text, numerous possibilities for interactions of individual phenomena on the biochemical level. The events and molecules reported to be relevant to the entrance of cells into mitosis or to metaphase include: the activities of protein kinases (Piras and Piras, 1975; Costa *et al.*, 1977; Costa and Ney, 1978); the phosphorylation of histone H1 (Gurley *et al.*, 1972, 1973, 1975; Marks *et al.*, 1973; Borun *et al.*, 1974; Bradbury *et al.*, 1974a,b); the phosphorylation of tubulin (Piras and Piras, 1975); the alteration of intracellular levels of cyclic AMP and cyclic GMP (Sheppard and Prescott, 1972; Nose and Katsuta, 1975; Al-Bader *et al.*, 1976; Kurz and Friedman, 1976; Zeilig *et al.*, 1976; DiPasquale *et al.*, 1978; Zeilig and Goldberg, 1977; Rochette-Egly and Castagna, 1979); the intracellular levels of $Mg^{2+}$ and $Ca^{2+}$ (Weisenberg, 1972; Timourian *et al.*, 1974; Mazia, 1974, 1975; Fuller *et al.*, 1975; Hayashi and Matsumuro, 1975; Olmsted and Borisy, 1975; Rosenfeld *et al.*, 1976; Zackroff and Weisenberg, 1977; Keller and Rebhun, 1978; Weisenberg and Piazza, 1978); the activity of ornithine decarboxylase (Friedman *et al.*, 1972; Heby *et al.*, 1975, 1976; McCann *et al.*, 1975), and the activities of several $Ca^{2+}$-dependent proteins (Cheung, 1970, 1971; Mazia *et al.*, 1972; Petzelt, 1974; Mazia, 1975; Petzelt and Auel, 1977; Welsh *et al.*, 1978).

*a. Protein Phosphorylation and Mitosis.* The phosphorylation of histone Hl in late $G_2$ and prometaphase has been suggested as a prerequisite for or inducer of the condensation of chromatin into chromosomes

(Marks *et al.*, 1973; Borun *et al.*, 1974; Bradbury *et al.*, 1974b; Gurley *et al.*, 1975), although this role for histone Hl may not be applicable to all eukaryotic cells (Gorovsky and Keevert, 1975). The activities of protein kinases which may be involved have been shown to increase through $G_2$ and peak in mitosis (Piras and Piras, 1975; Costa *et al.*, 1977; Costa and Ney, 1978). Recent data in support of this hypothesis are the observations that specific amino acid residues which are phosphorylated at mitosis are unphosphorylated at other cell cycle stages (Kurochkin *et al.*, 1977) and that histone Hl, phosphorylated by a kinase associated with actively dividing cells, more effectively cross-links calf thymus DNA *in vitro* (Matthews and Bradbury, 1978). Recently, histone Hl has been localized to the DNA region between the nucleosomes of chromatin and shown to form cross-links with histone H2A and H2A-ubiquitin (Goldknopf and Busch, 1977; Bonner and Stedman, 1979). It is conceivable that the site-specific phosphorylation of Hl alters its interaction with other nucleosomal histones to control the degree of compactness of chromatin. In fact, condensed chromatin in interphase cells maintains a basic nucleosome structure without apparent spacer DNA (Olins and Olins, 1979). However, the mechanism by which this phosphorylation could result in the extreme degree of condensation typifying metaphase chromosomes is unclear. At best, the evidence for such an influential role for histone Hl in chromatin–chromosome transformations is indirect. A natural extension of this proposed role for histone Hl is that its dephosphorylation at metaphase-specific sites would induce chromosome decondensation. The process could be progressive in that as more phosphate groups are removed, $G_1$-specific sites reported by Kurochlein *et al.* (1977) would be phosphorylated. Control of this phenomenon would then reside in regulation of the protein kinase(s). As yet, there is little evidence on the possible regulation of these enzymes at mitosis and studies with purified protein kinases may illuminate this point.

The phosphorylation of HeLa cell tubulin *in vivo* is also maximal at metaphase and *in vitro* studies suggested only a weak (10–20%) activation of microtubule or cytosol protein kinase preparations by cAMP (Piras and Piras, 1975). Protein kinase activity has been shown to be tightly bound to and copurify with brain tubulin from a variety of sources and these preparations display wide variations in sensitivity to stimulation by cAMP (Goodman *et al.*, 1970; Soifer, 1973; Eipper, 1974). Rat brain tubulin and protein kinase are resolvable (Eipper, 1974) and, in this form, the kinase is not cAMP dependent which could reflect the loss of a regulatory subunit during the extensive purification procedures (Tao, 1972; Miyamoto *et al.*, 1973; Eipper, 1974). These studies suggest that phosphorylation of tubulin can be involved in polymerization or microtubule rearrangements in the formation of the mitotic apparatus at metaphase. Recently, reduced sulfhydryl groups on tubulin have been reported to be necessary *in vitro* for polymerization (Rebhun *et al.*, 1978). The role of SH groups with regard to *in vivo* events is unknown.

b. *Cyclic Nucleotides and Mitosis.* Low endogenous levels of cyclic AMP are found at mitosis, being minimal in metaphase (Burger *et al.*, 1972; Sheppard and Prescott, 1972; Zeilig *et al.*, 1976; Zeilig and Goldberg, 1977). Exogenously adding cyclic AMP or dibutyryl cyclic AMP to elevate levels of endogenous nucleotides blocked cells in $G_2$ of the cell cycle (Remington and Klevecz, 1973; Nose and Katsuto, 1975; Kurz and Friedman, 1976; Dell'orco *et al.*, 1977). However, once cells are in metaphase, elevated cyclic AMP levels did not inhibit the progression of cells into $G_1$ but actually stimulated it (Zeilig *et al.*, 1976). In contrast, cyclic GMP levels peak in late $G_2$ and mitosis, being maximal at metaphase (Zeilig *et al.*, 1976). In a lower eukaryote, elevated cyclic GMP levels at mitosis were linked to increased activity of ornithine decarboxylase which may be of significance in cytokinesis (Sedory and Mitchell, 1977). As cells enter $G_1$, cyclic AMP levels begin to increase as cyclic GMP levels decrease (Zeilig *et al.*, 1976). Perhaps it is the ratio, rather than the absolute amounts of cyclic AMP and GMP which is important for mitotic events. An analogous cell cycle-related oscillation in the activities of adenyl and guanyl cyclases could also occur, but poor synchrony at mitosis preclude making a more definitive statement (Rochette-Egly and Castagna, 1979). It is conceivable that phosphorylation events at mitosis are regulated in part by cyclic nucleotides via protein kinases. However, the isolation of a variant of the S49 mouse lymphoma cell line, lacking a cyclic AMP protein kinase activity (Daniel *et al.*, 1973), but maintaining a "normal" cell cycle (Coffino *et al.*, 1975), argues against an obligatory role for cyclic AMP-dependent protein kinase activities in cell cycle events.

### 2. *Mediators of Chromosome Movements*

Evidence in this and the following section concerning mediators of cytokinesis is derived largely from *in vitro* and inhibitor studies or from studies that involved perturbations of the normal physiological conditions of the cell. Accepting the limitations inherent in these types of approaches, the results of these investigations do provide some insights into biochemical events underlying the structural changes seen at mitosis.

a. *Divalent Metal Cations.* The intracellular concentrations of divalent metal cations, especially $Ca^{2+}$, have been suggested to act as regulators of mitosis, principally because of effects on microtubule systems

(Weisenberg, 1972; Mazia, 1974, 1975; Fuller *et al.*, 1975). The evidence for such a role for $Ca^{2+}$ ions is diverse, if somewhat indirect. It has been widely observed that elevated $Ca^{2+}$ ion concentrations can prevent tubulin polymerization and lead to microtubule depolymerization *in vitro* and *in vivo* (Weisenberg, 1972; Timourian *et al.*, 1974; Hayashi and Matsumura, 1975; Olmsted and Borisy, 1975; Fuller *et al.*, 1975; Rosenfeld *et al.*, 1976). In the presence of physiological levels of $Mg^{2+}$ ions, micromolar concentrations of $Ca^{2+}$ can prevent brain tubulin polymerization and induce depolymerization of microtubules *in vitro* (Rosenfeld *et al.*, 1976). Depolymerization of microtubules appears to be by an all or none mechanism with the release of soluble tubulin (Weisenberg and Piazza, 1978; Keller and Rebhun, 1978). In addition, *in vitro* studies suggested that $Mg^{2+}$ ions were tightly bound to bovine brain microtubules and were required for polymer bond formation (Zackroff and Weisenberg, 1977). Calcium ions have also been reported to be localized in mitotic spindles (Timourian *et al.*, 1974). Studies with lower eukaryotes have clearly shown that the *in vivo* application of $CaCl_2$ can cause fading of the mitotic apparatus, so that *in vitro* and *in vivo* effects of the ion seem analogous (Kiehard and Inoúe, 1976). Furthermore, the addition of calcium to metaphase HeLa cells accelerated the entrance of cells into $G_1$ (Zeilig *et al.*, 1976). It is tempting to speculate that the latter effect was due to a $Ca^{2+}$ ion–mitotic apparatus interaction and it is possible to hypothesize that a balance between $Ca^{2+}$ and $Mg^{2+}$ ion concentrations is involved in mitotic apparatus assembly and disassembly via effects on microtubules.

*b. $Ca^{2+}$ Ion Regulatory Systems.* Mitosis represents only a 45-minute period in the mammalian cell cycle and in order to support the view that the $Ca^{2+}$ ion concentration is critical in determining the progression of some mitotic events, mechanisms must exist within the cell to rapidly and specifically alter the intracellular concentration of this ion. Two widely distributed systems that may be capable of regulating $Ca^{2+}$ ion concentration *in vivo* and which are localized in microtubules and in the mitotic apparatus have been described; a $Ca^{2+}$-ATPase (Mazia *et al.*, 1972; Petzelt and von Ledebur-Villiger, 1973; Petzelt, 1974; Mazia, 1975; Harris, 1975; Petzelt and Auel, 1977) and a $Ca^{2+}$-dependent regulatory protein (CDR) that appears to be structurally similar to muscle cell tropinin-C (Cheung, 1970, 1971; Kakiuchi *et al.*, 1970; Smoake *et al.*, 1974; Wang *et al.*, 1974; Kakiuchi *et al.*, 1975; Waisman *et al.*, 1975; Stevens *et al.*, 1976; Welsh *et al.*, 1978). Theoretically, by binding $Ca^{2+}$ ions these proteins can sequester intracellular $Ca^{2+}$, thereby lowering the $Ca^{2+}$ ion concentration so that the mitotic apparatus forms and is stable. Chromosome movement to the poles appears to require depolymerization of kineticore-associated microtubules (Fuseler, 1975; Inoúe and Ritter, 1975) so that a role for the release of $Ca^{2+}$ ions by these protein components in karyokinesis may also be proposed. It has also been postulated that the calcium-dependent regulatory protein (CDR) may be functionally, as well as structurally homologous to tropinin-C which regulates muscle ATPase activity (Welsh et al., 1978). Brain CDR protein has been reported to activate a $Ca^{2+}$–$Mg^{2+}$–ATPase from erythrocytes (Jarrett and Pennsiton, 1978). If this were true for the mitotic CDR protein and $Ca^{2+}$–ATPase, the function of both systems would be interrelated in the coordination of events at mitosis.

 *c. Contractile Proteins.* No discussion of the molecular basis of chromosome segregation or cytokinesis would be complete without commenting upon the potential roles of actin, myosin, and microtubules in this process.

 As discussed in Section II,B, although the presence of actin, myosin, and tubulin in the mitotic apparatus and cleavage furrow is well documented, differences in the relative distribution of these molecules at mitosis have been reported which bear directly on presumptive roles in chromosome movements and cytokinesis. The contradictory reports concerning the presence of actin at specific cellular locations during mitosis, in particular, must be resolved. In addition, only one detailed study of the distribution of myosin during mitosis has been reported (Fujiwara and Pollard, 1976) and these results should be further substantiated by other studies. The resolution of these problems is important since inferences on how actin, myosin, and tubulin may be involved in effecting chromosome movements have been based upon their cellular distributions during mitosis. For example, on the basis of actin's absence from the interzone but presence on the chromosome spindles that run to the poles, and tubulin's presence in the interzone, Cande et al. (1977) hypothesized that some movements to the poles depended in part upon actin while spindle elongation depended upon microtubules but not actin. However, if the studies that demonstrated the presence of intense actin-dependent fluorescence in the interzone region prove correct, the existence of these two distinct mechanisms mediating chromosome movements would seem unlikely. Clearly, additional studies are required before a satisfactory explanation(s) for biochemical roles of contractile proteins in mitosis can be formulated. The independent demonstrations of a tropinin C-like protein (Welsh et al., 1978) and a $Ca^{2+}$–ATPase (Petzelt and Auel, 1977) associated with the mitotic apparatus, makes all the more intriguing the hypothesis that an actin–myosin interaction, analogous to that found in muscle cells, has a role in chromosome movements during the mitosis to $G_1$ transition. In lower eukaryotes a role for microtubules but not myosin in chromosome movements has been suggested. In starfish, anti-dynein antibody which effects microtubules and not anti-myosin antibody inhibited chromosome movements in isolated mitotic apparati (Sakai *et al.*, 1976) which suggested that cross-bridged microtubules could be involved. In sea urchins while anti-myosin antibody led to distortion of the mitotic apparatus, karyokinesis occurred (Mabuchi and Okuno, 1977). Thus, although present in the mitotic apparatus, myosin's role in chromosome movements is open to question.

### 3. Mediators of Cytokinesis

The processes that lead to the appearance of a cleavage furrow and cytoplasmic bridge and which ultimately culminate in the production of two, approximately equal-sized progeny cells remain to be clearly defined and analyzed. However, two systems can be discussed which seem at least to be of potential importance to cytokinesis; ornithine decarboxylase-mediated polyamine biosynthesis and contractile proteins.

*a. Ornithine Decarboxylase and Polyamines.* Ornithine decarboxylase mediated polyamine biosynthesis has been linked to cytokinesis in a number of studies of the cell cycle. When the synthesis of polyamines was blocked at mitosis by inhibition of ornithine decarboxylase (ODC), karyokinesis occurred, but cytokinesis did not and binucleate cells formed (Sunkara *et al.*, 1979). The addition of exogenous spermidine was particularly effective in reversing this inhibition of cytokinesis which was correlated to a disorganization of cellular microfilaments (Sunkara *et al.*, 1979). Of the three polyamines, putrescene, spermine, and spermidine, failure to complete cell division was correlated with a fall in the intracellular level of spermidine (Sunkara *et al.*, 1979). Further data supporting the hypothesis that ODC and polyamines are involved in cell division are that: (1) ODC activity displays a peak at or just before mitosis (Friedman *et al.*, 1972; Heby *et al.*, 1975, 1976; McCann *et al.*, 1975); (2) agents such as colchicine which disrupt microtubules and/or microfilaments also inhibit ODC activity (Chen *et al.*, 1976); (3) peaks in ODC activity appear to parallel peaks in cGMP levels in several systems (Zeilig *et al.*, 1976; Sedory and Mitchell, 1977); (4) ODC activity is responsive *in vivo* and *in vitro* to cyclic nucleotides levels (Beck *et al.*, 1972; Richman *et al.*, 1973; Bachrach, 1975; Byus and Russell, 1975); and (5) polyamines bind to tubulin during polymerization *in.vitro* (Behnke, 1975; Jacobs *et al.*, 1975). Note that the link between ODC activity, cyclic nucleotides, and intact microtubules could also place this system under the unifying sphere of influence of the $Ca^{2+}$-dependent proteins. In fact, both calcium and cAMP, as well as asparagine, are required for ODC activation in intact Chinese hamster ovary cells (Costa and Ney, 1978). Localization of ODC activity to the mitotic apparatus would further support its functional relevance to cell division.

   b.  *Contractile Proteins.*  Cytokinesis, as well as chromosome movements, have been hypothesized to involve actin, myosin, and/or microtubules. As discussed in Section II,B, all three have been reported to be located in the cleavage furrow–intracellular bridge regions of dividing cells. However, also as discussed previously (Sections II,B and III,2,A,c), discrepencies or limitations in the published studies of the distribution of these components during late mitosis clearly indicate that further speculation on their roles in mitotic mammalian cells will depend on additional studies of their distribution in dividing cells.

   In lower eukaryotes both actin and myosin have been implicated as important for cytokinesis. In sea urchins, the injection of anti-myosin antibodies into cleavage eggs was primarily seen to inhibit cytokinesis, with little or no effect on karyokinesis (Mabuchi and Okuno, 1977). A role for microfilaments, and thereby presumably actin, was suggested by the observations that cytochalasin B inhibition of cytokinesis was concomitant with the disassembly of microfilaments *in vivo* (Wessells *et al.,* 1971; Schroeder, 1973; Weber *et al.,* 1976). It should be mentioned that actin may interact, at least in lower eukaryotes, with molecules other than myosin to produce movement (Tilney *et al.,* 1973; Tilney, 1975). However, at this point the role(s) of actin/myosin interactions at mitosis remain to be clearly established.

### 4.  Other Late Mitotic Events

   The other major ultrastructural events of late mitosis are nuclear envelope reformation, chromosome decondensation, and nucleolar reassembly. The underlying mechanisms driving these phenomena are unclear. Both the reformation of the nuclear membrane and nucleoli involve preexisting components (Section II,A), and could involve to some degree the process of self-assembly once the proper components are brought into close proximity in late anaphase and telophase. It has been suggested that the extreme condensation of chromatin seen at mitosis, i.e., metaphase chromosomes, represents a maximum in a continuous cycle of chromatin structure that encompasses both interphase and mitosis (Mazia, 1963; Pederson, 1972; Pederson and Robbins, 1972; Nicolini *et al.,* 1975; Schor *et al.,* 1975; Kendall *et al.,* 1977; Rao *et al.,* 1977; Setterfield *et al.,* 1978). The factors mediating the overall cycle are unknown, although both histone Hl phosphorylation and a lowering of intracellular $Ca^{2+}$ ion concentration have been suggested as "triggers" for chromatin condensation at mitosis. While detailed studies, specifically of the structural elements of chromosomes compared to nuclear components in the mitosis to G$_1$ transition, have not been carried out, such studies could provide new insights into the biochemistry of the reestablishment of the interphase nucleus.

### 5. Biochemical Interactions and the Integration of Individual Mitotic Events

There are obviously numerous ways in which the biochemical events and specific molecules that have been implicated in late mitotic events may interact to coordinate these events. In fact, multiple circles of overlapping interactions can be constructed. For example, the protein kinases active in phosphorylation reactions associated with mitosis could prove to be cyclic nucleotide regulated. In this manner these two systems that are thought to be important in the induction of the mitotic state would be linked (see Section III,A,1,a and b). In addition, numerous studies indicate that CDR proteins from a variety of tissues have other functional roles that make these proteins an attractive moiety for the control of diverse biochemical effects associated with mitosis. For example, CDR protein activates 3',5'-cyclic nucleotide phosphodiesterases (Cheung, 1970; Brostrom and Wolf, 1974a,b), a glial cell adenyl cyclase (Brostrom *et al.*, 1975), a smooth muscle myosin light chain kinase (Dabrowska *et al.*, 1978), and an erythrocyte $Ca^{2+}-Mg^{2+}-ATPase$ (Jarrett and Pennsiton, 1978). Binding of CDR protein to phosphodiesterase (Lynch *et al.*, 1976) and adenyl cyclase (Teshima and Kakiuchi, 1974) occurred only in the presence of $Ca^{2+}$ ions. If the CDR protein of the mitotic apparatus had analogous effects on the cyclic nucleotide and/or protein kinase systems *in vivo*, then the activities of both of these systems could be coordinated with the two $Ca^{2+}$-dependent systems whose possible interactions have already been described (Section III,A,2,b). This would impose a considerable degree of unity to some of the diverse biochemical events associated with mitotic induction, chromosome movements, and cytokinesis. The last obvious example of a potentially important interaction is that between actin and myosin to affect chromosome or cell movements, with or without the added involvement of one or more of the $Ca^{2+}$-dependent systems associated with metaphase cells (Section III,A,2,b). It should be noted that even though reports locate the same groups of contractile elements at the sites of chromosome movement and cytokinesis, there need not be a common underlying mechanism for these two phenomena. It is also obvious from inhibitor studies and natural systems that chromosome movements and cytokinesis can occur independently, which might suggest some differences in the underlying mechanism(s) involved (Mitchison, 1971; Schroeder, 1973; Fournier and Pardee, 1975; Lustig *et al.*, 1977).

## B.   Relationship to Macromolecular Synthesis

### 1.   RNA and Protein Synthesis

Apart from the biochemical events that must underlie the structural changes which are observed as mitotic cells enter $G_1$, dramatic changes also occur in both RNA and protein metabolism at this time. It is well known that as cells progress into mitosis the biogenesis of messenger RNA (mRNA) and ribosomal RNA (rRNA) ceases, while that of 4S and 5S cytoplasmic RNA continues at near interphase levels (reviewed in Prescott, 1976). Not only is the synthesis of the nuclear precursors corresponding to mRNA and rRNA inhibited, but so is the processing of already synthesized precursor molecules. As cells traverse from metaphase into $G_1$, these inhibitions are reversed and, based on studies with actinomycin D, the restoration of mRNA and rRNA synthesis and processing are independent events (Simmons *et al.*, 1973). The noncoordinate control of biogenesis of the various cellular RNAs could be a general feature of eukaryotic systems since it has been observed in yeast and in resting cells stimulated to proliferate (Mauck and Green, 1973; Johnson *et al.*, 1974; Benz *et al.*, 1977; Mauck, 1977; Schulman *et al.*, 1977). This phenomenon reflects in part the existence of a different species of DNA-dependent RNA polymerase for each major class of cellular RNA (reviewed in Chambon, 1975). Thus, some of the products of RNA polymerase III continue to be made and processed in colcemid-arrested metaphase Chinese hamster ovary cells while those of RNA polymerase II are not (Zylber and Penman, 1971). Since metaphase HeLa cells have been shown to synchronously enter $G_1$ and renew RNA synthesis similar to untreated controls, even in the presence of inhibitors of protein synthesis, the simple loss of polymerase molecules at mitosis does not appear to be responsible for the inhibition of RNA synthesis at this time. Inhibitor studies further indicated that multiple RNA polymerase activities persisted unexpressed in mitotic cells and that, while the initial restoration of RNA synthesis is the mitosis to $G_1$ transition occurred independently of protein synthesis, continued RNA synthesis in early $G_1$ was dependent upon protein synthesis (Simmons *et al.*, 1973). This is consistent with the hypothesis that new enzymes and/or protein factors were required to maintain interphase rates of RNA synthesis.

The location of the persistent, but inactive, RNA polymerase activities of metaphase HeLa cells was shown to be associated with a purified metaphase chromosome preparation (Simmons *et al.*, 1973, 1974a). This observation has been confirmed in several laboratories (Gariglio *et al.*, 1974; Matsui *et al.*, 1977) and the various studies have indicated that, at least, RNA polymerase I and II are tightly bound to metaphase chromosomes. Studies with the inhibitor rifamycin AF/013 (Gariglio *et al.*, 1974) and the incorporation of $[\gamma\text{-}P^{32}]$-ATP (Simmons and Hodge, 1974) indicated that the enzymes were largely in a bound, preinitiated state. Furthermore, high resolution electron microscope autoradiography of synchronous populations traversing the interval from metaphase to $G_1$ has provided evidence on the time and site of the earliest detectable renewed RNA synthesis, as well as indicated the relatively transient and rapidly reversible nature of the inhibition of RNA synthesis during mitosis. In HeLa $S_3$ cells, significant levels of RNA synthesis were first detected by electron microscope autoradiography in early telophase prior to chromosome decondensation (Simmons *et al.*, 1973). In Chinese hamster ovary cells, resumption of RNA synthesis may even occur slightly earlier, although only low levels of silver grains were observed (Fakan and Nobis, 1978). Both studies indicated that modification of template structure by the general process of chromosome condensation may not be a sufficient explanation for the cessation of the transcription of mRNA and rRNA precursors at mitosis. In addition silver grains were observed around the periphery of the chromosome masses in both cases, after nuclear envelope reformation was completed in the HeLa cell study (Simmons *et al.*, 1973) but prior to the completion of envelope reformation in the Chinese hamster ovary cell study (Fakan and Nobis, 1978).

In addition to the dramatic changes in RNA metabolism observed during mitosis, total protein synthesis is reduced to 20–30% of interphase levels (Steward *et al.*, 1968; Hodge *et al.*, 1969; Fan and Penman, 1970; reviewed in Prescott, 1976), while the synthesis of specific nuclear non-histone proteins has been reported to continue at premitotic levels (Stein and Baserga, 1970). The decrease observed in the incorporation of radioactive amino acids into protein correlates well with the disaggregation of 70–80% of cytoplasmic polyribosomes observed by electron microscopy (Robbins and Gonatas, 1964; Scharff and Robbins, 1966). As cells complete mitosis and enter $G_1$, interphase levels of protein synthesis are rapidly restored in association with the reformation of cellular polyribosomes (Steward *et al.*, 1968; Hodge *et al.*, 1969). Furthermore, in part, the restoration of polyribosome structure and protein synthesis occurred in the absence of *de novo* RNA synthesis which indicated the use of premitotic components of the translation machinery (Hodge *et al.*,

1969). In HeLa cells, approximately 40–50% of the premitotic RNA, with some characteristics of mRNA, was reutilized on early $G_1$ cell polyribosomes in the presence of high levels of actinomycin D. The bulk of this reutilized mRNA appeared to be distributed with the 80S monomer in sucrose gradient analyses of cytoplasmic extracts of metaphase cells (Hodge *et al.*, 1969). The properties, distribution, and behavior of this class of mitotic RNA were distinct from nuclear species which appeared to either remain bound to nucleus-derived structures as chromosomes or were rapidly repackaged into reforming progeny cell nuclei (Fan and Penman, 1971; Neyfakh *et al.*, 1971; Phillip, 1972; Abramova and Neyfakh, 1973). It is unknown whether the translated class represents a specific or random collection of mRNA sequences.

Studies of the effect of low concentrations of cycloheximide on the level of polyribosomes in metaphase-arrested Chinese hamster ovary cells are consistent with the interpretation that the inhibition of protein synthesis observed at mitosis reflected a decrease in the overall level of initiation of translation. Further analysis is necessary to document this conclusion more directly. Phosphorylation of a specific ribosome-associated polypeptide in the metaphase cell could also play a role in this inhibition, since a specific-sized class of polypeptide(s), reported to be phosphorylated in metaphase HeLa cells, and is dephosphorylated as the cells enter $G_1$ (Rupp *et al.*, 1976).

## 2. *Potential for Expression of Regulatory Events*

The data summarized above clearly indicate the potential existence and importance of transcriptional, posttranscriptional, and translational controls in the regulation of macromolecular metabolism at mitosis. The precise nature of these regulatory mechanisms in the mitosis to $G_1$ transition remain unclear, but some, at least, must operate on a coarse, all-or-none level of control. With the exception of histone mRNAs and proteins, studies that sought to demonstrate finer levels of control of protein and mRNA metabolism during the cell cycle have met with mixed success. Differences in polyacrylamide gel electrophoresis patterns of proteins isolated at different stages of the cell cycle have been reported. Such studies have suggested the existence of both $G_1$- and $G_2$-specific proteins (Kolodny and Gross, 1966; Fox and Pardee, 1971; Salas and Green, 1971; Ley, 1975; Al-Bader *et al.*, 1978; Gates and Friedkin, 1978). However, at least on the basis of molecular weight, in most cases, the majority of the proteins detected were generally similar throughout the cell cycle. This is not surprising considering the needs of cells to perform many housekeeping functions and the level of sensitivity of the techniques employed in these investigations. In studies of specific proteins, cell cycle-specific synthesis or activity of a number of proteins including enzymes (Klevecz and Kapp, 1973; Heby *et al.*, 1976; Mamot *et al.*, 1976), immunoglobulins (Buell and Fahey, 1969; Lerner and Hodge, 1971; Watanabe *et al.*, 1973; Garatun-Tjeldstø *et al.*, 1976), albumin (Chen and Redman, 1977), and non-histone chromosomal proteins (Bhorjee and Pederson, 1972; Stein and Borun, 1972; Naha *et al.*, 1975) have been reported. Detailed analyses of the molecular basis of these presumptive incidences of cell cycle-specific translational events or protein activations are lacking. With regard to the mitosis to $G_1$ transition period, it is not known whether the restoration of interphase levels of protein synthesis involves any stage-specific translations.

Attempts to study the existence of cell cycle-specific mRNA transcription, distribution, or utilization have likewise met with limited success except in the case of histone mRNAs. Historically, investigations were burdened with severe technical limitations in the isolation and characterization of cellular RNA fractions. Early studies designed to demonstrate differences in sequence composition of total HnRNA and cytoplasmic RNA during the cell cycle were inconclusive (Miller, 1967; Bello, 1969; Papoulatos and Darnell, 1970). This was not surprising since the methods employed could not have detected small differences in sequence composition and were, by and large, restricted to the detection of stable hybrids between repetitive sequences of RNA and DNA. The latter fact is highly significant since mammalian cell mRNA is largely made up of nonrepetitive sequences of RNA. In HeLa cells an average of only 6% of the poly(A)-containing mRNA was found to represent repetitive sequences (Klein *et al.*, 1974), while in various mouse tissue, 80% or more of the poly(A)-containing mRNA of each abundance class was shown to represent nonrepetitive RNA sequences (Hastie and Bishop, 1976). It is conceivable that improved methods of nucleic acid hybridization, especially the use of cDNA probes, would permit more precise investigation of possible cell cycle-specific changes in the composition of mRNA. This approach has been successfully used in delineating small differences, on a percentage basis, in the poly(A)-containing mRNA in resting and growing cells (Williams and Penman, 1975). It is significant that even with semisynchronous populations that result when resting cells are stimulated to proliferate (Johnson *et al.*, 1974), small but potentially significant differences in mRNA composition between the two growth states were detected. Depending upon which mRNA abundance class the differences reflected, from 400 to 1400 mRNA sequences could be involved. These studies at the least suggest some unique mRNAs in the $G_1/G_0$ to S transition of resting cells stimulated to proliferate. It would, therefore, be of interest to extend this approach to the mitosis to $G_1$ transition and determine whether or not there would be any evidence for the selective survival of mRNA.

It should be noted that the many studies of mRNA biogenesis and protein synthesis in exponentially growing cells have yielded insights into these processes and into mRNA structure which could prove relevant to cell cycle controls at the mitosis to $G_1$ transition. It is important that future investigations of protein and mRNA metabolism make use of this information which includes the findings that mammalian cell mRNA is long-lived, that mRNA has unique 5'- and 3'-termini and internal methylations, that mRNA sequences are found unassociated with polyribosomes, that mRNA and its precursor(s) exist *in vivo* as ribonucleoprotein complexes, and that the initial interaction of mRNA and ribosomes is likely to be the most significant level of control of translation. These and other aspects of protein synthesis and mRNA biogenesis, structure, and function have been the subjects of recent reviews (Lodish, 1976; Revel, 1977; Shafritz, 1977; Revel and Groner, 1978).

## IV. RESEARCH IN PROGRESS CONCERNING THE MITOSIS TO $G_1$ TRANSITION

### A. Rationale and Basic Experimental Approach

Our approach to the study of the biochemistry of the reestablishment of the interphase state involves both nuclear and cytoplasmic events. As stated above, there is extensive reorganization of cellular structure associated with the reestablishment of interphase function. In order to gain more precise information about any of these events, it seemed reasonable that each process should be analyzed separately. We have previously pursued this approach with populations of HeLa $S_3$ cells first synchronized in metaphase by selective detachment and then incubated at 37°C to permit the cells to progress form metaphase, through anaphase and telophase, into $G_1$. Metabolic inhibitors were used to dissect events in populations of cells and electron microscope autoradiography was used to follow events in individual cells. In this manner we were able to provide evidence, for example, that synthesis of both nuclear heterogeneous RNA and preribosomal RNA approached interphase levels in the absence of a return to $G_1$ levels of protein synthesis, and that the restitution of RNA synthesis occurs as early as telophase at the periphery of the reforming nucleus in individual cells (Simmons *et al.,* 1973).

These results demonstrated that other events in the mitosis to $G_1$

transition should be capable of being investigated with such synchronous populations of HeLa $S_3$ cells. During this interval of approximately 90 minutes, the mitotic index falls from 90% or greater to less than 10%, the interphase cellular morphology is reestablished, and RNA and protein synthesis are restored to interphase levels. Incubation of populations of metaphase cells at 37°C for shorter periods of time can be used to obtain a population that represents a continuum of mitotic stages from late anaphase to early $G_1$, the majority of the latter still connected by an intracellular bridge. This range of late mitotic stages represents only a 20- to 25-minute interval in the 18-hour life cycle of this cultured cell line (Robbins and Gonatas, 1964). Thus, with such tightly synchronized populations other more precise questions concerning the biochemistry of the mitosis to $G_1$ transition should be approachable.

## B. Nuclear Reformation

### 1. Incorporation of Lipid Precursors and Nuclear Envelope Reformation

Currently, there is little biochemical understanding of the reformation of the nuclear envelope which apparently is, at least in part, carried through cell division at the telomeres of chromosomes. Since phospholipids are relatively easy membraneous components to identify, we have begun to examine the synthesis and fate of nuclear envelope phospholipids during the mitosis to $G_1$ transition (Henry *et al.*, 1977a; Henry and Hodge manuscript in preparation). Essentially, mitotic cells are being collected and permitted to enter $G_1$ in the presence of radioactive orthophosphate, choline, or glycerol. At intervals during the transition into $G_1$, whole cells and/or preparations of nuclei have been analyzed for incorporation of radioactivity.

Populations of metaphase cells progress at a typical rate into $G_1$ under lipid labeling conditions, and by 90 minutes essentially all cells have progressed to interphase (Fig. 1, inset). During this interval, the incorporation of radioactive choline into acid-insoluble material is linear, both in the presence and absence of an inhibitor of protein synthesis (Fig. 1). Extraction of whole $G_1$ cells to specifically prepare lipids demonstrated that greater than 95% of the incorporated radioactivity was in cellular lipids, and analysis by thin-layer chromatography revealed that greater than 95% of this radioactivity was phosphatidylcholine. When metaphase cells are incubated with [$^{32}$P]orthophosphate, at a minimum, the radioactivity extracted from the $G_1$ cells can be recovered in lysophosphatidylcholine, sphingomyelin, phosphatidylcholine, phosphatidylino-

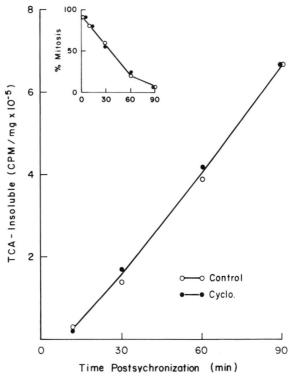

**Fig. 1.** Choline incorporation into cells progressing from metaphase into $G_1$. Preparations of metaphase cells were collected and suspended at a concentration of $1 \times 10^6$ cells/ml in prewarmed medium containing 1/100th choline and 10 $\mu$Ci/ml of [$^3$H]choline in the absence or presence of 60 $\mu$g/ml cycloheximide. Both cultures were maintained at 37°C to permit the cells to progress into $G_1$. At the indicated times an equal aliquot of each culture was used to determine the trichloroacetic acid insoluble radioactivity per milligram of cell protein. *Inset:* the mitotic index of control and inhibitor-treated cells was ascertained at successive times after synchronization by phase contrast microscopy. Untreated cells (O—O); cells incubated with 60 $\mu$g/ml cycloheximide (●—●).

sitol/serine, and phosphatidylethanolamine. Apparently, during this transition, most types of phospholipids are synthesized.

To visualize the cellular site of choline incorporation, we have examined individual cells by electron microscope autoradiography. Cells synchronized in metaphase were suspended in choline-minus medium containing [$^3$H]choline and were permitted by incubation to progress to the late stages of mitosis—anaphase, telophase and early $G_1$. Cells in these stages as well as a typical metaphase cell are depicted in Fig. 2 (A–D). A typical metaphase cell with electron-dense chromosomes displayed along

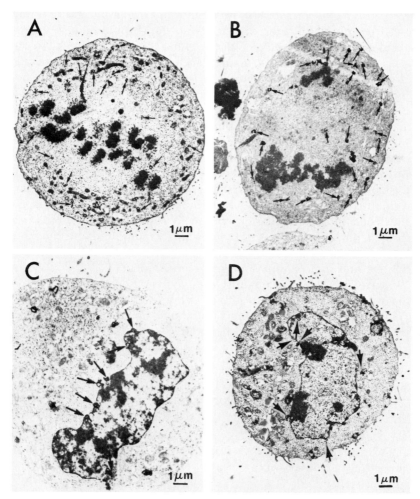

**Fig. 2.** Visualization of choline incorporation in late mitotic cells. A preparation of metaphase cells was exposed to [³H]choline at a concentration of 100 $\mu$Ci/ml as described in the legend to Fig. 1. Incubation was continued for 40 minutes at 37°C to maximize the opportunity to examine cells that had progressed to late stages of mitosis. After fixation and embedding in Epon, gold to silver sections were mounted on grids and electron microscope autoradiography using Ilford L₄ emulsion was performed as previously described (Simmons *et al.*, 1974b). A is a section through a metaphase cell; B is a section through an anaphase cell; C is a section through a telophase cell; and D is a section through an early G₁ cell. × 10,625. Details are described in text (Section III,B,1). Arrows indicate the location of silver grains.

the equatorial plate is overlaid with numerous cytoplasmic grains toward the cell periphery, at some distance from the chromosomes (Fig. 2A). Careful examination revealed that the grains are associated with known cytoplasmic membraneous structures, such as mitochondria and vesicles. In anaphase, before obvious envelope reformation, only small numbers of grains have been visualized at the site of the fusing chromosome masses (Fig. 2B). However, by early telophase, obvious sites of incorporation are observed at the periphery of the newly reformed nucleus before significant chromosome decondensation has begun, and before nucleoli have reformed (compare Figs. 2B and 2C). The marginated appearance of the telophase nucleus is consistent with nuclear envelope reformation by this time. The last stage in the metaphase to interphase transition is early $G_1$, and a typical cell in this stage is displayed in Fig. 2D. Chromatin decondensation is virtually complete and nucleoli have reformed, but the nucleus is still somewhat irregular in shape. Similar to the telophase labeling pattern, grains can be visualized over the nuclear periphery and some cytoplasmic grains are evident.

We have also begun to quantify these observations by counting total cytoplasmic grains, as well as grains within three half-distances of the chromosomal or nuclear periphery and the plasma membrane (Table I). Within this distance, one expects to find 85% or more of the silver grains

**Table I   Relative Grain Densities**[a]

| Cellular compartment | Stage in cell cycle | Number of grains | Relative area (%) | Relative grain density |
|---|---|---|---|---|
| Nuclear | Anaphase | 84 | 6.5 | 1.00 |
| envelope | Telophase | 227 | 3.8 | 4.62 |
| | $G_1$ | 231 | 4.7 | 3.80 |
| Plasma | Anaphase | 198 | 6.2 | 1.00 |
| membrane | Telophase | 243 | 6.8 | 1.12 |
| | $G_1$ | 202 | 7.8 | 0.81 |
| Cytoplasm | Anaphase | 519 | 79.0 | 1.00 |
| (other | Telophase | 536 | 79.2 | 1.03 |
| membranous structures) | $G_1$ | 322 | 58.4 | 0.84 |

[a] Measurements of cellular area and grain counts were made from autoradiographs of sections of 30 anaphase, 30 telophase, and 30 early $G_1$ cells printed $\times$ 12,500. The relative area of each cellular compartment was estimated from a uniformly dispersed grid of points by dividing the points per compartment by the total number of points. The relative density represents the grains per area for each compartment normalized to that of the nuclear envelope.

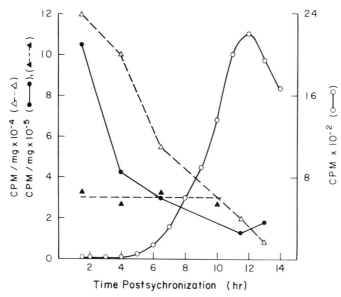

**Fig. 3.** Capacity to incorporate lipid precursors. Preparations of metaphase cells with a mitotic index of 95% were incubated at 37°C to permit the population to pass through $G_1$ and into S phase. For 90-minute intervals beginning with the collection of metaphase cells, aliquots of $2 \times 10^6$ cells were exposed to 20 $\mu$Ci of [³H]glycerol or to 20 $\mu$Ci of [³H]choline. The acid-insoluble incorporation per milligram of protein into whole cells or detergent-cleaned nuclei was ascertained. The same data are also obtained when incorporation is based on microgram of phospholipid. In addition, at the indicated times an aliquot of $4 \times 10^4$ cells was exposed to 0.5 $\mu$Ci of [methyl-³H]thymidine for 20 minutes and the cell associated acid-insoluble radioactivity determined. Incorporation of radioactive choline into whole cells (●—●); incorporation of radioactive choline into nuclei (△ - - △); incorporation of radioactive glycerol into nuclei (▲- - - -▲); incorporation of radioactive thymidine (O—O).

associated with the presence of radioactivity in a given cellular structure (Salpeter *et al.*, 1969). As seen in Table I, grains over cell sections from each of the three final stages in the transition from mitosis to G₁ have been scored. Consistent with the previous observations, we have found significant incorporation, at least by telophase, in the vicinity of the nuclear envelope as compared to other cellular membrane structures. Whether this represents a slow accumulation of radioactivity throughout late metaphase, anaphase, and telophase and/or a sudden accumulation after nuclear envelope reformation is not clear. Nevertheless, these results do indicate that incorporation is a fairly early event in the mitosis to G₁ transition.

Comparisons are also in progress of the incorporation of different lipid precursors into whole cells and into isolated, "detergent-cleaned" nuclei. The comparison appears to be of interest because choline, a head group on phospholipids, is thought to be readily available for exchange reactions, whereas the incorporation of glycerol would more likely represent lipid synthesis. As a first approximation, phospholipids were prepared from detergent-cleaned nuclei, which preserves about 50% of the nuclear lipids (Hodge *et al.*, 1977). Metaphase cells were incubated continuously at 37°C until the population had proceeded into S phase (Fig. 3). At selected 90-minute intervals beginning at metaphase, portions of the population were exposed to either radioactive choline or glycerol to measure relative incorporation. The capacity to incorporate choline was maximal in both whole cells and nuclei during the first 90-minute interval after collection of metaphase cells and this capacity steadily fell as cells progressed through the $G_1$ period. In contrast, with radioactive glycerol the level of incorporation in nuclei remained fairly uniform from the first 90-minute interval throughout early and late $G_1$.

Overall, these data suggest that lipid precursors are incorporated into the nuclear envelope during this transition period in the absence of newly synthesized protein components. In addition, it is possible to speculate that an exchange reaction could be of primary importance for envelope reformation in late mitosis.

## 2. Defined Structural Components and Nuclear Reformation

Following the fate of defined nuclear structural components should offer an additional approach to the study of nuclear reformation in the mitosis to $G_1$ transition. The use of well-defined, isolated nuclear subfractions such as ribonucleoprotein networks, nuclear pore–laminar complexes, and nuclear matrices may permit us to probe protein–protein interactions during nuclear reformation in late mitosis. The isolation of these nuclear fractions have the following features in common: (1) they appear to retain some nuclear structural integrity in the absence of the majority of the nucleic acids; (2) they are relatively difficult to visualize within the nucleus until after most of the nucleic acids have been removed; and (3) they are isolated by some combination of detergent treatment, salt extraction, and nucleic acid digestion. We have experience with one such preparation; a nuclear matrix fraction, whose isolation and characterization have previously been published (Hodge *et al.*, 1977). The final product is about the size and shape of the nucleus, contains about 12% of the protein of detergent-cleaned nuclei, and is composed of a peripheral component or dense lamina, material that spans the volume of the nucleus and a residual nucleolus (Figs. 4A and B).

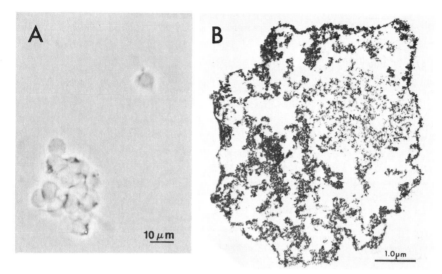

**Fig. 4.** Microscopic appearance of a nuclear matrix fraction. By phase microscopy the nuclear matrix is seen to be more or less spherical in shape, about two-thirds the size of the nucleus, and to lack visible intranuclear detail; A × 800. By electron microscopy this residual nuclear fraction is seen to be composed of a peripheral component or nuclear lamella, an internal portion that extends inwards from the lamella and a residual nucleolar component; B × 17,000.

By weight this nuclear matrix fraction is approximately 87% protein, 12% phospholipid, 1% DNA, and less than 1% RNA.

Generally, at least some matrix polypeptides survive nuclear dispersal at mitosis and are used as components of the G₁ nucleus (Fig. 5). However, the exact biochemical nature of these components is unclear, and better electrophoretic resolution and labeling of these polypeptides would make this observation more precise. In order to exploit further the nuclear matrix fraction to analyze the reassembly of the nucleus, it is necessary first to characterize its protein components in more detail and, thereby, select candidate polypeptides for use in our studies. We have evidence for the presence in matrix preparations of DNA-binding proteins, nucleolar proteins, nuclear periphery proteins, and phosphorylated acidic proteins (Henry *et al.*, 1977b; Henry and Hodge, manuscript in preparation).

*a. DNA-Binding Proteins.* We have observed that during the isolation of the matrix fraction, digestion of DNA in nuclei led to a rapid loss of the majority of DNA, although, even after prolonged incubation periods of up to 60 minutes, about 2–3% of the initial DNA remained associated with the matrix. This observation suggested that the matrix

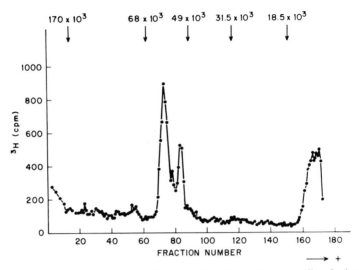

**Fig. 5.** Conservation of polypeptides through mitosis. S phase cells, obtained by a double thymidine blockade, were used to establish monolayer cultures in complete monolayer medium containing $^3$H-labeled mixed amino acids at a concentration of 2 $\mu$Ci/ml, and 8 hours later mitotic cells were collected. The population at a concentration of $5 \times 10^5$ cells/ml with a metaphase index of 95% was permitted to enter $G_1$ by incubation at 37°C in complete medium in the presence of cycloheximide at a concentration of 60 $\mu$g/ml. The nuclear matrix fraction was prepared and the constituent radioactive polypeptides were separated in a 7.5% SDS-polyacrylamide cylindrical gel.

fraction contained DNA-binding proteins. Comparisons of proteins solubilized by salt from nuclear matrices and detergent-cleaned nuclei have revealed that matrix preparations yield from 2- to 4-fold less DNA-binding proteins than nuclei. As judged by this criterion, the nuclear matrix should not be considered as an enriched fraction for this class of proteins. It is clear from the profiles of $^{35}$S-labeled extracts, applied to cellulose columns containing either bound single-stranded or native calf thymus DNA, that the DNA binding polypeptides obtained from nuclear proteins vary significantly from those obtained from the nuclear matrix (compare Figs. 6A and 6B). Detergent-cleaned nuclei yield binding proteins with a polypeptide profile which is fairly similar regardless of whether radioactive proteins are eluted either from single-stranded or double-stranded DNA columns (Fig. 6A). In contrast, the matrix fraction yields a far less heterogeneous profile with predominant polypeptides in the 55,000 and 38,000 MW ranges. Particularly striking is the polypeptide migrating to a region of approximately 55,000 MW after chromatography on double-stranded DNA (Fig. 6B). These results indicate that the nuclear matrix yields a relatively restricted class(es) of DNA-binding proteins as compared to whole nuclei. These matrix proteins would be prime candidates for additional studies of nuclear reformation, especially in view of the finding of polypeptides of similar molecular weights as components of the metaphase chromosome protein scaffold (Paulson and Laemmli, 1977.).

b. *Nucleolar, Peripheral, and Phosphorylated Proteins.* In order to assign polypeptides to topographical areas of the nuclear matrix, we have made comparisons between other nuclear fractions and the matrix fraction, and have used the technique of *in vitro* labeling with radioactive iodine.

The nuclear matrix has been resolved into at least 30 to 35 distinguishable polypeptides with apparent molecular weights in the 14,000 to 200,000 dalton range (Fig. 7). Polypeptides that migrate to the 49,000 to

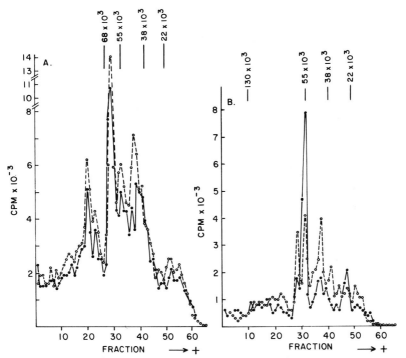

**Fig. 6.** Nuclear proteins (A) and nuclear matrix proteins (B) binding to single-stranded and double-stranded DNA. ³⁵S-labeled nuclear proteins from detergent-cleaned nuclei and ³⁵S-labeled nuclear matrix proteins, which bind to and elute from single-stranded or native calf thymus DNA-cellulose columns with 2 *M* NaCl, were precipitated with 10% trichloroacetic acid, disrupted into polypeptides, and separated electrophoretically in 6–20% gradient polyacrylamide gels containing 3% SDS (Hodge *et al.*, 1977). The radioactivity in 2-mm fractions was ascertained.

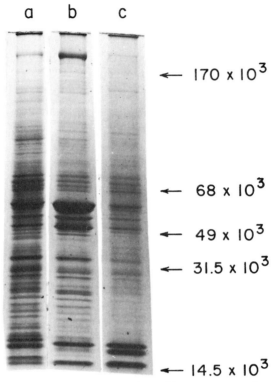

**Fig. 7.** Electrophoretic profile of the polypeptide components of a nucleolar and a nuclear matrix fraction. These subnuclear fractions were isolated and solubilized with SDS, and approximately 50 μg samples were analyzed by electrophoresis as indicated in the legend to Fig. 6. Gels were stained with Coomassie blue and calibrated for molecular weight with standard polypeptides. The sample order is: a, nucleoli; b, nucleoli treated with the final steps in matrix isolation; c, nuclear matrix fraction.

75,000 MW region of the gel are predominant in both nucleolar (Fig. 7a) and matrix (Fig. 7c) fractions, suggesting, as expected, some basic similarity between these two fractions. In fact, when nucleolar preparations are subjected to deoxyribonuclease digestion and dithiothreitol treatment, the final steps in matrix isolation, these polypeptides are even more pronounced (Fig. 7b). This suggests that polypeptides in this molecular weight range are major residual nucleolar components, and thus, candidates as markers to be used in experiments concerning nucleolar reformation in the mitosis to $G_1$ transition.

Iodination of nuclei before the final steps of matrix isolation should preferentially label peripheral components containing free tyrosine groups. The labeling profile of the polypeptides of matrices prepared from previously iodinated nuclei is displayed in Fig. 8A. Although a relatively low level of labeling seems to be associated with a number of the matrix peptides, there appears to be a preferential labeling in the 31,000 to 35,000 MW region which corresponds to minor components of the matrix as accessed by Coomassie blue stainability. However, when

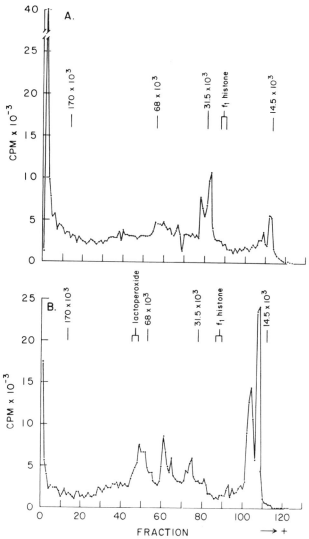

**Fig. 8.** *In vitro* iodination of nuclear matrix polypeptides. After labeling of detergent-clean nuclei with ¹²⁵I by the lactoperoxidase-catalyzed system, the nuclear matrix was isolated and solubilized (A). Solubilized matrix polypeptides were also labeled by the same procedure (B). Both samples were analyzed by electrophoresis as indicated in the legend to Fig. 6. Radioactivity in 1-mm fractions was ascertained.

matrices are prepared and solubilized with SDS before iodination, we find three major peaks of incorporation in the molecular weight region of 49,000 to 70,000 daltons which corresponds to the major polypeptide clusters seen in stained gels (Fig. 8B). In addition, extensive labeling of histones of 14,000 to 16,000 MW is also found. These results suggest that the 31,000 to 35,000 MW polypeptides represent surface proteins in the dense lamella. These putative peripheral subunits would represent a second class of candidate polypeptides with which to follow nuclear reformation in late mitosis.

In addition, we have found that the polypeptides of the matrix must be considered to be phosphonuclear proteins, and that the matrix fraction as isolated retains significant phosphorylating activity. These can easily be demonstrated by *in vitro* labeling of matrix polypeptides with gamma-labeled [$^{32}$P]ATP. It is evident that the phosphorylation pattern

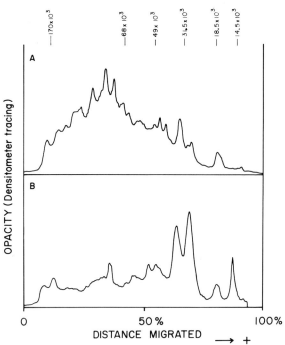

**Fig. 9.**   Phosphorylation of nuclear matrix polypeptides. Detergent-cleaned nuclei (A) and nuclear matrix fraction (B) were labeled in a reaction mixture containing [γ-$^{32}$P]ATP. After nuclear labeling, the nuclear matrix was isolated. Polypeptides from each preparation were analyzed by electrophoresis as indicated in the legend to Fig. 6 followed by autoradiography of each gel. A densitometer tracing of each autoradiograph is presented. All the radioactivity was in protein, as judged by pronase sensitivity.

observed for isolated matrices bears only a partial resemblance to that obtained when this structure is phosphorylated as part of the intact nucleus (compare Figs. 9A and 9B). When the whole nucleus is labeled followed by matrix isolation, approximately 24 bands are phosphorylated, with the most intensely labeled polypeptides migrating slower than 68,000 MW components (Fig. 9A). In contrast, when the isolated matrix fraction is labeled, radioactivity was concentrated in two peaks in the molecular weight range of about 28,000 to 31,000 daltons (Fig. 9B). It is possible to speculate that phosphorylation and/or dephosphorylation of specific proteins in the matrix fraction could be an important reaction in nuclear reformation. It has been suggested by others (Section III,A,1,a) that the phosphorylation of histone H1 is a major event in the condensation of chromatin at mitosis. Our results suggest that there exist many other nuclear structural phosphoproteins that could undergo similar changes in relation to chromatin condensation and decondensation at mitosis.

### 3. Future Directions

*a. Lipid Studies.* Additional experiments appear feasible with regard to the incorporation of lipid components. We have, for example, observed small numbers of grains by electron microscope autoradiography to be associated with the condensing chromosomes before complete membrane reformation. It is not clear whether or not this incorporation occurs only at sites of recognizable membrane. The use of higher specific activity lipid precursors followed by preparation of cells for electron microscope autoradiography using acetone dehydration to better preserve membrane phospholipids should make it possible to approach this question. In this manner, the sequence of events with regard to the incorporation of lipid precursors and membrane reformation could be investigated. Similarly, it would be of interest to visualize the site(s) and fate of premitotic membrane components through mitosis by electron microscope autoradiography.

It is conceivable that there is a unique profile of lipid incorporation during nuclear envelope reformation. By dual labeling, it should be possible to compare the incorporation of phospholipids in cells passing through the mitosis to G$_1$ transition to cells not undergoing this transition. For example, cells can be prelabeled to equilibrium followed by the harvest of metaphase cells in the presence of a second isotope. With better biochemical separation techniques, it should be possible to follow less abundant membrane lipids at this time.

*b. Protein Studies.* Continued analysis of specified polypeptides that coisolate with the matrix fraction should give more insight into nuclear reformation. Of particular interest are the DNA-binding proteins with approximate molecular weight of 55,000 and 38,000 daltons because similar size polypeptides have been reported to be part of the chromosome scaffolding. Such components conceivably could play a role as fixing points for DNA in the nucleus and could undergo significant alterations during nuclear reformation. Therefore, a detailed analysis of their localization, followed by an analysis of associated functional groups, and/or interactions could be revealing.

Matrix polypeptides appear to be highly phosphorylated and, thus, it is conceivable that the degree of phosphorylation of threonine or serine residues of one or several of these components could be correlated with nuclear reformation. Similarly, differences in methylation and acetylation may be associated with this event. It has been suggested that sulfhydryl groups (Section III,A,1,a) play a role in the process of cell division, and, it is reasonable to consider that protein–protein interactions are mediated by disulfide bond formation. Analyses of sulfhydryl groups and the use of sulfhydryl cross-linking compounds could explore these possibilities.

The raising of antibody to a specified polypeptide, either from matrix or the nuclear envelope, would provide a powerful reagent. This could be used to detail the synthesis and/or alterations of a given antigen. With immunoelectron microscopy, it is conceivable that the fate of a protein could be visualized in individual cells through the mitosis to $G_1$ transition.

Several procedures have been published for nuclear-envelope isolation and a well-defined preparation would permit additional analyses. In addition to studies using specific antibody, enzyme activities thought to be associated with the envelope could be assayed to determine whether new activities appear immediately after reformation or develope with time through $G_1$. Since it has recently been claimed that the architecture of HeLa $S_3$ nuclei can be analyzed by the binding of different lectins (Micheals *et al.*, 1977), it would be of interest to determine whether or not lectin binding could be used to investigate the architecture of the reforming nuclear envelope during the mitosis to $G_1$ transition.

## C. Cytoplasmic Reutilization of mRNP Particles

It would appear that multiple events will also have to be investigated in order to develop an understanding of the biochemistry of structural rearrangements that occur in the cytoplasm at the mitosis to $G_1$ transition. There is considerable lack of information concerning potential regulatory events at this time, especially those involved with the protein synthetic apparatus which reforms as the cell enters interphase. With regard to the metabolism of cytoplasmic mRNA at this time, as discussed above (Section III,B,1), there apparently is survival and reutilization of at least a portion of the premitotic mRNA in early $G_1$ progeny cells. Precise analyses of the molecular bases of these phenomena will depend upon our abilities (1) to define classes of translated (reutilized) and untranslated (nonutilized) cytoplasmic mRNA, and (2) to define the survival and function of mRNA fractions in terms of the current information derived from randomly growing cells concerning its metabolism and structure. Thus, we initially sought to define the reappearance of premitotic RNA on polyribosomes as mitotic cells enter $G_1$ in terms of poly(A)-containing sequences labeled after long term, as well as brief exposure periods to radioactive uridine (Simmons and Hodge, 1978; Simmons and Hodge, manuscript in preparation). Mitotic and $G_1$ cells were prepared from cells exposed to radioactivity for 9 hours or 1/2 hour prior to synchronization for metaphase, and poly(A)-containing radioactivity was selected from cytoplasmic fractions separated by sedimentation in sucrose gradients. In this manner, it is possible to fractionate the cytoplasm into polyribosomal, monoribosomal, and postribosomal fractions, and thereby select material for study. The optical density profile as seen in Figs. 10A and C display this fractionation and typify the polyribosome reformation that occurs as metaphase cells enter $G_1$. Metaphase cells yield a profile that predominantly contains 80S single ribosomes and few polyribosomes sedimenting faster than 80S, whereas cells after they have entered $G_1$ yield a typical interphase polyribosome profile sedimenting faster than the 80S monosomes with a concomitant decrease in monosomes.

To study the reappearance of poly(A)-containing mRNA on $G_1$ cell polyribosomes, cells were derived in the presence of actinomycin D to obviate any problems concerning reutilization of radioactivity, and greater than 90% of the radioactivity can be recovered in the $G_1$ cells. For metaphase cells, prelabeled for 9 hours prior to synchronization, approximately 40% of the premitotic poly(A)-containing RNA sediments to the region of residual polyribosomal structures, while approximately 60% sediments to the monoribosomal and postribosomal areas (Table II). However, in $G_1$ cells the levels of polyribosome-associated radioactivity increased to 68% which coincided with decreases in both the monoribosome and postribosome fractions (Table II). Generally, the shifts in RNA distribution represent one-third to one-half of the nonpolyribosome-associated material in the metaphase cell cytoplasm. Similarly, there is an analogous shift in the distribution of pulse-labeled (30 minutes) poly(A)-containing RNA, although the proportion of

**Table II    Distribution of Long-Term Labeled Cytoplasmic Poly(A)+RNA**[a]

| Cytoplasmic fraction | Incorporation (cpm/$10^7$ cells) | | Relative distribution (%) | |
|---|---|---|---|---|
| | M cell | $G_1$ cell | M cell | $G_1$ cell |
| Polysome | 31,110 | 48,625 | 40.9 | 67.6 |
| 80 S monosome | 26,245 | 9,385 | 34.5 | 13.1 |
| Preribosomes | 18,750 | 13,890 | 24.6 | 19.3 |

[a] HeLa $S_3$ cells were prelabeled in the presence of 0.4 $\mu$Ci/ml of [$^3$H]uridine after release from a second thymidine blockade. Approximately $4.4 \times 10^7$ synchronized cells with a mitotic index of 93% were divided in two equal portions. With one portion a cytoplasmic preparation (M cells) was prepared immediately. The second portion was incubated for 2 hours at 37°C to permit the mitotic cells to enter $G_1$, and then a cytoplasmic preparation was prepared. Both fractions were subjected to sedimentation in $7\frac{1}{2}$–45% sucrose gradients. Material sedimenting to greater than 80 S region with the polyribosomes, to the 80 to 60 S region, and to less than 60 S region of the gradients was obtained by pooling the appropriate fractions. The acid-insoluble poly(A)+RNA was determined in each pooled sample by oligo(dT) column chromatography followed by trichloroacetic acid precipitation. The data were adjusted on the basis of a 2 N DNA content.

radioactivity that sediments with $G_1$ polyribosomes is less compared to the long-term labeled material. In general, 20–30% of the pulse-labeled poly(A)-containing RNA synthesized before mitosis sedimented with polyribosomes in metaphase cells and this increased to 30–50% in $G_1$ cells. Of interest is that the greatest proportion of the radioactivity remained in the postribosomal regions of the gradients which was analogous to the original observation for pulse-labeled RNA not selected on the basis of poly(A) content (Hodge et al., 1969).

Originally, it had been suggested that the mitotic mRNA that regained a polyribosomal distribution in $G_1$ remained associated with single ribosomes during mitosis. Consistent with our new data concerning the shift of poly(A)-containing RNA to the polyribosomes, it now appears that mRNA exists as a free ribonucleoprotein particle in metaphase cells. To demonstrate this point, we have used the fact that ribosomes bound to mRNA are stable at high ionic strength, whereas ribosomes not bound dissociate into ribosomal subunits (Cooper et al., 1976). Thus, if metaphase 80S ribosomes were in stable association with mRNA, they should be resistant to dissociation to 60S and 40S subunits in the presence of high salt. For this analysis, the cytoplasmic fraction of a metaphase cell population, exposed to tritiated uridine for 30 minutes prior to synchronization under conditions that suppress ribosomal RNA synthesis, was analyzed by sedimentation in 10 m$M$ sodium chloride and by sedimentation in 500 m$M$ sodium chloride. As can be seen, the uv absorbing material in the 80S monomer region of the low salt gradient is absent in the high salt gradient and there is an increase in the 60S and 40S regions (compare Figs. 10A and 10B). Salt-resistant monosomes, when detected, never exceeded 5% of the amount seen in the presence of low salt. In addition, under these conditions a considerable amount of the premitotic pulse-labeled radioactivity continued to sediment through the 80 S monomer region. Shifts in the distribution of this radioactivity to lower density regions of the gradient at high ionic strength could reflect conformational changes in ribonucleoprotein particles (Zylber and Penman, 1970). Once the population of cells has entered G₁, however, salt-resistant monoribosomes can be recovered (compare Figs. 10C and

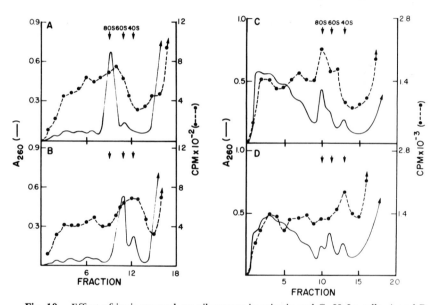

**Fig. 10.** Effect of ionic strength on ribosomes in mitotic and G₁ HeLa cells. A and B: Approximately $2.2 \times 10^7$ metaphase cells, pulse-labeled with 0.4 $\mu$Ci/ml [³H]uridine prior to synchronization, were collected in the presence of 2 $\mu$g/ml actinomycin D. A cytoplasmic fraction containing polyribosomes was prepared and dividied into two equal portions: one portion was subjected to sedimentation in a 10 to 40% (w/w) sucrose gradient at low ionic strength and the other portion to sedimentation in a 10 to 30% (w/w) sucrose gradient at high ionic strength according to Cooper *et al.* (1976). Gradients were monitored for absorption at 260 nm and the distribution of trichloracetic acid insoluble radioactivity determined. A, low ionic strength (0.01 *M* NaCl) gradient; B, high ionic strength (0.5 *M* NaCl) gradient. C and D: Metaphase cells labeled and collected as above were incubated for 2 hours at 37°C to allow the cells to enter G₁. Polyribosomes were prepared from $4.1 \times 10^7$ cells, divided into two equal portions and analyzed by the same sedimentation procedure. C, low ionic strength (0.01 *M* NaCl) gradient; D, high ionic strength (0.5 *M* NaCl) gradient).

10D). This stable fraction represents nearly half of the single ribosome material that can be demonstrated in low salt. In general, the presence of salt-resistant ribosomes in $G_1$ cells is consistent with increased levels of initiation of protein synthesis.

In order to continue an analysis of the fate of cytoplasmic mRNA, it appears necessary to define populations of mRNA that reassociate with, as well as populations of mRNA that fail to reassociate with, polyribosomes when the cells enter $G_1$. Permitting metaphase cells to enter $G_1$ in the presence of cycloheximide should permit this distinction. Low levels of this inhibitor reduces the rate of elongation while initiation of protein synthesis continues (Lodish, 1971), and this should maximize mobilization of messenger RNP particles into polyribosomes without recycling into a postribosomal fraction. Initial analysis of the kinetics of polyribosome reformation in the presence and absence of cycloheximide revealed that reformation was complete by 2 hours postsynchronization for metaphase and that cycloheximide treatment elevated polyribosome levels by 10–16%. To analyze the effect of this inhibitor of protein synthesis on the distribution of poly(A)-containing RNA, a metaphase cell population, pulse-labeled prior to synchronization in the absence of rRNA synthesis, was divided into two aliquots; one entered $G_1$ in the presence of actinomycin D alone while the other was exposed to actinomycin D plus cycloheximide. The content of poly(A)-containing RNA in the polyribosome, the monoribosome, and postribosome regions

**Table III    Effect of Cycloheximide on the Distribution of Pulse-Labeled Poly(A)+RNA in $G_1$ HeLa Cells[a]**

| Treatment | Region of the gradient | Incorporation (cpm/$10^7$ cells) | Relative distribution (%) |
|---|---|---|---|
| Actinomycin D | Polyribosomes | 2931 | 30.9 |
| | 80–60 S complex | 1952 | 20.6 |
| | Preribosomes | 4610 | 48.5 |
| Actinomycin D | Polyribosomes | 3870 | 39.7 |
| + | 80–60 S complex | 1632 | 16.7 |
| cycloheximide | Preribosomes | 4251 | 43.6 |

[a] HeLa cells were prelabeled in the presence of 1 $\mu$Ci/ml of [$^3$H]uridine for 30 minutes prior to collection of mitotic cells which were then incubated for 2 hours to allow the population to enter $G_1$. The poly(A)+RNA content of gradient fractions corresponding to the polyribosome region, 80–60 S region, and less than 60 S (preribosome) region was determined as described in Table II. The poly(A)+RNA content represents the acid-insoluble material corresponding to the oligo(dT) column fraction eluted in the absence of salt. The data are adjusted on the basis of a 2 $N$ DNA content. Actinomycin D and cycloheximide concentrations are as in Fig. 11.

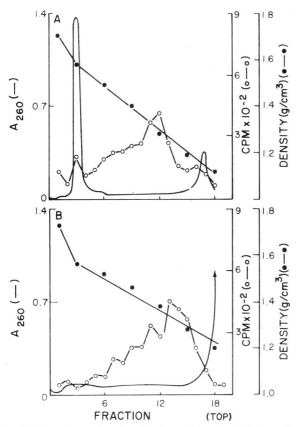

**Fig. 11.** Distribution of cytoplasmic mRNP particles derived from $G_1$ HeLa cells. Metaphase HeLa cells, pulse-labeled prior to synchronization as stated in the legend to Fig. 10 in the presence of 0.04 μg/ml actinomycin D, were collected and incubated at 37°C for 2 hours in the presence of 1 μg/ml each of cycloheximide and actinomycin D in order to maximize polyribosome reformation in the absence of renewed RNA synthesis. A cytoplasmic extract was prepared and made 30 mM with respect to $MgCl_2$ (Warner, 1960) so as to obtain a precipitate (polysomes, monosomes, and associated mRNP) and a supernatant fraction (free mRNP, tRNA, cytoplasmic protein). Both samples were then subjected to equilibrium centrifugation in $Cs_2SO_4/Me_2SO$ according to Greenberg (1977). Gradients were collected, monitored for optical density, and 0.3-ml fractions were collected and aliquots analyzed for acid-insoluble radioactivity. Densities were determined by weighing 0.1 ml of every third or fourth fraction. A, $MgCl_2$ precipitate. B, $MgCl_2$ supernatant fraction. (−), absorbance at 260 nm; (O—O), radioactivity; (●—●), density (g/cm³).

of the gradients following sedimentation was once again ascertained. Cycloheximide treatment led to a modest (9%) increase in the level of pulse-labeled, poly(A)-containing RNA in the polyribosome fraction of the $G_1$ cells (Table III). When the same analysis is performed for long-labeled material, modest increases (8–13%) were also found and from 20 to 30% of the radioactive RNA remained in the postribosomal fraction.

Using equilibrium centrifugation in cesium sulfate-dimethyl sulfoxide gradients, mRNA can be isolated in combination with its associated proteins (Greenberg, 1977). Typical results for this analysis of cytoplasmic fractions of $G_1$-inhibitor treated cells are shown in Figs. 11A and B. It is obvious that both polyribosome (Fig. 11A) and postribosome (Fig. 11B) fractions yield material that bands in the density range characteristic of mRNP particles. Thus, the mobilization of polyribosomes and mRNA with cycloheximide combined with the isolation of mRNP particles from metaphase cells and metaphase-derived $G_1$ cells should permit the isolation of mRNA that fails to reassociate with polyribosomes, as well as reutilized mRNA.

### Future Directions

With our ability to prepare populations of mRNP particles that survive mitosis and interact differently with the protein synthetic machinery in $G_1$ cells, it now becomes possible to approach experimentally such questions as whether there is selective survival and utilization of a class(es) of messenger, whether there is storage of mRNA in postpolysomal fractions for use at other times in the cell cycle or whether mRNP structure has a role in determining its survival and reutilization in $G_1$ cells. For example, data can be obtained on the nature of the posttranscriptional alterations of the mRNA found in each of the populations of mRNP particles. In this way we can explore whether 5'-terminal or 3'-terminal modifications of the mRNA are related to its fate in the mitosis to $G_1$ transition. Likewise, the proteins from the various populations of mRNP particles can be isolated and compared on the basis of electrophoretic mobilities in polyacrylamide gels to explore whether survival and/or reutilization is reflected in differences in messenger-associated proteins. Although the fate at this time in the cell cycle of a particular mRNA sequence cannot be followed, it may be possible with molecular probes to compare the sequence composition of the various mRNA preparations.

### V.  SUMMARY

The structural rearrangements and the alterations of macromolecular synthesis in mammalian cells that proceed through the mitosis to $G_1$

transition have been discussed. Several events associated with mitosis can be visualized microscopically and can be demonstrated biochemically. However, it is not clear which of these events are either obligatory for or merely temporally coincident with cell division. Nonetheless, the implication(s) of potential regulatory and/or inducing proteins have been considered. In addition, the locations of specific proteins thought to be elements of microtubules and microfilaments have been discussed and speculations have been advanced concerning cellular alterations in enzymatic activities such as Ca-dependent ATPase(s), protein kinase(s), and ornithine decarboxylase; and concerning optimal levels of cyclic nucleotides and divalent cations. Although plausible correlations between chromosome movements and cytoplasmic division (Section III,A and B) were drawn from this information, much still remains to be learned about these events. In addition, there is even less information about the basic biochemistry of, or regulatory implication of other structure–function relationships.

To dissect critical mitotic events and to evaluate their possible interrelationships in more detail, each event must be independently investigated. With this goal in mind, our current investigations concerning aspects of nuclear and cytoplasmic changes have been discussed (Section IV). For these studies a cultured human cell line, HeLa S$_3$, has been used to explore the mitosis to G$_1$ transition because large numbers of cells homogeneous for metaphase, which by incubation enter G$_1$, are obtained without the use of mitotic inhibitors.

Two approaches to the study of nuclear reformation are reported: (1) the use of lipid precursors to probe nuclear envelope reformation; and (2) the use of a defined nuclear subfraction, the nuclear matrix, to probe other nuclear phenomena. With regard to lipid incorporation, lipid precursors were incorporated into the nuclear envelope even in the absence of renewed protein synthesis and were visualized in the envelope at least by early telophase. Additional studies into whether or not there is a unique profile of incorporation of lipid precursors at this time are underway. Polypeptides that coisolate with a nuclear matrix fraction survive mitosis and appear in the G$_1$ nucleus. In order to exploit this approach, polypeptides are being assigned to components of the matrix, functional groups associated with the polypeptides ascertained, and DNA-binding proteins isolated. These marker polypeptides (proteins) should provide a means to elucidate more information concerning nuclear reformation. With regard to cytoplasmic events, the disaggregation of polyribosomes at metaphase resulted in the release of mRNP particles free of ribosomes. From cells that have entered G$_1$, a means to isolate reutilized mRNP particles and nonreutilized mRNP particles has been described. This should permit additional studies to determine whether or not there is a structural bases for reutilization of mRNP particles and, conceivably, whether or not there is selective survival of mRNA at this time.

The discussion of the literature and of current research only begins to probe selected events at a time in the cell cycle that is characterized by dynamic changes in many aspects of cell structure and function. However, an accumulation of additional information, both biochemical and morphological, should lead to an eventual integrated understanding of cytoskeletal and nuclear processes, which in turn should result in the elucidation of potential control points in cell division.

## REFERENCES

Aaronson, R. P., and Blobel, G. (1974). *J. Cell. Biol.* **62**, 746–754.
Abramova, N. B., and Neyfakh, A. A. (1973). *Exp. Cell Res.* **77**, 136–142.
Adolph, K. W., Cheng, S. M., and Laemmli, U. K. (1977a). *Cell* **12**, 805–816.
Adolph, K. W., Cheng, S. M., Paulson, J. R., and Laemmli, U. K. (1977b). *Proc. Natl. Acad. Sci. U.S.A.* **74**, 4937–4941.
Al-Bader, A. A., Rao, P. N., and Orengo, A. (1976). *Exp. Cell. Res.* **103**, 47–54.
Al-Bader, A. A., Orengo, A., and Rao, P. N. (1978). *Proc. Natl. Acad. Sci. U.S.A.* **75**, 6064–6068.
Bachrach, U. (1975). *Proc. Natl. Acad. Sci. U.S.A.* **72**, 3087–3091.
Back, F. (1976). *Int. Rev. Cytol.* **45**, 25–64.
Beck, W. T., Bellantone, R. A., and Canellakis, E. S. (1972). *Biochem. Biophys. Res. Commun.* **48**, 1649–1655.
Behnke, O. (1975). *Nature (London)* **257**, 709–710.
Bello, L. J. (1969). *Biochim. Biophys. Acta* **179**, 204–213.
Benz, E. W., Jr., Getz, M. J., Wells, D. J., and Moses, H. L. (1977). *Exp. Cell Res.* **108**, 157–165.
Berns, M. W., and Richards, S. M. (1977). *J. Cell. Biol.* **75**, 977–982.
Berridge, M. J. (1976). *Soc. Exp. Biol. Symp.* **30**, 219–231.
Bhorjee, J. S., and Pederson, T. (1972). *Proc. Natl. Acad. Sci. U.S.A.* **69**, 3345–3349.
Bonner, W. M., and Stedman, J. D. (1979). *Proc. Natl. Acad. Sci. U.S.A.* **76**, 2190–2194.
Bornens, M. (1977). *Nature (London)* **270**, 80–83.
Borun, T. W., Paik, W. K., and Marks, D. (1974). *In* "Control Processes in Neoplasia" (M. A. Mehlman and R. W. Hanson, eds.), pp. 187–219. Academic Press, New York.
Bradbury, E. M., Inglis, R. J., Matthews, H. R., and Sarner, N. (1973). *Eur. J. Biochem.* **33**, 131–139.
Bradbury, E. M., Inglis, R. J., and Matthews, H. R. (1974a). *Nature (London)* **247**, 257–261.
Bradbury, E. M., Inglis, R. J., and Matthews, H. R. (1974b). *Nature (London)* **249**, 553–556.
Brostrom, C. O., and Wolff, D. J. (1974a). *Arch. Biochem. Biophys.* **165**, 715–727.
Brostrom, C. O., and Wolff, D. J. (1974b). *Arch. Biochem. Biophys.* **172**, 301–311.
Brostrom, C. O., Huang, V. C., Breckenridge, B. M. L., and Wolff, D. J. (1975). *Proc. Natl. Acad. Sci. U.S.A.* **75**, 64–68.
Buell, D. N., and Fahey, J. L. (1969). *Science* **164**, 1524–1525.
Burger, M. M., Bombik, B. M., Breckenridge, B. M., and Sheppard, J. R. (1972). *Nature (London) New Biol.* **239**, 161–163.

Byrus, C. V., and Russell, D. H. (1975). *Science* **187**, 650–652.

Cande, W. Z., Lazarides, E., and McIntosh, J. R. (1977). *J. Cell Biol.* **72**, 552–568.

Chambon, P. (1975). *Annu. Rev. Biochem.* **44**, 613–638.

Chen, K., Heller, J. S., and Canellakis, E. S. (1976). *Biochem. Biophys. Res. Commun.* **68**, 401–407.

Chen, L. L., and Redman, C. M. (1977). *Biochim. Biophys. Acta* **479**, 53–68.

Cheung, W. Y. (1970). *Biochem. Biophys. Res. Commun.* **33**, 533–538.

Cheung, W. Y. (1971). *J. Biol. Chem.* **246**, 2859–2869.

Cobbs, C. S., Jr., and Shelton, K. R. (1978). *Arch. Biochem. Biophys.* **189**, 323–335.

Coffino, P., Gray, J. W., and Thomkins, G. M. (1975). *Proc. Natl. Acad. Sci. U.S.A.* **75**, 878–882.

Comings, D., and Okada, T. (1970a). *Exp. Cell Res.* **62**, 293–302.

Comings, D., and Okada, T. (1970b). *Exp. Cell Res.* **63**, 461–473.

Cooper, H. L., Berger, S. L., and Brawerman, R. (1976). *J. Biol. Chem.* **251**, 4891–4900.

Costa, M., and Ney, J. S. (1978). *Biochem. Biophys. Res. Commun.* **83**, 1158–1164.

Costa, M., Fuller, D. J. M., Russell, D. H., and Gerner, E. W. (1977). *Biochim. Biophys. Acta* **479**, 416–426.

Dabrowska, R., Sherry, J. M. F., Aromatorio, D. K., and Hartshshorne, D. J. (1978). *Biochem.* **17**, 253–258.

Daniel, V., Litwack, G., and Tomkins, G. M. (1973). *Proc. Natl. Acad. Sci. U.S.A.* **70**, 76–79.

deCapoa, A. D., Ferraro, M., Archidia, N., Pellicci, F., Rocchi, M., and Rocchi, A. (1976). *Human Genet.* **34**, 13–16.

Dell'orco, R. T., Martin, T. J., and Douglas, W. H. J. (1977). *In Vitro* **13**, 55–62.

DiPasquale, A., White, D., and McGuire, J. (1978). *Exp. Cell Res.* **116**, 317–323.

Eipper, B. A. (1974). *J. Biol. Chem.* **249**, 1398–1406.

Ely, S., D'Arcy, A., and Jost, E. (1978). *Exp. Cell Res.* **116**, 325–331.

Erlandson, R. A., and de Harven, E. (1971). *J. Cell Sci.* **8**, 353–397.

Fakan, S., and Nobis, P. (1978). *Exp. Cell Res.* **113**, 327–337.

Fan, H., and Penman, S. (1970). *J. Mol. Biol.* **50**, 655–670.

Fan, H., and Penman, S. (1971). *J. Mol. Biol.* **59**, 27–42.

Fournier, R. E., and Pardee, A. B. (1975). *Proc. Natl. Acad. Sci. U.S.A.* **72**, 869–873.

Fox, T. O., and Pardee, A. B. (1971). *J. Biol. Chem.* **246**, 6159–6165.

Friedman, S. J., Bellantone, R. A., and Canellakis, E. S. (1972). *Biochim. Biophys. Acta* **261**, 188–193.

Fujiwara, K., and Pollard, T. D. (1976). *J. Cell. Biol.* **71**, 848–875.

Fuller, G. M., Ellison, J., McGill, M., and Brinkley, B. R. (1975). *J. Cell. Biol.* **67**, 126a.

Fuseler, J. (1975). *J. Cell. Biol.* **67**, 789–800.

Garatun-Tjeldstø, O., Pryme, I. F., Weltman, J. K., and Dowben, R. M. (1976). *J. Cell Biol.* **68**, 232–239.

Gariglio, P., Buss, J., and Green, M. H. (1974). *FEBS Lett.* **44**, 330–333.

Gates, B. J., and Friedkin, M. (1978). *Proc. Natl. Acad. Sci. U.S.A.* **75**, 4959–4961.

Goldknopf, I. L., and Busch, H. (1977). *Proc. Natl. Acad. Sci. U.S.A.* **74**, 864–868.

Goodman, D. B. P., Rasmussen, H., DiBella, F., and Guthrow, C. E. (1970). *Proc. Natl. Acad. Sci. U.S.A.* **67**, 652–659.

Gorovsky, M. A., and Keevert, J. B. (1975). *Proc. Natl. Acad. Sci. U.S.A.* **75**, 2672–2676.

Greenberg, J. R. (1977). *J. Mol. Biol.* **108**, 403–416.

Gurley, L. R., Walters, R. A., and Tobey, R. A. (1972). *Arch. Biochem. Biophys.* **154**, 212–218.

Gurley, L. R., Walters, R. A., and Tobey, R. A. (1973). *Biochem. Biophys. Res. Commun.* **50**, 744–750.

Gurley, L. R., Walters, R. A., and Tobey, R. A. (1975). *J. Biol. Chem.* **250**, 3936–3944.

Harris, P. (1962). *J. Cell Biol.* **14**, 475–487.

Harris, P. (1975). *Exp. Cell Res.* **94,** 409–425.

Hastie, N. D., and Bishop, J. O. (1976). *Cell* **9,** 761–774.

Hayashi, M., and Matsumura, F. (1975). *FEBS Lett.* **58,** 222–225.

Heby, O., Marton, L. J., Zardi, L., Russell, D. H., and Baserga, R. (1975). *Exp. Cell Res.* **90,** 8–14.

Heby, O., Gray, J. W., Lindl, P. A., Marton, L. J., and Wilson, C. B. (1976). *Biochem. Biophys. Res. Commun.* **71,** 99–105.

Henderson, A. S., Warburton, D., and Atwood, K. C. (1972). *Proc. Natl. Acad. Sci. U.S.A.* **71,** 3394–3398.

Henry, S., Heywood, P., and Hodge, L. D. (1977a). *J. Cell Biol.* **75,** 9a.

Henry, S., Heywood, P., and Hodge, L. D. (1977b). *J. Cell Biol.* **75,** 413a.

Herman, I. M., and Pollard, T. D. (1978). *Exp. Cell Res.* **114,** 15–25.

Herman, I. M., and Pollard, T. D. (1979). *J. Cell Biol.* **80,** 509–520.

Hodge, L. D., Robbins, E., and Scharff, M. D. (1969). *J. Cell Biol.* **40,** 497–507.

Hodge, L. D., Mancini, P., Davis, F. M., and Heywood, P. (1977). *J. Cell Biol.* **72,** 194–208.

Hughes, S. H., Stubblefield, E., Payvar, F., Engel, J. D., Dodgson, J. B., Spector, D., Cordell, B., Schmike, R. T., and Varmus, H. E. (1979). *Proc. Natl. Acad. Sci. U.S.A.* **76,** 1348–1352.

Hughes, S. H. *et al.* (1979). *Proc. Natl. Acad. Sci. U.S.A.* **76,** 1348–1352.

Inoúe, S., and Ritter, H. (1975). *In* "Molecules and Cell Movement" (S. Inoúe and R. E. Stevens, eds.), pp. 3–30. Raven, New York.

Jacobs, M., Bennett, P. M., and Dickens, M. J. (1975). *Nature (London)* **257,** 707–709.

Jarrett, H. W., and Penniston, J. T. (1978). *J. Biol. Chem.* **253,** 4676–4682.

Johnson, L. F., Abelson, H. T., Green, H., and Penman, S. (1974). *Cell* **1,** 95–100.

Kakiuchi, S., Yamazaka, R., and Nakajima, H. (1970). *Proc. Jpn. Acad.* **46,** 587–592.

Kakiuchi, S., Yamazaki, R., Teshima, Y., Venishi, K., and Miyamoto, E. (1975). *Biochem. J.* **146,** 109–120.

Keller, T., and Rebhun, L. I. (1978). *J. Cell. Biol.* **79,** 304a.

Kendall, F., Swenson, R., Borun, T., Rowinski, J., and Nicolini, C. (1977). *Science* **196,** 1106–1109.

Kiehard, D. P., and Inoue, S. (1976). *J. Cell Biol.* **70,** 230a.

Klein, W. H., Murphy, W., Attardi, G., Britten, R. J., and Davidson, E. H. (1974). *Proc. Natl. Acad. Sci. U.S.A.* **71,** 1785–1789.

Klevecz, R. R., and Kapp, L. N. (1973). *J. Cell Biol.* **58,** 564–573.

Kolodny, G. M., and Gross, P. R. (1966). *Exp. Cell Res.* **56,** 117–121.

Kurochlein, S. N., Trakht, I. N., Severin, E. S., and Cole, R. D. (1977). *FEBS Lett.* **84,** 163–166.

Kurz, J. B., and Friedman, D. L. (1976). *J. Cyclic Nucleotide Res.* **2,** 405–415.

Lerner, R. A., and Hodge, L. D. (1971). *J. Cell. Physiol.* **77,** 265–275.

Levi-Montalcini, R., Caramia, F., and Angeletti, P. (1969). *Brain Res.* **12,** 54–73.

Ley, K. D. (1975). *J. Cell Biol.* **66,** 95–101.

Lodish, H. (1976). *Annu. Rev. Biochem.* **45,** 38–62.

Lodish, H. (1971). *J. Biol. Chem.* **246,** 7131–7139.

Lustig, S., Kosower, N. S., Pluznik, D. H., and Kosower, E. M. (1977). *Proc. Natl. Acad. Sci. U.S.A.* **74,** 2884–2888.

Lynch, T. J., Tallant, E. A., and Cheung, W. Y. (1976). *Biochem. Biophys. Res. Commun.* **68,** 616–625.

McCann, P. P., Tardif, C., Mamont, P. S., and Schuber, F. (1975). *Biochem. Biophys. Res. Commun.* **64,** 336–341.

Mabuchi, I., and Okuno, M. (1977). *J. Cell Biol.* **74,** 251–263.

Mamot, P. S., Bohlen, P., McCann, P. P., Bey, P., Schuber, F., and Tardif, C. (1976). *Proc. Natl. Acad. Sci. U.S.A.* **73**, 1626-1630.

Marks, D. B., Paik, W. K., and Borun, T. W. (1973). *J. Biol. Chem.* **248**, 5660-5667.

Matsui, S.-I., Weinfeld, H., and Sandberg, A. A. (1977). *J. Cell Biol.* **75**, 121a.

Matthews, H. R., and Bradbury, E. M. (1978). *Exp. Cell Res.* **111**, 343-351.

Mauck, J. C. (1977). *Biochemistry* **16**, 793-797.

Mauck, J. C., and Green, H. (1973). *Proc. Natl. Acad. Sci. U.S.A.* **70**, 2819-2822.

Maul, G. G. (1977). *J. Cell Biol.* **74**, 492-500.

Mazia, D. (1963). *J. Cell Comp. Physiol.* **62**, Suppl. 1, 123-140.

Mazia, D. (1974). *In* "Cell Cycle Controls" (G. M. Padilla, I. L. Cameron and A. Zimmerman, eds.), pp. 265-272. Academic Press, New York.

Mazia, D. (1975). *Ann. N. Y. Acad. Sci.* **253**, 7-13.

Mazia, D., Petzelt, C., Williams, R. O., and Meza, I. (1972). *Exp. Cell Res.* **70**, 325-332.

Micheals, G. A., Whitlock, S. D., Horowitz, P., and Levin, P. (1977). *Biochem. Biophys. Res. Commun.* **75**, 480-486.

Miller, A. O. A. (1967). *Arch. Biochem. Biophys.* **122**, 270-279.

Mitchison, J. M. (1971). "Biology of the Cell Cycle." Cambridge Univ. Press, London and New York.

Miyamoto, E., Petzol, D. G. L., Kuo, J. F., and Greengard, P. (1973). *J. Biol. Chem.* **248**, 179-189.

Mobley, W., Server, A., Ishii, D., Riopelle, R., and Shooter, E. (1977). *New Engl. J. Med.* **297**, 1096-1104.

Naha, P. M., Meyer, A. L., and Hewitt, K. (1975). *Nature (London)* **258**, 49-53.

Neyfakh, A. A., Abramova, N. B., and Bagrova, A. M. (1971). *Exp. Cell Res.* **65**, 345-352.

Nicolini, C. K., Ajiro, K., Borun, T. W., and Baserga, R. (1975). *J. Biol. Chem.* **250**, 3381-3385.

Nose, K., and Katsuta, H. (1975). *Biochem. Biophys. Res. Commun.* **64**, 983-988.

Olins, A. L., and Olins, D. E. (1979). *J. Cell Biol.* **81**, 260-265.

Olmsted, J., and Borisy, G. G. (1975). *Biochemistry* **14**, 2996-3005.

Pagoulatous, G. N., and Darnell, J. E. (1970). *J. Cell Biol.* **44**, 476-483.

Paulson, J. R., and Laemmli, U. K. (1977). *Cell* **12**, 817-828.

Pederson, T. (1972). *Proc. Natl. Acad. Sci. U.S.A.* **69**, 2224-2228.

Pederson, T., and Robbins, E. (1972). *J. Cell Biol.* **55**, 322-327.

Petzelt, C. (1974). *J. Cell Biol.* **63**, 267a.

Petzelt, C., and Auel, D. (1977). *Proc. Natl. Acad. Sci. U.S.A.* **74**, 1610-1613.

Petzelt, C., and von Ledebur-Villiger, M. (1973). *Exp. Cell Res.* **81**, 87-94.

Phillip, S. G. (1972). *J. Cell Biol.* **53**, 611-620.

Piras, R., and Piras, M. M. (1975). *Proc. Natl. Acad. Sci. U.S.A.* **72**, 1161-1165.

Prescott, D. M. (1976). *Adv. Genet.* **18**, 99-177.

Rao, P. N., Hittleman, W. N., and Wilson, B. A. (1975). *Exp. Cell Res.* **90**, 40-46.

Rao, P. N., Wilson, B. A., and Puck, T. T. (1977). *J. Cell Physiol.* **91**, 131-142.

Rao, P. N., Wilson, B. A., and Sunkara, P. S. (1978). *Proc. Natl. Acad. Sci. U.S.A.* **75**, 5043-5047.

Rasmussen, H. (1970). *Science* **170**, 404-412.

Rebhun, L. I., Wang, R., and Carbine, L. (1978). *J. Cell Biol.* **79**, 296a.

Remington, J. A., and Klevecz, R. R. (1973). *Biochem. Biophys. Res. Commun.* **50**, 140-146.

Revel, M. (1977). *In* "Molecular Mechanisms of Protein Biogenesis" (H. Weissbach and S. Pestka, eds.), pp. 245-321. Academic Press, New York.

Revel, M., and Groner, Y. (1978). *Annu. Rev. Biochem.* **47**, 1079-1126.

Richman, R., Dobbins, C., Voina, S., Underwood, L., Mahaffee, D., Gittelman, H. J., Van Wyk, J., and Ney, R. L. (1973). *J. Clin. Invest.* **52**, 2007–2015.

Riddle, V. C., Dubrow, R., and Pardee, A. B. (1979). *Proc. Natl. Acad. Sci. U.S.A.* **76**, 1298–1302.

Robbins, E., and Gonatas, N. K. (1964). *J. Cell Biol.* **21**, 429–463.

Rochette-Egly, C., and Castagna, M. (1979). *Biochem. Biophys. Res. Commun.* **86**, 937–944.

Rosenfeld, A. C., Zackroff, R. V., and Weisenberg, R. C. (1976). *FEBS Lett.* **65**, 144–147.

Rupp, R. G., Humphrey, R. M., and Shaeffer, J. R. (1976). *Biochim. Biophys. Acta* **418**, 81–92.

Sabatini, M. T., de Iraldi, P. A., and DeRobertis, E. (1965). *Exp. Neurol.* **22**, 370–383.

Sakai, H., Mabuchi, I., Shimoda, S., Kuriyama, R., Ogawa, K., and Mohri, H. (1976). *Dev. Growth Diff.* **18**, 211–219.

Salas, J., and Green, H. (1971). *Nature (London) New Biol.* **229**, 165–169.

Salpeter, M. M., Bachmann, L., and Salpeter, E. E. (1969). *J. Cell Biol.* **41**, 1–20.

Sanger, J. W. (1975a). *Proc. Natl. Acad. Sci. U.S.A.* **72**, 1913–1916.

Sanger, J. W. (1975b). *Proc. Natl. Acad. Sci. U.S.A.* **72**, 2451–2455.

Sanger, J. W., and Sanger, J. M. (1976). *In* "Cell Motility" (R. Goldman, T. D. Pollard and J. Rosenbaum, eds.), pp. 1295–1317. Cold Spring Harbor Lab., Cold Spring Harbor, New York.

Scharff, M. D., and Robbins, E. (1966). *Science* **151**, 992–995.

Schor, S. L., Johnson, R. T., and Waldren, C. A. (1975). *J. Cell Sci.* **17**, 539–565.

Schroeder, T. E. (1973). *Proc. Natl. Acad. Sci. U.S.A.* **70**, 1688–1692.

Schroeder, T. E. (1976). *In* "Cell Motility" (R. Goldman, T. W. Pollard and J. Rosenbaum, eds.), pp. 265–279. Cold Spring Harbor Lab., Cold Spring Harbor, New York.

Schulman, R. W., Sripati, C. E., and Warner, J. R. (1977). *J. Biol. Chem.* **252**, 1344–1349.

Sedory, M. J., and Mitchell, J. L. A. (1977). *Exp. Cell Res.* **107**, 105–110.

Setterfield, G., Sheinin, R., Dardick, I., Kiss, G., and Dubsky, M. (1978). *J. Cell Biol.* **77**, 246–263.

Shafritz, D. A. (1977). *In* "Molecular Mechanisms of Protein Biogenesis" (H. Weissbach and S. Prestka, eds.), pp. 555–601. Academic Press, New York.

Sheppard, J. R., and Prescott, D. M. (1972). *Exp. Cell Res.* **75**, 293–296.

Sieber-Blum, M., and Burger, M. M. (1977). *Biochem. Biophys. Res. Commun.* **74**, 1–8.

Simmons, T., and Hodge, L. D. (1974). *J. Cell Biol.* **63**, 633a.

Simmons, T., and Hodge, L. D. (1978). *J. Cell Biol.* **79**, 10a.

Simmons, T., Heywood, P., and Hodge, L. (1973). *J. Cell Biol.* **59**, 150–164.

Simmons, T., Heywood, P., Taube, S., and Hodge, L. D. (1974a). *In* "Cell Cycle Controls" (M. Padilla, T. L. Cameron, and A. M. Zimmerman, eds.), pp. 289–308. Academic Press, New York.

Simmons, T., Heywood, P., and Hodge, L. D. (1974b). *J. Mol. Biol.* **89**, 423–433.

Smoake, J. A., Song, S.-Y., and Cheung, W. Y. (1974). *Biochem. Biophys. Acta* **341**, 402–411.

Soifer, D. (1973). *J. Gen. Physiol.* **61**, 265.

Sperling, K., and Rao, P. N. (1974). *Chromosoma* **45**, 121–131.

Stein, G. S., and Baserga, R. (1970). *Biochem. Biophys. Res. Commun.* **41**, 715–722.

Stein, G. S., and Borun, T. W. (1972). *J. Cell Biol.* **52**, 292–307.

Stevens, F. C., Walsh, M., Ho, H. C., Teo, T. S., and Wang, J. H. (1976). *J. Biol. Chem.* **251**, 4495–4500.

Steward, D. L., Shaeffer, J. R., and Humphrey, R. M. (1968). *Science* **161**, 791–793.

Sunkara, P. S., Rao, P. N., Nishioka, K., and Brinkley, B. R. (1979). *Exp. Cell Res.* **119**, 63–68.

Tao, M. (1972). *Biochem. Biophys. Res. Commun.* **46**, 56–61.

Teshima, Y., and Kakiuchi, S. C. (1974). *Biochem. Biophys. Res. Commun.* **56,** 489–495.

Tilney, L. G. (1975). *J. Cell. Biol.* **64,** 289–310.

Tilney, L. G., Hatano, S., Ishikawa, H., and Mooseker, M. S. (1973). *J. Cell. Biol.* **59,** 109–126.

Timourian, H., Jotz, M., and Clothier, C. E. (1974). *Exp. Cell Res.* **83,** 380–386.

Tyson, J., Garcia-Herdugo, G., and Sachsenmaier, W. (1979). *Exp. Cell Res.* **119,** 87–98.

Waisman, D., Stevens, F. C., and Wang, J. H. (1975). *Biochem. Biophys. Res. Commun.* **65,** 975–982.

Wang, J. H., Teo, T. S., Ho, H. C., and Stevens, F. C. (1974). *Adv. Cyclic Nucleotide Res.* **5,** 179–194.

Warner, J. R. (1966). *J. Mol. Biol.* **19,** 383–398.

Wasserman, W. J., and Smith, L. D. (1978). *J. Cell Biol.* **79,** R15–22.

Watanabe, S., Yagi, Y., and Pressman, D. (1973). *J. Immunol.* **111,** 797–804.

Weber, K., Rathke, P. C., Osborn, M., and Franke, W. W. (1976). *Exp. Cell Res.* **102,** 285–297.

Weisenberg, R. C. (1972). *Science* **177,** 1104–1105.

Weisenberg, R., and Piazza, C. (1978). *J. Cell Biol.* **79,** 303a.

Welsh, J., Dedman, J. R., Brinkley, B. R., and Means, A. R. (1978). *Proc. Natl. Acad. Sci. U.S.A.* **75,** 1867–1871.

Wessells, N. K., Spooner, B. S., Ash, J. F., Bradley, M. O., Luduena, M. A., Taylor, E. L., Wrenn, J. T., and Yamada, K. M. (1971). *Science* **171,** 135–143.

Williams, T. G., and Penman, S. (1975). *Cell* **6,** 197–206.

Wunderlich, F. R., Berezney, R., and Kleinig, H. (1976). *In* "Biological Membranes" (D. Chapman and D. F. H. Wallach, eds.), Vol. 3, pp. 241–333. Academic Press, New York.

Zackroff, R. V., and Weisenberg, R. C. (1977). *J. Cell. Biol.* **75,** 292a.

Zeilig, C. E., and Goldberg, N. D. (1977). *Proc. Natl. Acad. Sci. U.S.A.* **74,** 1052–1056.

Zeilig, C. E., Johnson, R. A., Sutherland, E. W., and Friedman, D. L. (1976). *J. Cell Biol.* **71,** 515–534.

Zentgaf, H., Falk, H., and Franke, W. W. (1975). *Cytobiologie* **11,** 10–29.

Zylber, E. A., and Penman, S. (1971). *Science* **172,** 947–949.

# 4

# Temperature-Sensitive Mutants in the Study of Cell Cycle Progression in Mammalian Cells

ROSE SHEININ

## I. INTRODUCTION

The general concepts of progression through the somatic cell cycle of eukaryotes are now well formulated and documented (cf. Dirksen *et al.*, 1979). Many of the biochemical processes which proceed in the succes-

NUCLEAR–CYTOPLASMIC INTERACTIONS
IN THE CELL CYCLE

sive stages have been revealed; some have been well characterized (cf. Prescott, 1976a; Pardee *et al.*, 1978; Sheinin *et al.*, 1978b). We are still, however, a long way from having a precise temporal map of the many biochemical and morphogenetic processes that move cells through any given phase, or of the regulatory mechanisms that result in exact cell duplication at the end of cytokinesis.

Many experimental approaches have been used to elucidate individual and coupled biochemical pathways which are the basis of cellular physiology. Perhaps one of the most powerful has been that of biochemical genetics, introduced in 1941 by Beadle and Tatum (cf. Beadle, 1945). It is therefore not surprising that the analytical tools of biochemistry and genetics were applied to the study of higher eukaryotic cells as soon as the essential technology for their growth and manipulation *in vitro* was developed. This chapter addresses itself to the progress made using temperature-sensitive mammalian cells in analyzing cell cycle progression and in revealing the nature and extent of interweaving and overlapping of control mechanisms between the different stages of the cell cycle.

## II. TEMPERATURE-SENSITIVE MUTANTS OF MAMMALIAN CELLS

It is obvious that with few exceptions, a mutation in a gene coding for any protein which is essential for progression through the cell cycle will be lethal, in the sense that the cell carrying such a mutation will be unable to propogate normally. This has directed several laboratories toward the isolation and characterization of conditional lethal mutants, and in particular those which are temperature sensitive. The reasoning and methodology employed are reviewed elsewhere (cf. Siminovitch *et al.*, 1973; Basilico, 1977, 1978; Meiss *et al.*, 1978).

For the present purposes it is perhaps adequate to recall that temperature-sensitive (*ts*) mutants carry missense mutations (Edgar and Leilausis, 1964) which are masked at a permissive temperature (*pt*), but are expressed at the nonpermissive temperature (*npt*). Thus, cells that carry a *ts* mutation in a cell cycle function will grow normally at the *pt;* whereas at the *npt* growth is restricted *in toto* or in part. In general, cells recover from the effects of a *ts* mutation, provided incubation at the *npt* is kept to a time frame which precludes unbalanced growth (Sparkuhl and Sheinin, 1980; see also Section V,F). Mutant *ts* cells can therefore be carried essentially normally at the *pt* and can be experimentally manipulated to express the *ts* mutation by altering the temperature.

In interpreting the various experiments presented below it would be useful to know whether the protein affected by a particular *ts* mutation is inherently unstable at the *npt,* whether its synthesis is temperature sensitive, and whether it is protected against inactivation at the *npt,* transiently or otherwise, by association with other cellular constituents. Such information is only recently becoming available, as is indicated below. A catalogue of currently available *ts* mammalian cells, updated from one collected earlier, (Sheinin *et al.,* 1978b) is presented in Tables I, V, and VI.

## III. METHODS EMPLOYED IN CELL CYCLE STUDIES

### A. Operational Definitions of Cell Cycle Stages

The somatic cell duplication cycle was operationally redefined by Howard and Pelc in 1953, using mitosis (M), cytokinesis, and DNA synthesis (S) as major diagnostic landmarks (Hartwell, 1978). The temporal gaps separating cell division from S phase and DNA synthesis from mitosis were designated $G_1$ and $G_2$, thereby completing the nomenclature for the various stages of the ongoing cycle. Modifications to this general concept were introduced to accommodate the so-called $G_0$ (Lajtha, 1963) or A state (Smith and Martin, 1973), in which cells may exist for long periods of time outside the cycle of proliferation.

Cells in S phase and mitosis are readily identified; the former because they synthesize DNA which can be measured by chemical methods, by radioautography, by incorporation of radioactive precursor molecules, and by cytochemical procedures defined most recently in the technique of flow microfluorimetry (FMF) (Crissman *et al.,* 1975). Cells in mitosis are identified by their characteristic chromosome morphology, movement, and segregation to newly evolving daughter cells, using microscopic and cytochemical techniques (cf. Dirksen *et al.,* 1979). Because well-defined morphological, biochemical, and cytochemical criteria for cells in $G_1$ and $G_2$ are still not generally available, these are indirectly classified in relation to the other cell cycle stages (Prescott, 1976a). Thus $G_1$ cells are those characterized by a postdivision and pre-S phase physiology; $G_2$ cells are those with a post-DNA-synthetic and premitotic physiology.

Most at risk in assessment of their cell cycle localization are cells at or near the $G_1/S$ interface and the $S/G_2$ traverse. In general, techniques (like FMF or cytomicrophotmetry) which depend upon measurement of DNA content, cannot distinguish between cells in late $G_1$, at the $G_1/S$ traverse, or early in S phase, nor do such methods discriminate between cells very late in S, those at the S/$G_2$ traverse, those in early $G_2$, or indeed anywhere in $G_2$.

Noncycling cells, in $G_0$ or A state, are also indirectly characterized at the present time. They remain in a quiescent state, sometimes for very long periods of time, retaining full viability. They can be mobilized into partial or full cell cycle activity by a variety of methods (cf. Baserga, 1978), some of which are described in the next section.

## B.   Synchronization of Cell Populations

For many physiological and biochemical studies of cell cycle progression it is useful to work with mutant (or nonmutant) cells brought to a specific stage of the cell cycle. A number of procedures have been used to effect such synchronization. It is important to recognize that most of these do not produce physiologically or biochemically homogeneous cell populations. Only those methods which have been widely applied in the study of *ts* mammalian cells are discussed here.

The most satisfactory methods for synchronizing mammalian cells are clearly those which do not interrupt, or impinge upon, normal cell cycle progression. Only one such is available, that for collection of cells in M by mitotic shakeoff (Terasima and Tolmach, 1963). This is applicable only to cells which grow on a solid substratum to which they attach during $G_1$ in a flattened, epithelioid-like configuration. As they progress to the $G_2$/M interface and into M, the cytoskeletal organization within the cytoplasm is altered. The cells round up, they release most of their anchorage links to the growth surface and can, therefore, readily be detached by gentle shaking. In some instances the free cell population recovered contains > 95% of mitotic cells. However with other cell types, the procedure is less successful.

Cells reversibly arrested in metaphase can be obtained by treating growing cultures with colchicine, vinblastine, or other antimitotic agents under controlled conditions (Dustin, 1978). Removal of drug gives rise to a highly synchronized population of cells which moves through one or more successive cell duplication cycles (cf. Prescott, 1976a).

Untransformed fibroblasts, which are subject to contact inhibition of growth, enter an $G_0$ state when grown to confluence on a solid surface (Green and Todaro, 1967; Stanners *et al.,* 1979). These can be forced into one duplication cycle by gentle protease treatment *in situ* (cf. Reich *et al.,* 1975), or they can be released for many rounds of multiplication by trypsinization and subculture at low concentration (Green and Todaro, 1967).

Cells can be brought into $G_1$ arrest by starvation of serum (Green and

Todaro 1967), isoleucine or other amino acids (Tobey and Ley, 1971), or by treatment with excess thymidine (cf. Prescott, 1976a). These manipulations give rise to different stages of $G_1$ arrest (Pardee *et al.*, 1978; Moats-Staats *et al.*, 1980; Ashihara *et al.*, 1978b), all of which can be released by reversing the treatment regimen. Cells released from $G_0$ and $G_1$ arrest move through a pre-DNA-synthetic state and into DNA synthesis. Entry of the population into S phase and subsequent progression is usually highly asynchronous (see Fig. 5).

The most widely used method for bringing cells into early S phase is treatment with hydroxyurea (Walters *et al.*, 1976) or 5-fluoro-2'deoxy-uridine (FdUrd; Prescott, 1976a). Subsequent removal of drug provides a cell population which moves immediately through S and appears to cycle normally once or twice (see Figs. 3 and 4). In some instances investigators have combined a double thymidine block which arrests cells in $G_1$, with subsequent hydroxyurea treatment to improve synchronization in early S (e.g., Roufa, 1978).

It should be noted that all of the methods used to collect and/or synchronize cells in $G_1$ or S phase interrupt normal cell cycle progression. The physiological and biochemical mechanisms affected are not well understood; however, it is clear that more than a single metabolic sequence must be involved. Reversal of these pleiotropic effects does not usually occur at the same time, thereby giving rise to parasynchronous cell populations (for example, see Fig. 5).

Additional problems attend the use of antimetabolites like hydroxyurea and FdUrd to bring cells to early S phase arrest. These interfere with polymerization of the polydeoxyribonucleotide chain either directly or by depleting the pool of deoxyribonucleoside triphosphate precursors. They give rise to accumulation of newly made, low molecular weight, single-strand DNA (Martin *et al.*, 1977; Chan and Walker, 1975) and, in addition, cause a major structural rearrangement of the chromatin (Sheinin *et al.*, 1980). Notwithstanding such caveats these synchronization procedures have been useful in characterizing *ts* mammalian cells.

## C. Operational Definitions for Characterizing *ts* Cell Cycle Mutants

Definition of mammalian cell cycle mutants is still quite arbitrary because the physiological and biochemical diagnostic landmarks (Hartwell, 1978) of cell cycle progression are few. In general, they permit only the grossest of assignment of *ts* function to one or other of the phases of the cell division continuum. The parameters used for such assignment are

**Table I  Catalogue of Temperature-Sensitive Mammalian Cells**

| Putative cell cycle arrest point | Mutant designation | Cell of origin | Reference |
|---|---|---|---|
| $G_0$ | cs 4 D3 | CHO | Crane and Thomas (1976) |
| $G_1$ | ts AF8 | BHK-21/13 | Meiss and Basilico (1972) |
| | ts 11, ts 13 | | Talavera and Basilico (1977) |
| | CH-K12, K18, K27 | Chinese hamster, WgIA | Roscoe et al. (1973a) |
| | ts K/34C | | Tenner et al. (1977) |
| | ts 3/1, 4/2, 4/3 | | Melero (1979) |
| | ts BF-113 | Chinese hamster, CCL 39 | Scheffler and Buttin (1973) |
| | $G_1^-$-4, $G_1^-$-5 | Chinese hamster lung (CHL-V79) | Liskay (1978); Liskay and Prescott (1978) |
| | | | Liskay and Prescott (1978) |
| | cs CH$^R$ E5 | CHO | Ling (1977) |
| | ts AMA$^R$-1 | | Ingles (1978) |
| | ts 13A, ts 15C | | Gupta et al. (1980) |
| | B54 | Mouse, CAK | Farber and Liskay (1974) |
| | A31, A8 | BALB/c-3T3 | Naha et al. (1975); Naha (1979) |
| $G_1$/S traverse | ts HJ4 | BHK-21/13 | Talavera and Basilico (1977) |
| | ts 154 | CHL-V79 (HT-1 Clone) | Roufa et al. (1979) |
| | ts 2 | BALB/c-3T3 | Slater and Ozer (1976) |
| S | ts AlS9 | Mouse L | Thompson et al. (1970) |
| | ts C1 | | Thompson et al. (1971) |
| | ts BN 2 | BHK-21/13 | Nishimoto and Basilico (1978) |

110

| Stage | Cell type | Mutant | Reference |
|---|---|---|---|
| M | Murine lymphoma (L5178Y) | ML-*ts* 2 | Shiomi and Sato (1976) |
| | Hamster (HM-1) | ML-*ts* 39 | Sato and Hama-Inaba (1978) |
| | | *ts* 546 | Wang (1974) |
| | | *ts* 655 | Wang (1976) |
| | | *ts* 687 | Wissinger and Wang (1978) |
| Cytokinesis | Syrian hamster | NW-1 | Smith and Wigglesworth (1972) |
| | Chinese hamster | *ts* 111 | Hatzfield and Buttin (1975) |
| | CHO-K1 | MSl-1 | Thompson and Lindl (1976) |
| | | *ts* 13B11 | Marunouchi and Nakano (1980) |
| | | *ts* Cl.B59 | Nakano *et al.* (1978) |
| Random block | Mouse FM3A | AARS[ts] | Thompson *et al.* (1973, 1975, 1977) |
| | CHO | *ts* 18, *ts* 14 | Roufa and Reed (1975) |
| | CHL-V79, HT-1 | *ts* 155-46 | Roufa and Haralson (1975) |
| | | *ts* 422E | Meiss and Basilico (1972) |
| Not assigned | BHK-21/13 | *ts* 1 | Wittes and Ozer (1973) |
| | BALB/c-3T3 | ML-*ts* 1 | Sato and Shiomi (1974) |
| | BSC-1 | *ts* BTNI, *ts* BTN3 | Nishimoto and Basilico (1978) |
| | L5178Y | *ts* BN series | Nishimoto and Basilico (1978) |
| | BHK-21/13 | *ts* T22, *ts* T23 | Meiss and Basilico (1972) |
| | | *ts* BCH, *ts* BCB, | Meiss and Basilico (1972) |
| | | *ts* BCL | |
| | | *ts* AF6 | |
| | WgIA | *ts* 5/2 | Meiss and Basilico (1972) |
| | CHO | *cs* 11-29, *cs* 11-32 | Melero (1979) |
| | Rat skeletal muscle | E3, H6 | Ohlsson-Wilhelm *et al.* (1976) |
| | HM-1 | *ts* 542 | Loomis *et al.* (1973) |
| | | | Wang and Sheridan (1974) |

those developed in an expansive study of *ts* mutants of yeast (Hartwell, 1978; Pringle, 1978). Thus *ts* cell cycle mutants of mammalian cells are defined as cells that arrest at a specific point in the cell cycle as a result of temperature inactivation of the affected gene product. The *execution point* for a given function, defined by experiments in which cells are upshifted from the *pt* to the *npt,* is that point in the cycle (or indeed in any metabolic pathway) beyond which temperature upshift will no longer affect progression or completion of the process underway. The execution point is not always identical with the *arrest point,* which defines the *terminal phenotype* of the temperature-inactivated cell. These various parameters have been used to arrive at a tentative assignment of the *ts* cell cycle function of the mutant cells listed in Table I. In addition they have been employed to develop *temporal maps of cell cycle events* presented in subsequent sections.

## IV.  GENETIC ANALYSIS OF *ts* MAMMALIAN CELLS

All *ts* mammalian cells analyzed to date carry a recessive mutation (see Table II). This property was revealed in experiments in which the *ts* cell

**Table II   Complementation of *ts* Mutants by Wild-Type Mammalian Cells**

| Mutant cell | Wild-type cell | Reference |
|---|---|---|
| *ts* AlS9 | Mouse L | Thompson *et al.* (1970) |
| | BALB/c-3T3 | H. L. Ozer (personal communication) |
| *ts* Cl | Mouse L | Thompson *et al.* (1971) |
| | Human | Giles and Ruddle (1976) |
| | BALB/c-3T3 | H. L. Ozer (personal communication) |
| *ts* AF8, *ts* AF6, *ts* T22, *ts* T23, *ts* BCH | BHK-21/13 | Meiss and Basilico (1972) |
| *ts* CH-K12 | BHK-TK | Smith and Wigglesworth (1973) |
| | Mouse, 3TP1 | Marin and Labella (1977) |
| *ts* Hl | CHO | Thompson *et al.* (1973) |
| AARS*ts* | CHO | Thompson *et al.* (1975, 1977) |
| B54 | Mouse CAK | Liskay (1974) |
| *ts* 2 | BALB/c-3T3 | Slater and Ozer (1976) |
| | Human | Jha and Ozer (1977) |
| *ts* 025 | Human | Giles *et al.* (1977) |
| *ts* K/34C | Hamster, human | Tenner *et al.* (1977) |
| *ts* 11, *ts* 13, *ts* HJ4 | BHK-21/13 | Talavera and Basilico (1977) |
| *ts* AMA$^R$-1 | CHO | Ingles (1978) |
| *ts* 422E | Mouse (3T3, 3T6) | Toniolo and Basilico (1974) |
| | HeLa | Bramwell (1977) |

**Table III  Complementation between *ts* Mammalian Cells**

| Mutant cells | | Reference |
|---|---|---|
| *ts* CH-K12 | × *ts* CH-K18 | Smith and Wigglesworth (1974) |
| | × *ts* AF8 | Dubbs and Kit (1976); Ashihara *et al.* (1978b) |
| | × *ts* 4/1, 4/2, 4/3, 5/2 | Melero (1979) |
| | × *ts* K27 | Melero (1979) |
| *ts* AF8 | × B54 | Liskay and Meiss (1977) |
| | × *ts* 422E, *ts* AF6 | Meiss and Basilico (1972) |
| | × *ts* BN2, BN7, BN39, BN51, BN63, BTN1, BTN3 | Nishimoto and Basilico (1978) |
| | × G$_1^+$-4, G$_1^+$-5 | Liskay and Prescott (1978) |
| *ts* 13 | × *ts* BN2,7,39,51,63,67,75,119, *ts* BTN1, *ts* BTN2 | Nishimoto and Basilico (1978) |
| | × *ts* 11, HJ4, *ts* AF8 | Talavera and Basilico (1977) |
| *ts* AlS9 | × *ts* Cl | R. Mankovitz (personal communication) |
| | | H. L. Ozer (personal communication) |
| | × *ts* 2 | H. L. Ozer (personal communication) |
| *ts* B54 | × *ts* AF8 | Liskay and Meiss (1977) |
| | × G$_1^+$-4, G$_1^+$-5 | Liskay and Prescott (1978) |
| *ts* 546 | × *ts* 687 | Wissinger and Wang (1978) |

was fused with one carrying a normal allele for the *ts* gene. The heterokaryon or hybrid cell so formed exhibited wild-type properties when incubated at a temperature nonpermissive for the *ts* parent. The prognosis for genetic identification of *ts* gene products has brightened recently as a result of four important developments. The first produced methods for successful interspecies fusion of mammalian cells (cf. Harris, 1974). The second is rapidly moving toward total genetic mapping of the human genome (Ruddle and Creagan, 1975; Shows and McAlpine, 1978). The third is the demonstration that individual *ts* mutaions of the cells under discussion here can be corrected by human genes on specific chromosomes (see Table IV). The fourth is the effective transfer of chromosomes and of specific DNA fragments from donor to recipient cell (Graham, 1977; Ruddle *et al.*, 1979; Wigler *et al.*, 1979; Lewis *et al.*, 1980).

At the present time classical genetic mapping of *ts* mammalian cells is rudimentary. Complementation analysis has permitted identification of a number of complementation groups within G$_1$, S, and M (see Table III). It has also revealed noncomplementing cells which may be *ts* within a single cistron or in different cistrons or genes which must cooperate in effecting a given process of cell cycle progression. Considerable interest focuses on the recently emerging observation that a number of *ts* mutants of mammalian cells are complemented at the *npt* in hybrid cells

Table IV   Human Chromosome Correction of Mammalian Cell *ts* Lesions

| Chromosome number | *ts* Lesion | Reference |
|:---:|:---:|:---:|
| 3 | *ts* AF8 | Ming *et al.* (1976) |
| 14, 14$^q$ | *ts* CH-K12 | Ming *et al.* (1979) |
| X | BALB/c-3T3- *ts* 2 | Jha and Ozer (1977) |
| X | *ts* Cl | Giles and Ruddle (1976) |
| 5 | *ts* 025 | Giles *et al.* (1977) |
| X | *ts* BHK cells | Schwartz *et al.* (1979) |

carrying the human X chromosome (Table IV). Many of these are non-complementing (Schwartz *et al.*, 1979). This may indicate a highly mutable genetic locus. It may point to mutation in a regulatory process of cell cycle progression encoded in a gene of the X chromosome, perhaps in a cell surface molecule (Miller *et al.*, 1978).

## V.  TEMPORAL MAPPING OF THE CELL CYCLE USING *ts* MAMMALIAN CELLS

We have now learned a great deal about some of the physiological, biochemical, and morphogenetic events of mitotic division among eukaryotic organisms (Pardee *et al.*, 1978; Simchen, 1978; Dirksen *et al.*, 1979). However we remain a long way from fully appreciating the nature and sequence of the many molecular processes that make up the totality of cell cycle progression. The studies with *ts* mammalian cells have begun to yield information which permits tentative development of temporal maps of pathways within and between given stages of the cell division cycle.

### A.   Studies with *ts* Cells Which Arrest in $G_1$ (or $G_0$)

Of the many *ts* mammalian cells isolated to date, a majority have tentatively been designated as *ts* in a function required for traverse of $G_1$. Limited genetic analysis has provided evidence for at least 12 genes which determine such progression (see Table III; Melero, 1979; Naha, 1979; Naha *et al.*, 1975). It seems likely that these differ from the six $G_1^+$ genes shown by Liskay and Prescott (1978) to permit establishment of a $G_1$ state among variants and mutants of the V79-8 subclone of V79 CHL cells which are genetically $G^-$ (Liskay, 1974, 1977). Thus *ts* B54 and *ts* AF8 cells are complemented by those V79-8 $G_1^+$ mutants which grow with a measurable $G_1$ period at 39°C, but are $G_1^-$ at 33°C (Liskay and Prescott, 1978).

The candidates for the products of genes which determine $G_1$ progression abound. Even so, few of the many biochemical events which proceed during $G_1$ have been identified (Pardee *et al.*, 1978). Little is known with certainty about their sequential relationship or their causal link-up (if any) with other cell cycle events. For the purposes of the present discussion six of the major diagnostic biochemical landmarks have been tentatively aligned on the temporal map of $G_1$ progression shown in Fig. 1. They include: (1) Cytokinesis which by definition generates a $G_1$ cell; (2) reorganization of the cytoskeletal system to produce the characteristic morphology of interphase cells; (3) initiation of cell dupli-

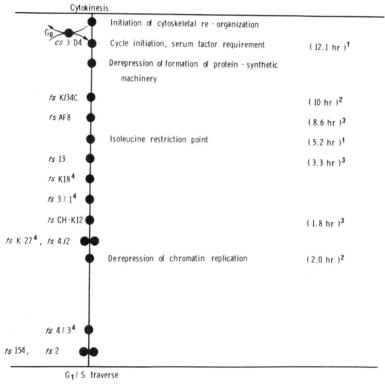

**Fig. 1.** Temporal map for $G_1$ progression. The temporal arrangement is not meant to indicate precise times within $G_1$ (S in Fig. 7 and M in Fig. 9), but rather sequence relative to major events designated on the right side of the diagram. The alignment of expression of a gene encoded in a given *ts* mutant is based on data discussed in the text and in the references cited as superscripts to the bracketed interval (in hours) before the $G_1$/S traverse at which the gene product is expressed. Key to superscript numbers: (1) Yen and Pardee (1978); (2) Landy-Otsuka and Scheffler (1978); (3) Ashihara *et al.* (1978a); (4) Melero (1979).

cation triggered by growth factor(s) in serum; (4) derepression of formation of the protein synthetic machinery; (5) derepression of chromatin replication (see Fig. 7); and (6) traverse of the $G_1/S$ interface. The map anticipates the results of the studies performed with mammalian cells which appear to be *ts* in a function of $G_1$ phase.

### 1.  ts CH-K12 Hamster Cells

CH-K12 cells, isolated from the WgIA clone of Chinese hamster DON cells (Roscoe *et al.*, 1973a), were shown to be *ts* in a $G_1$ phase function (Roscoe *et al.*, 1973b; Smith and Wigglesworth, 1973, 1974; Marin and Labella, 1977) expressed approximately 1.6 hours before the $G_1/S$ traverse (Ashihara *et al.*, 1978a) and beyond the restriction points for isoleucine and serum starvation (Pardee *et al.*, 1978; see Fig. 1).

CH-K12 cells brought into a $G_1$ arrest state by depletion of serum at 36.5°C progress into S phase $\simeq 6$ hours after serum supplementation, and begin to divide $\simeq 15$ hours thereafter (Dubbs and Kit, 1976). If such cells are incubated with serum at the *npt*, temperature inactivation of the *ts* CH-K12 gene product blocks derepression of the enzymes of DNA biosynthesis (e.g., thymidine kinase and deoxycytidylate deaminase) which occurs normally as cells traverse the $G_1/S$ interface (Kit and Jorgensen, 1976). The execution point for such derepression cannot be separated from that for expression of the *ts* CH-K12 defect. These observations have been interpreted as indicating that the *ts* CH-K12 gene product may be specifically involved in the transcription or post-transcriptional processing associated with derepression of thymidine kinase, of deoxycytidylate deaminase and, perhaps, by extension, of all of the enzyme proteins of DNA replication (Sheinin, 1967; Littlefield, 1977). This would place the *ts* CH-K12 execution point in advance of derepression of the machinery of chromatin replication (Fig. 1).

This conclusion is in accord with the observation that temperature-inactivated CH-K12 cells are deficient in S phase factors. This has been demonstrated in two experimental systems. Thus chromatin replication is not reactivated when mature inactive chick erythrocytes are fused with CH-K12 cells and the heterokaryons so obtained are incubated at the nonpermissive temperature (Dubbs and Kit, 1976). Reactivation is obtained with analogous heterokaryons incubated at the *pt*, or with heterokaryons of chick erythrocytes and WgIA cells maintained at low or high temperature. Floros *et al.* (1978b) have tested extracts of control and temperature-inactivated CH-K12 cells for their capacity to stimulate DNA synthesis by isolated nuclei of the naturally inert oocyte of *Xenopus leavis*. The former did initiate DNA replication, the latter did not. (See Section V,D,2,c for further discussion.)

These experiments indicate that temperature-inactivated *ts* CH-K12 cells are deficient in a protein involved in traverse of the $G_1/S$ interface. Focussing on reactivation of chick erythrocytes, the data suggest that the *ts* CH-K12 gene product may participate in derepression of chromatin replication. This involves decondensation of the super-condensed chick erythrocyte chromatin and subsequent transcription as prerequisites for initiation of DNA synthesis (cf. Harris, 1974). Analogous processes are thought to be expressed late in $G_1$ of normally cycling cells (cf. Setterfield *et al.*, 1978, Sheinin *et al.*, 1980 and Section D2).

The suggestion that the *ts* CH-K12 gene product may be involved with derepression of chromatin replication gains interest when one examines the complex terminal phenotype of *ts* CH-K12 cells incubated for extended periods at the *npt*. Histone synthesis (Rieber and Bacalao, 1974b) and phosphorylation (Pochron and Baserga, 1979) become uncoupled from temperature-inactivated DNA replication, and the formation of some phosphorylated, nonhistone chromosomal proteins is suppressed at the *npt* (Rieber and Bacalao, 1974a). The significance of these observations for *ts* CH-K12 cells, and vis à vis other *ts* mutants is discussed in Section V,D,2.

In addition to effecting changes in chromosomal proteins, inactivation of the *ts* CH-K12 gene product also appears to lead to a major regulatory upset reflected in the accumulation of at least three cytoplasmic proteins, not usually detected in wild-type cells (Rieber and Bacalao, 1974b). Melero and Fincham (1978) have studied these proteins, designated A, B and C, and shown them to be of 64,000, 79,000, and 94,000 MW, respectively. They concluded that expression of the *ts* CH-K12 defect results in disruption of regulation of their synthesis, leading to overproduction. Such is not observed with the *ts* 4/3 mutant of WgIA cells which also arrests in $G_1$ phase at the *npt* (Melero, 1979), or with *ts* 5/6 mutant in a non-S phase function (Melero and Fincham, 1978). In hybrid cells derived by the fusion of *ts* CH-K12 and *ts* 5/6 cells growth, DNA replication and synthesis of proteins A, B, and C approaches wild-type activity at 35° and 40.5°C. Melero and Smith (1978) have demonstrated, *in vivo* and *in vitro*, that synthesis of proteins A, B, and C results from transcription of distinct heat-stable messenger RNA (mRNA) molecules, the translation of which is regulated by the *ts* CH-K12 gene product.

The *ts* CH-K12 lesion can be corrected in human × CH-K12 hybrids which carry a normal or modified human chromosome 14 (Ming *et al.*, 1979), thereby excluding thymidine kinase, which maps on human chromosome 17 (McDougall *et al.*, 1973). These findings are in accord with the demonstration that expression of the *ts* CH-K12 defect precedes derepression of synthesis of thymidine kinase (Kit and Jorgensen,

1976). Recently five genes on human chromosome 14 have been identified (Ferguson-Smith and Westerveld, 1978; Aitken and Ferguson-Smith, 1978; Denney *et al.*, 1978; Owerbach *et al.*, 1979). More extensive studies are bringing genetic identification of the *ts* CH-K12 gene within long-range view.

### 2. ts AF8 Hamster Cells

The *ts* AF8 cell (Meiss and Basilico, 1972), one of a growing number of mutants of BHK-21/13 cells which arrest in a $G_1$ state upon temperature inactivation (Burstin *et al.*, 1974), has an execution point approximately 8.6 hours prior to the $G_1/S$ interface (Burstin *et al.*, 1974; Kane *et al.*, 1976; Chang and Baserga, 1977; Ashihara *et al.*, 1978a). This assignment on the temporal map of $G_1$ progression (Fig. 1) is in accord with a number of observations. Thus, *ts* AF8 cells brought into early S phase by treatment with hydroxyurea proceed apparently normally through the ongoing S phase, whether released at the *pt* or *npt* (Burstin *et al.*, 1974). *ts* AF8 cells arrested in a $G_1$-state by serum deprivation are unable to move into DNA synthesis at the *npt*, whereas cells starved of isoleucine move essentially without impediment through DNA synthesis at 39.5°C. On the basis of such studies Burstin *et al.* (1974) placed the *ts* AF8 execution point in late $G_1$ between the restriction points for serum starvation and isoleucine starvation. It is beyond the $G_0$ or pseudo-$G_1$ state established when *ts* AF8 cells are allowed to grow to confluence at 33°C. If such cells are subcultured at this *pt* they will, of course, move through many turns of the duplication cycle. If plated at 39.5°C, no progression occurs unless the cells are allowed recovery to S phase entry at 33°C.

The terminal $G_1$-state of temperature-inactivated *ts* AF8 cells is not the same as that which is established in normal mouse or hamster cells which grow to confluence *in vitro* on a solid surface (Stanners *et al.*, 1979). The latter are forced into extensive cellular DNA synthesis by infection with polyomavirus (cf. Sheinin, 1967), whereas virus-infected, temperature-inactivated *ts* AF8 cells exhibit only a small and transient stimulation of DNA synthesis (Burstin and Basilico, 1975) presumably mediated by T antigen which is synthesized at the *npt* (Rossini *et al.*, 1979a).

*ts* AF8 cells can be transformed by polyomavirus at the *pt*. If transformed cells (Py *ts* AF8) are incubated at the *npt* they continue to synthesize DNA and to divide for $\simeq$ 30 hours longer than observed with the parental, non-transformed *ts* AF8 cells (Burstin and Basilico, 1975). This observation suggests that the presence of the polyoma genome can, for a short period, overcome the *ts* AF8 defect. The data are compatible with a model which places the *ts* AF8 site before the $G_1$ target for the polyoma

T antigen, thereby permitting some stimulation of cell DNA synthesis in both Py-infected *ts* AF8 and Py *ts* AF8 cells incubated at the *npt*.

The above conclusion is of interest in the light of observations made with two other DNA viruses—human adenovirus-2 (Nishimoto *et al.*, 1975, 1977; Rossini *et al.*, 1979a) and herpes simplex virus Type I (HSV-I, Yanagi *et al.*, 1978). Adenovirus-infected *ts* AF8 cells are unable to replicate virus at the *npt* due to restriction of an early event prior to commitment to viral DNA synthesis. This may be related to temperature inactivation of synthesis of RNA which sets in several hours after *ts* AF8 cells are incubated at the *npt* (Burstin *et al.*, 1974; see below). Whereas inhibition of cellular DNA synthesis in uninfected *ts* AF8 cells incubated at 39.5°C begins within 3–6 hours, that in adenovirus-infected cells sets in only after ≃ 20 hours. This suggests that the adenovirus genome also carries information which can transiently overcome the effects of expression of the *ts* AF8 defect, a conclusion confirmed by the demonstration that cells rendered quiescent by serum starvation at the *pt* are forced into S phase synthesis by adenovirus infection at the *npt* (Rossini *et al.*, 1979a).

*ts* AF8 cells infected at low multiplicity with HSV-I and incubated at 39.5°C are unable to support virus multiplication or viral DNA synthesis (Yanagi *et al.*, 1978). Under conditions of high multiplicity of infection, nearly normal virus yields are obtained at the *npt*. This is an unexpected finding in light of the emerging nature of the *ts* AF8 defect discussed below. It suggests heretofore unsuspected interactions between the genetic information provided by viral and cellular genomes.

Since the *ts* AF8 execution point is well before that at which derepression of the DNA synthetic machinery occurs, it is not surprising that mature chick erythrocyte nuclei cannot be reactivated at the *npt*, in preparations of chick red blood cells fused with intact *ts* AF8 cells brought into $G_1$ arrest by isoleucine starvation (Tsutsui *et al.*, 1978) or with their cytoplasts (Lipsich *et al.*, 1979; see also Section V,D,2,c). The absence of active S phase factors in temperature-inactivated *ts* AF8 cells is also indicated by the fact that the cytoplasm of cells incubated at the *npt* is unable to stimulate DNA synthesis in isolated frog nuclei (Floros *et al.*, 1978b).

Using *in vitro* assays with isolated nuclei, Rossini and Baserga (1978) demonstrated that *ts* AF8 cells released from serum deprivation at the *pt* initiated transcription during $G_1$ traverse, as did control cells. However, RNA synthesis did not proceed in nuclei of cells incubated at 40°C, almost exclusively because of temperature-inactivation of α-amanitin-sensitive RNA polymerase II. The α-amanitin-resistant RNA polymerase I associated with the nucleolus was unaffected, even after cells were incubated for 24 hours at the *npt*. Rossini *et al.* (1979b, 1980) have extended these studies to demonstrate that loss of RNA polymerase II in temperature-inactivated *ts* AF8 cells is associated with loss of the α-amanitin binding subunit, under conditions in which the function of other enzymes (e.g., hexokinase) is unaffected. In this connection it may be significant that *ts* AF8 cells and CHO-*ts* Ama$^R$-1 cells do not complement at 39°C (C. J. Ingles, M. Shales, and J. Bergsagel, personal communication), whereas complementation at the *npt* has been observed between *ts* AF8 and members of many classes of *ts* $G_1$ mutant cells (Table III). CHO-*ts* AMA$^R$-1 cells are temperature-sensitive for growth and α-amanitin resistance, due to a *ts* α-amanitin binding subunit of RNA polymerase II (Ingles, 1978).

These observations suggest that the *ts* AF8 gene may encode information for the structure of the α-amanitin-sensitive protein subunit of RNA polymerase II. Final genetic validation of this conclusion has been brought closer by the demonstration that the *ts* defect can be corrected in *ts* AF8 × human hybrid cells which carry human chromosome 3 (Ming *et al.*, 1976), already extensively mapped (Bootsma and Ruddle, 1978; Shows *et al.*, 1978; Wijnen *et al.*, 1978; Naylor *et al.*, 1979).

### 3. *ts 154 Hamster Cells*

The *ts* 154 derivative of the HT-1 clone of V79 CHL cells has tentatively been classified amongst those mutants which arrest in $G_1$ at the *npt* (Roufa, *et al.*, 1979), even though such arrest occurs only after two complete division cycles have transpired at 39°C. This observation is compatible with any of the following models: (1) The *ts* 154 gene product is marginally mutated and is therefore only slowly inactivated at the *npt;* (2) the synthesis of the *ts* 154 protein is *ts;* (3) the *ts* 154 protein is present in relatively large amounts in the cell and is therefore functionally diluted out slowly at the *npt;* and (4) the *ts* gene product once formed at the *pt,* is well protected against temperature inactivation.

Fully temperature-inactivated *ts* 154 cells resume DNA synthesis almost immediately when down-shifted to 33°C and attain a rate 25–30% higher than control levels after 10 hours. Cell division begins approximately 5 hours later at the normal rate, by a cycloheximide-sensitive process. These data are compatible with a terminal phenotype of accumulation at the $G_1$/S interface (see Fig. 1). It has been noted by Roufa *et al.* (1979) that the behavior of *ts* 154 cells at the *npt* resembles that of *dna* A *ts* mutants of *Escherichia coli* (cf. Wickner, 1978). They suggest that a membrane protein, perhaps a nuclear membrane protein, specifically involved with initiation of DNA replication may be the *ts* 154 gene product.

### 4. *CH$^R$ E5 Chinese Hamster Ovary Cells*

CH$^R$ E5 cells are characteristic of a large class of mutants isolated from CHO cells which are cold-sensitive in growth and are therefore designated as GRO$^{cs}$ (Ling, 1977). They proliferate at 39°C but become increasingly defective in cell division as the temperature is lowered. At 33.5°C they are unable to form colonies unless plated at high cell density. This mutant cell class was isolated at 39°C by a procedure which used colchicine to select for cells blocked in mitotic division. The mutants obtained are indeed resistant to colchicine, hence the CH$^R$ nomenclature. In addition they arrest at the low temperature in a G$_1$-like state (Ling, 1977). Extensive studies have revealed that the CH$^R$, GRO$^{cs}$ mutants exhibit pleiotropic drug resistance (Ling and Thompson, 1974) due to an alteration in a major plasma membrane glycoprotein (Ling, 1977; Juliano and Ling, 1976; Juliano *et al.*, 1976).

### 5. *cs3 D4 CHO Cells*

The GRO$^{cs}$ mutants have some properties in common with the *cs*3 D4 mutant isolated by Crane and Thomas (1976) from CHO cells. These also grow well at 39°C, they arrest in a pre-DNA synthetic phase, and may remain viable at the *npt* for extended periods, perhaps in G$_0$ arrest. It is postulated that the biochemical basis for the terminal phenotype of these GRO$^{cs}$ mutant cells lies in aberrant synthesis and/or function of a major plasma membrane glycoprotein as noted above for CH$^R$ E5 cells.

### 6. *ts K/34C Hamster Cells*

*ts* K/34C cells were isolated from Wg1A hamster cells by Tenner *et al.* (1977) and shown to arrest in a G$_1$ state upon incubation at 40°C $\simeq$ 6 hours before the G$_1$/S traverse (see Fig. 1). A second arrest point was, however, observed in G$_2$.

The primary *ts* defect was first detected as rapid and extensive inhibition of incorporation of sugars (fucose, mannose, and glucosamine) into macromolecular form, which set in within 2 hours *pts,* well before any effect was observed with respect to synthesis of protein and RNA, and loss of cell integrity. It could be dissociated from inhibition of DNA synthesis which was temporally linked to expression of the *ts* K/34C gene product.

Tenner *et al.* demonstrated that temperature inactivation did not result from excessive turnover of preformed glycoprotein. Synthesis of sugar–phosphate intermediates, neutral lipids, and most phospholipids was not temperature sensitive. The *ts* lesion was shown to affect generalized glycoprotein biosynthesis, but not at the level of glycosyl-transferase activity nor at the level of polypeptide formation (Landy-Otsuka and Scheffler, 1978).

The specificity of the *ts* defect was emphasized by the demonstration that the replication of mengovirus, which parasitizes the host cell machinery for RNA and protein synthesis (Perez Bercoff, 1979), proceeds normally in *ts* K/34C cells incubated at 40°C for 16 hours. In contrast, the multiplication of vesicular stomatitis and sindbis viruses, which depend upon the cellular glycoprotein-synthetic apparatus (Toneguzzo and Ghosh, 1978; Bonatti *et al.*, 1979), is totally prevented in temperature-inactivated cells.

In pursuing these findings Tenner and Scheffler (1979) were able to identify the *ts* function in *ts* K/34C cells as that reaction in which the complex core oligosaccharide is transferred from its dolichol–lipid carrier to the amino group of the acceptor asparagine residue of the growing polypeptide chains destined to become glycoproteins (Parodi and Leloir, 1979). Normal formation of core oligosaccharide–lipid intermediate and of nascent polypeptides on free and bound polysomes in cells incubated at 40°C was demonstrated in both *in vivo* and *in vitro* experiments. In addition, a four-fold accumulation of oligosaccharide-dolichol intermediate was observed at the *npt* under conditions which lead to turnover in wild-type cells. Tenner and Scheffler have suggested that the *ts* K/34C defect may involve faulty insertion of the nascent polypeptides into membrane sites of glycoslation. Such a defect may explain the apparently unrelated observation that the pathway for synthesis of phosphatidylglycerol and cardiolipin is *ts* in *ts* K/34C cells (Tenner *et al.*, 1977).

Although the *ts* K/34C gene product has still not been identified, the execution point for its action in $G_1$ can be placed between the restriction point for serum starvation and the induction of ornithine decarboxylase at approximately 10 and 2 hours before the $G_1/S$ traverse, respectively (Landy-Otsuka and Scheffler, 1978; see Fig. 1). It clearly precedes expression of the major $G_1$ process involved with $G_1/S$ traverse, i.e., derepression of the chromatin replication machinery as monitored in these studies by the synthesis of ornithine decarboxylase, a key enzyme of polyamine biosynthesis (Pardee *et al.*, 1978).

The *ts* K/34C mutation is of additional interest because, like $CH^R$ E5 discussed above, it has a primary *ts* mutation which affects membrane biogenesis and/or function. It is tempting to postulate that arrest of such cells at the *npt* in $G_1$ and $G_2$ may result from defective plasma membrane metabolism. Such may indeed be a terminal biochemical phenotype which characterizes many $G_1$ mutants.

### 7. ts 13 Hamster Cells

ts 13 cells, isolated from BHK-21/13 cells by Talavera and Basilico (1977), exhibit an execution point beyond the restriction points for serum starvation and isoleucine starvation, approximately 3 hours before the $G_1/S$ traverse (Floros et al., 1978a; Fig. 1). Since this execution point precedes the time at which chromatin replication is derepressed, it is not surprising that the cytoplasm of temperature-inactivated ts 13 cells does not induce DNA synthesis in nuclei isolated from frog oocytes (Floros et al., 1978b), or that chick erythrocyte nuclei are not reactivated in heterokaryons incubated at the npt (Floros et al., 1978a; see Section V,D,2,c).

It has been observed that when polyoma-infected ts 13 cells are incubated at the npt they are not induced to make cellular DNA; nor do they synthesize viral DNA (Rossini et al., 1979a). This can be contrasted with the effects of infection by adenovirus-2, which does permit ts 13 cells to synthesize cellular DNA at the npt, in the normal semiconservative mode. This observation makes it possible to use adenovirus-2 expression as a probe for identification of the ts 13 gene product.

### 8. BALB/c-3T3 ts 2 Cells

The ts 2 mutant of BALB/c-3T3 fibroblasts isolated and given preliminary characterization by Slater and Ozer (1976) has now been analyzed further (Sheinin and Lewis, 1980; Sheinin et al., 1980). When ts 2 cells are upshifted to 38.5°C they continue to synthesize RNA and protein at near control levels for 10–20 hours, whereas DNA synthesis begins to decline within 2 hours pts reaching a plateau at approximately 20% of control rate at 16–20 hours.

ts 2 cells arrested in the $G_1$-state by starvation of isoleucine at 34°C resume DNA synthesis after 8–9 hours, upon restoration of the amino acid. They move through a first S phase and cell division and into a second DNA-synthetic period over the subsequent 15 hours. Cells released from isoleucine starvation at 38.5°C maintain their low level of DNA synthesis for 4–5 hours and then exhibit further temperature inactivation well before the $G_1/S$ traverse (Sheinin et al., 1980). ts 2 cells brought into early S phase arrest by treatment with hydroxyurea at 34°C complete DNA synthesis with apparently normal kinetics, whether released at the pt or at 38.5°C. In the former instance cells continue to move through cell division and into a second DNA-synthetic period under the experimental conditions employed. The cells incubated at the *npt* progress through cell division, but are unable to initiate a second cycle of DNA replication. These observations are compatible with a model which places expression of the *ts* 2 defect in $G_1$.

This conclusion derives support from studies in which *ts* 2 cells were treated with hydroxyurea at 38.5°C for 12 hours. Such cells, when re-leased at 34°C, moved into DNA synthesis somewhat more slowly than the comparable cells drug-treated at 34°C, and progressed through cell division. Cells released at 38.5°C exhibited a pattern of continued DNA synthesis and exit from S similar to that shown by cells pretreated with hydroxyurea at 34°C. However, no cell division was detected, presumably because of the lethal effects of combined and sequential treatment with hydroxyurea and temperature inactivation. Since hydroxyurea protects *ts* 2 cells from temperature inactivation, it is perhaps possible to place the *ts* 2 site on the temporal map shown in Fig. 1 near the $G_1$/S traverse.

The conclusion that *ts* 2 cells are *ts* in a late $G_1$ function is in accord with the kinetics of recovery upon downshift of *ts* 2 cells subjected to temperature-inactivation for 25 hours observed by Slater and Ozer (1976). It is also compatible with observations made with cells depleted of serum at 33°C and then released for varying periods at this *pt* prior to upshift to 38°C. Upshift between 0 to 12 hours resulted in significant inhibition of DNA synthesis measured at the 21 to 25-hour interval, the extent depending upon the duration of recovery at 33°C.

It is suggested in Section V,D that true *dna*$^{ts}$ mutants should exhibit coupled temperature inhibition of DNA and histone synthesis at the *npt*, and coordinated ultrastructural reorganization of the chromatin. It is therefore of considerable interest that neither of these phenomena is observed in temperature-inactivated *ts* 2 cells. The synthesis of chromatin-bound histones (and of other chromosomal proteins) pro-ceeds almost at control levels during 16–40 hours *pts*, under conditions in which little or no replication of chromatin DNA occurs (Sheinin and Lewis, 1980). The heterochromatin of the nucleoplasm does not undergo permanent decondensation (Sheinin *et al.*, 1980).

Biochemical attention now focuses on a reaction of late $G_1$ as the functional expression of the *ts* 2 defect. Genetic attention is placed on the human X chromosome which corrects this lesion (Jha and Ozer, 1977).

## B. Studies with Cells Which Arrest in S Phase

The major biochemical event of S phase is chromatin replication. This simple statement belies the complexity of individual reactions which con-stitute the separate but coordinated processes of DNA replication, replication of the histones, and of the very large number of other chromosomal proteins which contribute to the structure and function of the chromatin (cf. Sheinin *et al.*, 1978b). Biochemical dissection has permitted identification and isolation of many of the proteins which may be involved and in characterization of the specific reactions catalyzed. However, we remain a long way from arriving at a complete picture of all the integrated events of S phase, and their interaction with other cell cycle processes. Mammalian cells *ts* in DNA replication, or other reactions of S phase progression, will be extremely useful in this endeavor.

A number of criteria may be applied in designating a mutant as *ts* in S phase. In increasing cumulative order of desirability they are the following: (1) To demonstrate that cells arrest at the *npt* at some stage during the DNA synthetic phase of the cell cycle, using physiological or biochemical methods; (2) to demonstrate that a specific biochemical process of DNA replication is temperature-inactivated at the npt; (3) to demonstrate that a specific enzyme–protein of DNA replication is temperature sensitive *in vivo* and *in vitro;* (4) to demonstrate that the temperature-inactivation of this protein is subject to genetic control by the affected *ts* locus.

Evidence permits us tentatively to assign only three of the many *ts* mutants isolated so far to S phase on the basis of some, but not all, of the criteria listed. These are *ts* A1S9 (Sheinin, 1976a) and *ts* Cl (Guttman and Sheinin, 1979) mouse L cells and *ts* BN2 BHK-21/13 cells (Hand *et al.*, 1980).

### 1. ts A1S9 Mouse L Cells

*ts* A1S9 cells, isolated from mouse L cells by Thompson *et al.* (1970) have been shown to be mutant in a protein involved in the conversion of newly replicated, single-stranded DNA of $\leq 5 \times 10^6$ MW to chromosomal DNA (Sheinin, 1976a,b). By definition this is classified as an S phase mutant. Evidence indicates that most known enzymes of DNA replication are not involved in expression of the *ts* A1S9 defect. Thus synthesis of "Okazaki fragments" and their conversion to the higher molecular weight intermediates which accumulate in temperature-inactivated *ts* A1S9 cells (Sheinin, 1976a) proceeds at control rates (C. S. Schwartz and R. Sheinin, unpublished). This implies that the major DNA polymerase and the major polydeoxyribonucleotide ligase (Sheinin *et al.*, 1978b) are unlikely to be *ts.* This conclusion is supported by the finding that the replication of polyoma DNA, which is rigorously dependent upon these and other host enzymes of DNA synthesis (Sheinin, 1967), proceeds normally in *ts* A1S9 cells incubated for at least 54 hours at the *npt* (Sheinin, 1976b). Experiments with the

DNA polymerase $\alpha$ recovered in the cytoplasm, and with the nucleus-associated $\beta$ and $\gamma$ enzymes, gave no indication for temperature sensitivity of these proteins (R. Sheinin, unpublished). A similar conclusion was drawn by Dr. Marvin Gold (personal communication) who worked with polymerases partially purified by column chromatography. The mitochondrial DNA polymerase $\gamma$ appears not to be encoded in the *ts* A1S9 locus, since mitochondrial DNA synthesis is unaffected, in quantity or quality, for long periods during which nuclear DNA replication undergoes extensive inactivation (Sheinin *et al.,* 1977).

That polydeoxyribonucleotide ligase I (Soderhall and Lindahl, 1976) is not the *ts* A1S9 gene product is indicated by the studies noted above on the kinetics of transfer of "Okazaki fragments" to the growing polydeoxyribonucleotide chain. It seems also to be precluded by the experiments of Cleaver (1972) and by M. McBurney and R. Sheinin (unpublished) which revealed that temperature-inactivated *ts* A1S9 cells are fully able to repair their DNA damaged by prior exposure to ultraviolet light and X-rays, respectively. Furthermore, when *ts* A1S9 cells are incubated at the *npt* for an interval exceeding the equivalent of two generation periods, ongoing but temperature-inactivated, semiconservative replication is totally suppressed, and is supplanted by repair replication which can attain a rate of dThd incorporation approaching 20% of control levels (Sheinin and Guttman, 1977; see Section V,D,1). All repair replication requires the participation of polydeoxyribonucleotide ligase I (cf. Hanawalt *et al.,* 1979).

These experiments make it unlikely that polydeoxyribonucleotide ligase I is the *ts* A1S9 gene product. They do not, however, exclude a second enzyme detected in mammalian cells (Soderhall and Lindahl, 1976). This protein of unknown function is designated as polydeoxyribonucleotide ligase II. To address this problem J. Robertson and R. Sheinin (unpublished) examined the polydeoxyribonucleotide ligase activity in crude lysates and in partially-purified nuclei and cytoplasmic extracts of wild-type mouse L cells, and *ts* A1S9 cells incubated at 34°C or 38.5°C for periods long enough to permit temperature inactivation of the affected gene product. They detected no differences in the ligase activities in these various preparations.

One additional enzyme, poly(ADP)-ribose polymerase (Hayaishi and Ueda, 1977), has been tested as a possible *ts* A1S9 gene product recently postulated to play a regulatory role in S phase entry and/or progression (Berger *et al.,* 1978). Our studies (Savard *et al.,* 1980) reveal that the poly(ADP)-ribose polymerase activity (measured in isolated nuclei) of temperature-inactivated *ts* A1S9 cells remains at control levels, comparable to those of wild-type cells for over 40 hours. (see also Section V,D,1).

Although it has not yet proved possible to identify the *ts* A1S9 gene product we have been able to confirm its S phase action in a number of different experiments. Sheinin and Guttman (1977) demonstrated that when *ts* A1S9 cells are upshifted to the *npt*, normal semiconservative replication continues at a control level for 6–8 hours, after which it is inactivated with a $T_{\frac{1}{2}} \simeq 4$ hours to a minimum rate $\simeq$ 1–5% of control, during the equivalent of approximately two generation periods. This temperature inactivation is seen at the level of *de novo* synthesis of chromatin-bound DNA (Sheinin *et al.*, 1978a; Sheinin and Lewis, 1980). It does not result in turnover of preformed DNA, or that made at the *npt* (Sheinin, 1976a; Sheinin *et al.*, 1978a).

A second approach was to ask whether that DNA synthesis effected at the *npt* was performed by S phase cells, or by cells at other or all stages of the cell cycle. We chose to examine this by velocity sedimentation analysis of appropriately labeled wild-type (WT-4) and *ts* A1S9 cells, using the procedure of Macdonald and Miller (1970). Cells were prelabeled in their DNA with $^{14}$C-dThd as noted in the legend of Fig. 2, and then incubated for 20 hours at 34°C or 38.5°C. The cultures were pulse-labeled at the appropriate temperature with $^{3}$H-dThd to mark newly made DNA. The cells were then analyzed by velocity sedimentation in a bovine serum albumin gradient to assess cell cycle staging based on size.

The sedimentation profiles for cell distribution are shown in Fig. 2, with bar indications of cell cycle assignment calculated on the basis of the sedimentation velocities. The DNA-synthetic capacities are reported as the ratio of $^{3}$H-DNA formation at the *pt* or the *npt* to the $^{14}$C-DNA content (the distribution of which is coincident with total cell number). The data obtained with WT-4 and *ts* A1S9 cells are essentially the same. They indicate that only S-phase cells synthesize DNA during establishment of the *ts* A1S9 defect.

A third series of experiments was performed to determine whether *ts* A1S9 cells brought into S phase arrest with hydroxyurea would cycle upon drug removal. Control studies were done with WT-4 cells. As noted in Fig. 3, WT-4 cells treated with hydroxyurea at 34°C or 38.5°C progressed through one full cycle of DNA synthesis and cell division, and then into a second and/or third, whether released at 34°C or 38.5°C. Fig. 4a and b reveal that *ts* A1S9 cells blocked with hydroxyurea at 34°C cycled in control pattern when released at the *pt*. If incubated at the *npt*, the *ts* A1S9 cells continued to synthesize DNA at control rate for $\simeq$ 4 hours, after which temperature inactivation set in resulting in early exit from S and no progression to cell division. The results obtained with *ts* A1S9 cells treated with hydroxyurea at 38.5°C are shown in Fig. 4c and d. Clearly such drug treatment did not prevent temperature inactivation

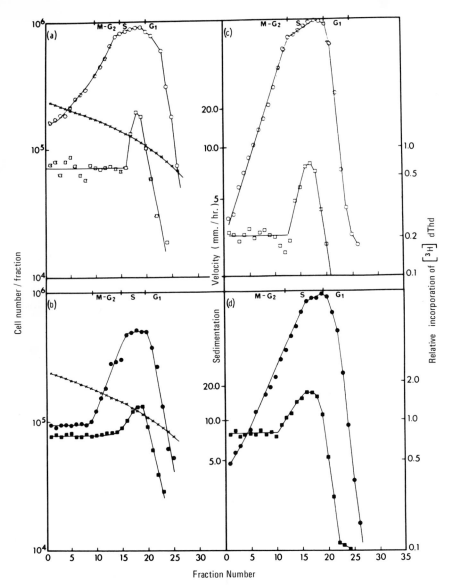

**Fig. 2.** Cell cycle analysis of DNA-synthesizing mouse L cells, by velocity sedimentation at unit gravity. WT-4 and *ts* A1s9 cells were grown at 34°C to early logarithmic phase, in medium containing ¹⁴C-dThd to uniformly label normal DNA. The cells were harvested, washed, resuspended in dThd-free medium, incubated at 34° or 38.5°C for 20 hours, and then pulse-labeled for 10 minutes with ³H-dThd (500 μCi/ml) to mark newly made DNA. For experimental details see Guttman and Sheinin (1979). The cells were chilled, harvested, and analyzed by velocity sedimentation at unit gravity in a gradient of bovine serum albumen (McDonald and Miller, 1970). (a) and (b) WT-4 ³H-labeled cells at 34° and 38.5°C, respectively. (c) and (d) *ts* A1S9 ³H-labeled cells at 34° and 38.5°C, respectively. Circles indicate cell numbers; squares the ratio of ³H-dThd/¹⁴C-dThd incorporated into DNA.

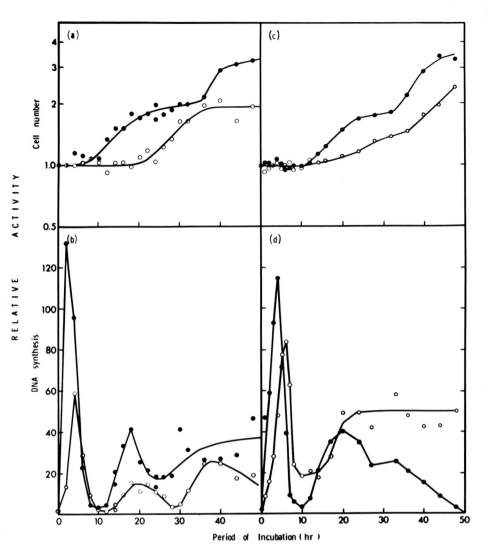

**Fig. 3.** Recovery of WT-4 cells from hydroxyurea treatment at 34° and 38.5°C. WT-4 cells were prelabeled in their DNA by growth at 34°C through four to five generations in medium containing $^{14}$C-dThd (0.01 $\mu$Ci/ml). The harvested cells were resuspended in prewarmed medium containing 1 m$M$ hydroxyurea and incubated for 12 hours at 34° or 38.5°C. They were than removed from the drug-containing medium and resuspended in appropriately prewarmed medium. Half of each culture was maintained at the temperature of drug treatment, half at the other temperature. At the intervals noted, cell number and capacity for incorporation of $^{3}$H-dThd (10 $\mu$Ci/ml for 1 hour) into DNA were measured. For experimental details see Guttman and Sheinin (1979). Recovery calculations based on preformed $^{14}$C-labeled DNA were applied to normalize $^{3}$H-labeled DNA synthesis. O—O, ●—●; cells allowed to recover at 34° and 38.5°C, respectively. (a), (b) and (c), (d); cells incubated with hydroxyurea at 34° and 38.5°C, respectively.

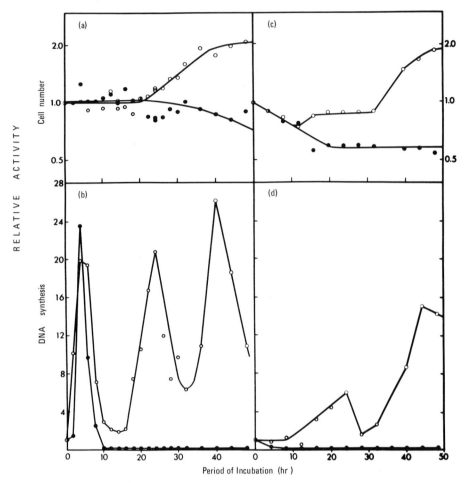

RELATIVE ACTIVITY

**Fig. 4.** Recovery of *ts* A1S9 cells from hydroxyurea treatment at 34° and 38.5°C. See legend of Fig. 3 for details.

of the *ts* A1S9 gene product as indicated by the absence of DNA synthesis and subsequent cycling in cells released at the *npt*. Release from the high-temperature drug treatment at 34°C permitted *ts* A1S9 cells to recover, but only after an extensive lag period, commensurate with that exhibited by cells temperature inactivated for 16–24 hours, in the absence of hydroxyurea (Sheinin, 1976a).

The final set of experiments in this series was performed with *ts* A1S9 and WT-4 cells arrested in a $G_1$ state by starvation of isoleucine. The control patterns of recovery observed with WT-4 cells are shown in Fig. 5. These cells, whether amino acid starved at 34°C or 38.5°C recovered

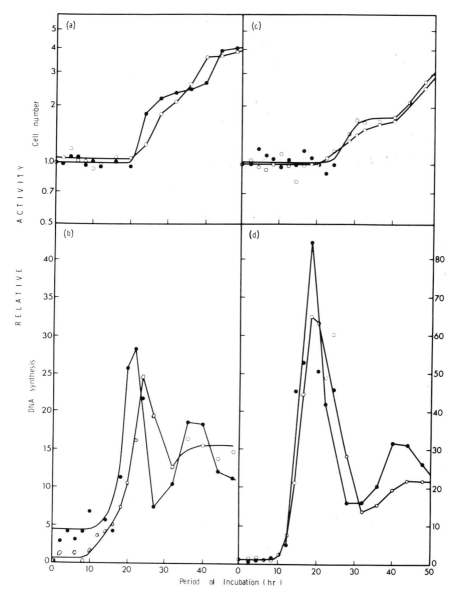

**Fig. 5.** Recovery of WT-4 cells from isoleucine starvation at 34° and 38.5°C. WT-4 cells, prelabeled in their DNA by growth at 34°C as described in the legend of Fig. 3, were incubated at 34° or 38.5°C for 20 hours in medium lacking isoleucine. They were harvested and resuspended in appropriately prewarmed complete medium. Half of each culture was maintained at the temperature of isoleucine starvation; half was shifted to the other temperature. At the intervals noted cell number and capacity for $^3$H-dThd incorporation into DNA was measured. For experimental details see Guttman and Sheinin (1979). Normalization was accomplished as noted in Fig. 3 legend. O—O, ●—●; cells allowed to recover at 34° and 38.5°C, respectively. (a), (b) and (c), (d); cells starved of isoleucine at 34° and 38.5°C, respectively.

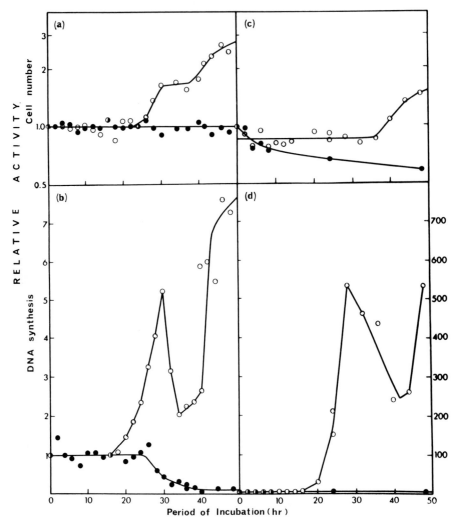

**Fig. 6.** Recovery of *ts* A1S9 cells from isoleucine starvation at 34° and 38.5°C. See legend of Fig. 5 for details.

when incubated at either temperature in the presence of isoleucine. After a lag period of ≃ 10 hours, they moved through S, through division, and into subsequent cycles of multiplication.

As shown in Fig. 6a and b, *ts* A1S9 cells deprived of isoleucine at 34°C and subsequently allowed to recover at this *pt* behaved like the wild-type cells. *ts* A1S9 cells released at 38.5°C did not synthesize DNA or divide. As shown in Fig. 6c and d, *ts* A1S9 cells starved of isoleucine at 38.5°C and incubated further at this *npt* were unable to progress into cell cycle activity. Cells subsequently incubated with amino acid at 34°C did eventually recover, with kinetics of DNA synthesis and cell division comparable to those exhibited by cells temperature-inactivated for 20–24 hour and then downshifted to the *pt* (Sheinin, 1976a).

These experiments yield results which are compatible with a model which brings *ts* A1S9 cells into early S phase arrest upon full expression of the *ts* A1S9 defect. This conclusion is in accord with the FMF data (cited in Setterfield *et al.*, 1978) which reveals accumulation at the *npt* of cells with an unreplicated or early S phase-DNA complement. On the basis of these various studies the *ts* A1S9 arrest point has been placed early on the temporal map for S phase progression, before the hydroxyurea restriction point (Fig. 7). It should be noted that the *ts* A1S9 gene product appears to act at a relatively late step in the duplication of replication units of chromosomal DNA (Sheinin *et al.*, 1978b). However, because DNA replication in higher eukaryotes is so tightly and firmly regulated throughout the DNA-synthetic period, temperature inactivation of the *ts* A1S9 protein yields cells with an early S phase terminal phenotype. This is associated with a number of interesting pleiotropic effects which are discussed in Section V,D.

### 2. *ts C1 Mouse L Cells*

The *ts* C1 mouse L cells isolated by Thompson *et al.* (1971) appear to arrest at the *npt* throughout S phase (Guttman and Sheinin, 1979). This conclusion is based upon the following evidence. Upon upshift to 38.5°C, semiconservative DNA replication is affected almost immediately and decays with a $T_{\frac{1}{2}} \simeq 4$ hour (Sheinin and Guttman, 1977). Radioautographic analysis indicates that the number of cells synthesizing DNA remains at the control level for 6–8 hours. However, the DNA-

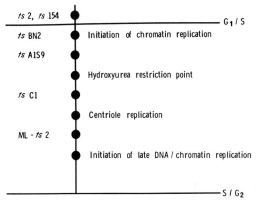

**Fig. 7.** Temporal map for S phase progression. See comments in Fig. 1 legend.

synthetic activity per cell declines rapidly (Setterfield *et al.*, 1978; Guttman and Sheinin, 1979). *ts* C1 cells brought into early S phase arrest by hydroxyurea treatment at 34°C recover in the control pattern of cell cycle progression at the *pt* (noted in Figs. 3 and 4a and b). If released from hydroxyurea arrest at 38.5°C, the cells continue DNA synthesis at the control rate for 1–2 hour and then exit rapidly from S and are unable to progress to cell division. *ts* C1 cells subjected to hydroxyurea at the *npt* and then released at the same temperature remain unable to synthesize DNA (Guttman and Sheinin, 1979). If released at the *pt*, they exhibit an initial burst of DNA-synthetic activity which is not associated with a detectable increase in cell number, as seen in Fig. 8. Entry into S phase and subsequent cell cycling begins 10–12 hours after temperature downshift.

Guttman and Sheinin (1979) have shown that *ts* C1 cells deprived of isoleucine at 34°C or 38.5°C give patterns of recovery of cell cycling activity analogous to those exhibited by *ts* A1S9 cells (Fig. 6), upon restoration of the amino acid. In addition they demonstrated that *ts* C1 cells amino acid starved at 34°C and allowed to recover at this *pt* to late $G_1$ or early S will continue progression through S phase if upshifted to 38.5°C, but only for the short survival time of the *ts* gene product at the *npt*. No cell division is detected.

Velocity sedimentation analysis has also been used to assess the cell cycle arrest point of temperature-inactivated *ts* C1 cells, using the procedure described in the legend of Fig. 2. The results obtained with *ts* C1 cells incubated throughout at 34°C were entirely analogous to those given by wild-type cells (Fig. 2a and b), and by *ts* A1S9 cells incubated under test conditions at the *pt* (Fig. 2a). The data derived with *ts* C1 cells post-labeled with $^3$H-dThd at 38.5°C resembled those obtained with temperature-inactivated *ts* A1S9 cells (Fig. 2d). However, the curve of $^3$H-labeled DNA cells was broader, suggesting that DNA synthesis had occurred in cells at various stages of S phase progression. Such a conclusion also emerges from FMF analyses (Setterfield *et al.*, 1978; Guttman and Sheinin, 1979). Because *ts* C1 cells treated with hydroxyurea at 34°C enter DNA synthesis immediately upon drug removal at 38.5°C, albeit not in the normal course of cell cycle progression, it is suggested that the *ts* C1 gene product may act beyond the hydroxyurea restriction point (see Fig. 7).

The available evidence indicates that the *ts* C1 gene probably does not carry information for the nuclear DNA polymerases (Guttman and Sheinin, 1979). Neither does it encode the mitochondrial DNA polymerase-$\alpha$, since mitochondrial DNA synthesis proceeds normally under conditions that preclude normal nuclear DNA replication

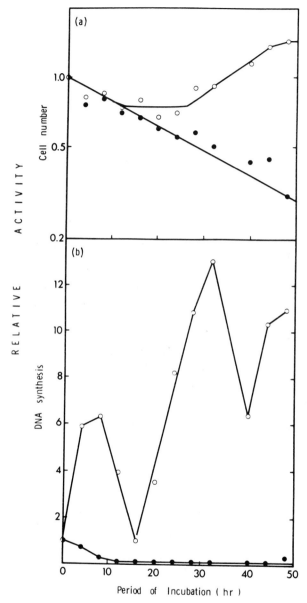

**Fig. 8.** Recovery of *ts* C1 cells from Hydroxyurea treatment at 38.5°C. See legend of Fig. 3 for details.

(Sheinin and Guttman, 1977; Guttman and Sheinin, 1979). Nor does it carry information for the poly ADP-ribose polymerase. Thus nuclei isolated from *ts* C1 cells incubated for up to 3 days at the *npt* exhibit control levels of enzyme activity (Savard *et al.*, 1980). Pleiotropic expressions associated with temperature inactivation of the *ts* C1 gene product are discussed under section D.

### 3.  *ts BN2 Hamster Cells*

*ts* BN2 cells, isolated from BHK-21/13 cells by Nishimoto and Basilico (1978) undergo rapid temperature inactivation of their DNA-synthetic activity starting about 2 hours *pts*. If *ts* BN2 cells are arrested in a $G_1$-state by isoleucine starvation at 33.5°C. they are unable to carry out DNA synthesis at 38.5°C. unless first allowed to progress to within 2–4 hours of the $G_1$/S traverse at the *pt* (Nishimoto *et al.*, 1978). Even under these conditions normal S phase progression at the *npt* appears to be impeded.

These observations are compatible with a model (see Fig. 7) that suggests that the *ts* BN2 product is required for entry into and maintenance of DNA synthesis. It is in accord with the evidence, which indicates that *ts* BN2 cells are able to effect apparently normal elongation and ligation of newly-synthesized DNA at the *npt*. It also concurs with recent studies using DNA fibre radioautography which suggest that iniation of DNA synthesis on replicating units is affected in temperature-inactivated *ts* BN2 cells (Hand *et al.*, 1980).

Nishimoto *et al.* (1978) have examined S-phase progression at the *npt* of *ts* BN2 cells brought to an early stage of DNA synthesis at 33.5°C. by sequential starvation of isoleucine and treatment with hydroxyurea. Such cells did indeed continue DNA synthesis at a very high rate; however, after ≃ 2 hours at 39.5°C, such synthesis was rapidly inactivated. FMF studies indicated accumulation of DNA by all cells for approximately 3 hours after release at the *npt*. Therefore, there appeared to be increased accumulation by some cells and loss by others. At no time did any cell acquire a $G_2$ complement of DNA, thereby explaining why FMF analysis of cultures of *ts* BN2 cells incubated at 39.5°C appeared to indicate arrest with a $G_1$ complement of DNA.

The evidence presented by Nishimoto *et al.* (1978) leave no doubt that *ts* BN2 cells are unable to maintain DNA synthesis for longer than 2–3 hours at 39.5°C. However, their data also suggest that cells cannot traverse the $G_1$/S interface at the *npt*. In addition, they show that if cells are incubated at the *npt* for any interval during progression through $G_1$, entry into DNA synthesis is delayed by slightly more than this period of time. These observations may be interpreted as indicating that the *ts* BN2 protein is synthesized and accumulates during the latter portion of

$G_1$ and early S, whereas it is active only in S. The temperature-inactivation data suggest that this protein must be present at a certain concentration (perhaps in stoichiometric amount relative to sites on the chromatin) for DNA synthesis to proceed. On the basis of these various studies expression of $ts$ BN2 is placed early in S phase, perhaps at the $G_1$/S traverse (see Fig. 7).

It is not immediately clear why the DNA of HSV-I should not be replicated in temperature-inactivated $ts$ BN2 cells (Yanagi *et al.*, 1978), since the genome of this virus is thought to encode information·for all of the enzymes and proteins necessary for HSV-DNA replication. Further studies of HSV infection of $ts$ BN2 cells may therefore reveal factors required for initiation of both cellular and viral DNA.

## C. Studies with *ts* Mammalian Cells Which Arrest in Mitosis

A number of $ts$ mammalian cells have been isolated which exhibit a terminal phenotype which has been identified with one or other of the diagnostic landmarks of mitosis, using light microscopy. Mitosis and cytokinesis, like the other stages of the cell cycle, are a continuum of a number of independent metabolic pathways, the progress of which is regulated by interactions between mitotic events, and between these and processes expressed in other stages of the cell cycle. These include the dynamic equilibrium between the microtubules of the interphase cytoskeleton, the pool of tubulin subunits and the microtubules of the mitotic spindle (cf. Dirksen *et al.*, 1979; Dustin, 1978). These encompass the analogous reactions that tie together the cytoskeletal microfilaments of interphase cells and the actin-containing contractile ring formed during cytokinesis (Gabiani, 1979; Aubin *et al.*, 1980). They also must embrace the reactions of formation, and mobilization into action, of the centrioles (Tucker *et al.*, 1979), the microtubule-associated proteins (Kirschner, 1978), and other as yet unrecognized proteins which participate in chromosome segregation. Even though the following mutants have not yet been studied at the biochemical level, they are of interest in understanding mitotic progression and its interrelationship with other cell cycle stages.

### 1. ts 655 Hamster Cells

$ts$ 655 cells (Wang, 1976) divide apparently normally at the $npt$ (39°C) for 12 hours *pts,* or perhaps the equivalent of one-half generation ($T_D$ at 33°C = 30 hours). Thereafter cells arrest with a mid-prophase terminal phenotype, characterized initially by condensation of the chromatin and disassembly of the nuclear membrane. This chromatin condensation progresses over the next 60–70 hour interval; however, distinguishable chromosomes are not formed. Instead, the continuously condensing chromatin coalesces into increasingly large masses, which never become re-enveloped by nuclear membrane. These observations suggest a *ts* defect in the microtubule system which aligns the condensing chromosomes into a metaphase configuration in preparation for what would normally be subsequent association with, and movement along, the mitotic spindle fibers.

The *ts* 665 defect appears not to affect at least some of the prerequisite processes for metaphase which undoubtedly occur during $G_2$. Thus disassembly of the cytoskeletal microtubules, which is thought to provide a pool of tubulin for subsequent mitotic spindle formation, appears to proceed normally, as indicated by the fact that all of the temperature-inactivated cells undergo a change from the fibroblastoid to the rounded morphology, in parallel with accumulation of abnormal prophase chromatin. Nevertheless progression into and beyond metaphase is completely blocked at the *npt*. As indicated by the model shown in Fig. 9, the *ts* 665 protein appears to function after those prophase events which set in motion chromosome condensation and the disassembly of the nuclear membrane.

The molecular basis for the *ts* 665 phenotype remains undefined. However, it is undoubtedly related to the fairly specific temperature inactivation of incorporation of acidic amino acids into protein, which occurs within 2 hours after upshift to 39°C under conditions which leave incorporation of other amino acids and synthesis of RNA and DNA relatively unaffected (Wang, 1976).

### 2. *ts 546 Hamster Cells*

*ts* 546 hamster cells (Wang, 1974; Wang and Yin, 1976) do form metaphase plates at the *npt* (cited in Wissinger and Wang, 1978); however, subsequent progression through metaphase appears to abort quickly and the cells acquire well-defined "metaphase chromosomes" in a pattern characteristic of those present in cells treated with compounds which specifically prevent formation of, or induce disassembly of the microtubules of the mitotic spindle (cf. Dustin, 1978). With prolonged incubation at 39°C, the "metaphase-chromosomes" become clumped and are subsequently enveloped by nuclear membrane giving rise to mono-, bi-, and multinucleate cells (Wang and Yin, 1976).

These observations can be explained by a model which places expression of the *ts* 546 defect at early anaphase on the mitotic progression map shown in Fig. 9. Thus temperature-inactivated *ts* 546 cells can effect

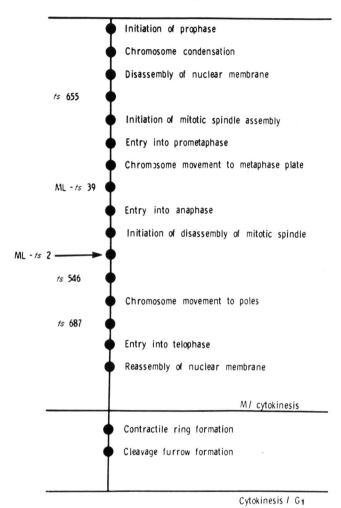

**Fig. 9.** Temporal map for progression through mitosis and cytokinesis. See comments in Fig. 1 legend. Arrow denotes signal triggered by primary expression of the ML-*ts* 2 gene product during S phase (Fig. 7).

dissolution of the cytoskeletal microtubules as evidenced by the change from fibroblastoid to rounded-up morphology, thereby releasing tubulin subunits for assembly into the spindle transiently detected as metaphase plate formation. Subsequent movement of the chromosomes appears to be abnormal; however, information for subsequent events of telophase progression seems to be transmitted. This results in the envelopment of coalesced clumps of chromosomes in reformed nuclear membrane.

### 3.  ts 687 Hamster Cells

ts 687 hamster cells appear to be defective in a protein which participates in anaphase chromosome movement toward the poles (Wissinger and Wang, 1978). After 16 hours at the npt (36°C) they arrest with a pleiotropic terminal phenotype of cells already programmed to move through various subsequent stages of mitotic termination. Thus after progressively longer intervals at the npt, the following are observed to accumulate in an apparently sequential pattern: (1) Cells in late anaphase with excessively condensed chromosomes at the poles, still in association with a thin band of microtubules arrayed along the length of the cell; (2) mono-, bi-, and increasingly multinucleated cells produced during aberrant telophase or faulty cytokinesis; (3) cells with incompletely formed cleavage furrows, perhaps resulting from the presence of residual mitotic spindle material or from an abnormally formed contractile ring.

On the basis of these observations expression of the ts 687 gene is placed in anaphase after those events which set in motion subsequent progression into and through telophase; which in turn permit or promote movement into and through cytokinesis (see Fig. 9). Such physiological mapping of the ts 546 and ts 687 functions is in accord with the evidence indicating complementation between the two genetic loci (Wissinger and Wang, 1978).

Approximately 24 hours after maximum mitotic arrest, temperature-inactivated ts 687 cells undergo a change in morphology which differs from that exhibited by ts 546 cells (Wang, 1974) or by ts 665 cells (Wang, 1976). The former lose their normal fibroblastoid character and become elongated rather than rounded. This may reflect two separate but related phenomena. Thus ts 687 cells must continue to synthesize protoplasm at the npt and therefore increase in volume in the absence of normal cell division. In addition, progression proceeds, albeit abnormally, through and beyond telophase and well into cytokinesis beyond the point at which one would have expected dissolution of the spindle microtubules and reassembly of the cytoskeletal microtubules, thereby promoting resumption of normal cellular morphology.

### 4.  ML-ts 2 Murine Lymphoma Cells

The ts 2 mutant of mouse lymphoma cells (referred to herein as ML-ts 2) are able to divide twice after upshift to the npt (39°C) and remain viable for at least 4 days, as judged by their ability to resume growth (after an extended recovery phase) at 33°C (Shiomi and Sato, 1976).

RNA and protein synthesis appear to be unaffected for at least 30 hours, whereas DNA synthesis falls gradually as temperature-inactivated cells fail to enter S. Cultures incubated at the *npt* accumulate cells with aberrant mitotic figures and with multiple nuclei.

Temperature-induced mitotic arrest was first observed $\simeq$ 6 hours *pts,* suggesting that the execution point for subsequent expression of the ML-*ts* 2 terminal phenotype may be in S phase (Fig. 7). Support for this conclusion was obtained using ML-*ts* 2 cells partially synchronized at 33°C by treatment with excess thymidine, followed by colcemid to arrest them at metaphase. The cells were released from drug block to accumulate with a majority of the population in $G_1$ or $G_2$. When the $G_2$ cells were upshifted to 39°C, they proceeded normally through the first mitotic division, and arrested in the succeeding M phase. The upshifted $G_1$ population was unable to complete even the first mitosis normally. The execution point for primary expression of the ML-*ts* 2 defect is therefore placed early to mid-S phase, whereas the terminal phenotype of temperature-arrested cells is that of cells blocked in mitosis beyond the stage at which progression into late telophase and/or cytokinesis is already programmed (Shiomi and Sato, 1978).

Cells arrested in metaphase by temperature inactivation of the ML-*ts* 2 gene product recover when downshifted to the *pt,* which is not so for those which have progressed to multinucleation. These findings suggest that mitotic spindle microtubule formation and all preceding events occur normally at the *npt.* This is in accord with the observation that colcemid treatment of temperature-inactivated ML-*ts* 2 cells effectively blocked subsequent micronucleation at the *npt.*

It has been suggested by Shiomi and Sato (1978) that the ML-*ts* 2 defect which is initially expressed prior to mid-S phase, may affect normal replication of the centriole, thereby setting in train the post-metaphase, temperature-sensitive events observed in ML-*ts*-2 cells. Manifestation of the terminal phenotype has therefore been placed early in anaphase after expression of the signal for subsequent nuclear membrane reassembly in telophase (Fig. 9). Because ML-*ts* 2 cells accumulate aberrant mitotic figures at the *npt,* whereas the chromosomes of *ts* 546 cells become aggregated and fused after metaphase plate formation, the ML-*ts* 2 expression site has been placed earlier in anaphase to reflect the possible continued presence of the prophase factors causing chromatin condensation and coalescence.

### 5. ML-ts 39 Murine Lymphoma Cells

ML-*ts* 39 mouse lymphoma cells incubated at the non-permissive temperature accumulate in metaphase and then proceed to micronucleation (Sato and Hama-Inaba, 1978). Since metaphase cells do appear, albeit transiently, the ML-*ts* 39 defect is located on the physiological map close to the metaphase/anaphase traverse, beyond the function which signals subsequent nuclear membrane reformation in telophase (Fig. 9).

## D. Pleiotropic Expression of *ts* Mutations Which Affect DNA Synthesis

### 1. Semiconservative to Nonconservative DNA Replication

It has now been demonstrated that semiconservative DNA replication proceeds in *ts* A1S9, *ts* Cl (Sheinin and Guttman, 1977) and *ts* 2 cells (Sheinin et al., 1980) incubated at the *npt* for an interval equivalent to about two generation periods, even though DNA synthesis is severely temperature-inactivated. In addition, that DNA which is made at the *npt* by these cells is effected by the normal discontinuous mode (Sheinin, 1976a; Sheinin *et al.,* 1978b, 1980; Guttman and Sheinin, 1979), even though this is interrupted or aborted by temperature inactivation of the specific gene product. Similar observations have been made with respect to discontinuous DNA synthesis in *ts* BN2 cells incubated at the *npt* (Nishimoto *et al.,* 1978).

Our studies reveal that if *ts* A1S9, *ts* Cl or *ts* 2 cells are incubated at the *npt* for an interval longer than that required for two turns of the cell duplication cycle normal, semi-conservative DNA replication is totally suppressed (Sheinin and Guttman, 1977; Sheinin *et al.,* 1980). At this time nonconservative DNA replication becomes apparent and increases to about 20% of control DNA-synthetic activity (assessed on the basis of incorporation of precursor into DNA) after 4–5 days. This movement of *ts* Cl and *ts* 2 cells into repair replication may be accompanied by a significant increase in the activity of poly (ADP)-ribose polymerase (Savard *et al.,* 1980).

These findings suggest that enhancement of repair replication may occur in all *ts* mammalian cells which exhibit a terminal phenotype of arrested semiconservative DNA synthesis. *ts* A1S9, *ts* Cl and *ts* 2 cells thus provide an important experimental system in which to study DNA repair replication in higher eukaryotes, the regulatory mechanisms which determine whether cells effect semiconservative or repair replication, and the possible role of poly (ADP)-ribose polymerase in these processes.

### 2. Chromatin Replication and Segregation Studied with ts Mammalian Cells

Interest in *ts* mammalian cells lies not only in what they can tell us about the specific reaction determined by the affected *ts* gene product.

They may provide even more exciting information about the regulatory links which operate within and between phases of the proliferation cycle.

   a.   *Regulation of Synthesis of DNA, Histones, and Other Chromosomal Proteins in Mammalian Cells* ts *in* $G_1$ *and S Functions.*   The evidence linking histone synthesis to DNA replication in the duplication cycle of eukaryotic cells is extensive and convincing (cf. Borun, 1975). Coupling of these two synthetic processes appears to be firmer in the forward direction, in that histone synthesis proceeds normally only during the DNA-synthetic phase of the cell cycle and is usually initiated at or near the $G_1$/S interface (Sheinin *et al.*, 1978b; Cremisi, 1979). In contrast, cessation of DNA replication is often not very tightly coupled with termination of histone synthesis (Nadeau *et al.*, 1978). In addition there are interesting examples in nature of total uncoupling, e.g., during meiotic recombination events in plants (Borun, 1975; Stern and Hotta, 1977), during embryogenesis in sea urchin (DiLiegro *et al.*, 1978) and amphibians (Woodland and Adamson, 1977), and in nucleated erythrocytes (Mura *et al.*, 1978). The discovery of the nucleosome subunit structure of chromatin has made the questions asked previously about the origin or necessity of coupled DNA and histone duplication redundant. It is now obvious that normal replication of the chromatin during S phase requires coordinated and highly regulated replication of the DNA, the histones, the non-histone chromosomal proteins, and other chromosomal constituents (Sheinin *et al.*, 1978b; Cremisi, 1979).

   Although we have considerable insight into those processes that generate new DNA and new histones, we know very little about the events that trigger chromatin replication, nor do we understand how coupled DNA–histone synthesis is regulated. A preliminary approach to this problem was made using *ts* A1S9 and *ts* Cl mouse L cells and BALB/c-3T3 *ts* 2 fibroblasts described in Sections V,B, and V,A,8, respectively. The results show that in the two mouse L cell mutants which are *ts* in an S phase function, temperature inactivation of DNA replication is followed by termination of synthesis of histone, of other basic proteins, and of the non-histone chromosomal proteins (NHCP) (Sheinin *et al.*, 1978a; Sheinin and Lewis, 1980). In contrast, the continued synthesis of histones and the other chromosomal proteins is uncoupled from temperature-inactivated DNA synthesis in the *ts* 2 cells and *ts* CH-K12 cells (Rieber and Bacalao, 1974b), which are known to be mutant in a gene of $G_1$ expression (see Section V,A,1).

   It is of interest that cessation of histone synthesis is not tightly tied to temperature inactivation of DNA replication in *ts* A1S9 and *ts* Cl cells, but rather to a signal emitted at termination of the aborting S phase in progress. These temporal relationships are shown in Fig. 10 which schematically presents the data obtained in these studies. It includes for

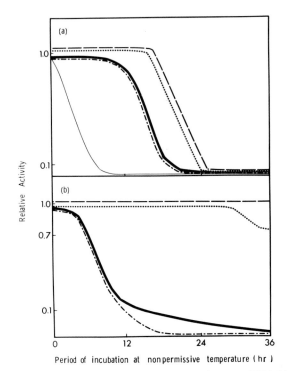

**Fig. 10.** Schematic representation of the relationship between DNA, histone synthesis, and chromatin decondensation in temperature-inactivated mammalian cells which exhibit *ts* DNA Synthesis. (a) DNA*ts* cells: *ts* A1S9, *ts* C1; (b) $G_1{}^{ts}$ cells: BALB/c-3T3 *ts* 2, *ts* CH-K12, CHO-*ts* 13A, CHO-*ts* 15C. Incorporation of ³H-dThd into TCA-insoluble material by *ts* A1S9, *ts* 2, CHO-*ts* 13A, CHO-*ts* 15C (—) and *ts* C1 cells (–), respectively. Percentage of cells making DNA measured by radioautagraphy (-·-·-·). Incorporation of [³H]lysine + [³H]arginine into chromatin-bound histones (......). Disaggregation of condensed heterochromatin (- - -).

comparison the results given by *ts* 2 cells, which in turn are similar to those reported for *ts* CH-K12 cells (Rieber and Bacalao, 1974b), and obtained in recent studies with CHO-*ts* 13A and CHO-*ts* 15C cells (R. Sheinin, unpublished), which also exhibit temperature-sensitive DNA synthesis (Gupta *et al.,* 1980) probably because they arrest in a $G_1$ state (P. R. Srinivasan, personal communication).

These observations suggest that derepression of chromatin replication, which occurs late in $G_1$, activates a number of independent metabolic pathways; however, these are subject to multiple interconnecting regulatory links that ensure normal chromatin replication and segregation to daughter cells. Four of these several processes are presented in the model shown in Fig. 11. The DNA pathway gives rise to the enzymatic machinery of DNA replication; the second pathway provides for synthesis of the histones and the structural high mobility group (HMG) proteins (Goodwin *et al.*, 1978); the third for the formation of the NHCP; and the fourth for the chromatin modeling events which are recognized as a cycle of changing chromatin ultrastructure discussed below.

The finding that histone synthesis proceeds for very long periods after temperature-inactivation of the gene products of *ts* 2, *ts* CH-K12, CHO-*ts* 13A, and CHO-*ts* 15C cells suggests that histone synthesis may be switched on late during $G_1$ by a regulatory process which can be dissociated from initiation of DNA replication. Evidence for major histone synthesis in the absence of DNA replication comes from experiments with developing oocytes of sea urchin (DiLiegro *et al.*, 1978) and amphibians (Woodland and Adams, 1977), which go through an extended $G_1$ period during early embryogenesis. It also derives from studies with Friend leukemia cells forced into a prolonged $G_1$ state during induction of differentiation by *n*-butyrate or dimethyl sulfoxide (Zlatanova and Swetly, 1978). Significant and uncoordinated histone synthesis has been reported in BHK-21 cells brought into $G_1$ arrest by isoleucine starvation (Tarnowka *et al.*, 1978).

On the basis of the experiments discussed here, it is postulated that

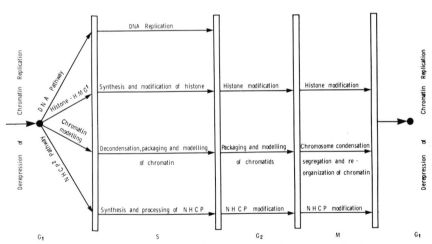

**Fig. 11.** Model for cell cycle progression as it affects chromatin replication and segregation. Histone-HMG[1] denotes the pathway of derepression of synthesis of those basic proteins which contribute to maintenance of primary nucleosomal structure and perhaps to higher order chromatin structure. NHCP[2] refers to the nonstructural chromatin proteins (including some HMG proteins) which may participate primarily in chromatin modeling for transcription.

mutants of mammalian cells which are truly *ts* in DNA replication should exhibit coordinate cessation of histone synthesis upon expression of the *dna^ts* defect. Such may not be the case with cells which are *ts* in a late $G_1$ function, or in non-S phase functions normally affecting turnover and modification of specific histones (cf. Newrock *et al.*, 1977; Gurley *et al.*, 1978). Clearly this is a tentative hypothesis because the number of *dna^ts* mammalian cells available for study has been limited.

Little is known about the mechanism of coordinate stimulation of histone and DNA replication at the beginning of S phase. Although histone mRNA may be present throughout the cell cycle (Melli *et al.*, 1977), its synthesis is greatly enhanced and its translation is set in motion in association with initiation of DNA replication at the $G_1$/S traverse (Borun, 1975; Stein *et al.*, 1975; Melli *et al.*, 1977). The trigger for these events remains unidentified (Kedes, 1979), but may involve post-transcriptional processing and transport of nuclear histone mRNA.

Information concerning the mechanism of the linked termination of DNA and histone synthesis as chromatin replication nears completion is equally sparse (Kedes, 1979). It is clear, however, that cessation of DNA replication, whether it occurs normally during S phase or is induced by viral or chemical inhibitors of DNA synthesis (Borun, 1975; Stahl and Gallwitz, 1977; Tallman *et al.*, 1977; Hand and Kasupski, 1978), is rapidly followed by coupled inhibition of histone synthesis. It is suggested that the latter occurs as a result of dissociation of polysomes carrying histone mRNA and degradation of at least some of this mRNA (Perry and Kelley, 1973; Kedes, 1979; Stahl and Gallwitz, 1977). Once again, the trigger for this series of events remains unrecognized. It seems likely that it acts prior to termination of DNA replication since the evidence indicates little or no pool of free histones in or out of S phase; nor is there an accumulation of excess histone on chromatin during normal replication (Borun, 1975; Elgin and Weintraub, 1975; R. Sheinin, unpublished).

   *b.   Relationships between Synthesis of Chromatin-DNA and Protein and Chromatin Structure.*   It has long been postulated that synthesis of the DNA of eukaryotic cells requires that the chromatin with which it is compacted and condensed to subserve mitotic division, become uncondensed and accessible to the enzymatic machinery of replication (discussed in Setterfield *et al.*, 1978, Sheinin *et al.*, 1980). It is therefore of considerable interest to discover that expression of the *ts* A1S9 and *ts* Cl lesions sets in motion a process that results ultimately in a major structural reorganization of the chromatin.

   Chromatin is organized within the interphase nucleus in a number of different morphotypes. Of particular interest in the present context are the nucleolus, the heterochromatin (which is recognized as highly condensed, electron-dense material associated with the nuclear membrane and the nucleolus, but is also dispersed throughout the nucleoplasm), and the euchromatin or the uncondensed chromatin of the nucleoplasm. The heterochromatin exists as two subtypes, termed constitutive and facultative. The former appears to be present at all times as highly condensed material in association with the nuclear membrane and the nucleolus. The latter appears to change reversibly, going from a state of condensation to one of decondensation and back again, in association with chromatin replication (Back, 1976). The likelihood is that all of the chromatin changes its morphological configuration at some time during the total process of the chromatin modeling pathway noted in Fig. 11, which is, of course, linked to the other pathways shown.

Our studies with the $dna^{ts,}$ $ts$ A1S9, and $ts$ Cl cells revealed that temperature inactivation of the gene products results in a major structural reorganization of the chromatin which is first detected 18–20 hours $pts,$ using the tools of electron microscopy or of light microscopy (Dardick $et$ $al.,$ 1978a,b; Setterfield $et$ $al.,$ 1978; L. Naismith, I. Dardick and R. Sheinin, unpublished). This change is manifest in decondensation of the facultative heterochromatin which is complete $\simeq$ 28–32 hours $pts.$

It is clear from the kinetics of temperature inactivation schematically shown in Fig. 10, that the process of chromatin decondensation may be temporally linked to the inability of temperature-inactivated $ts$ A1S9 and $ts$ Cl cells to synthesize, and subsequently to modify, the histones and other basic structural proteins of the chromatin, which are essential to the evolution of higher order structure (Fig. 11; Georgiev $et$ $al.,$ 1978; Bradbury, 1979). It is not tied to inhibition of DNA replication suggesting that although linked, the two processes are biochemically distinct. A similar conclusion is emerging from studies in which wild-type WT-4 cells were directly inhibited in DNA replication by hydroxyurea, cytosine arabinoside, FdUrd, and mitomycin C, at 34°C. These showed the same pattern of chromatin structural reorganization observed with temperature-inactivated $ts$ A1S9 and $ts$ Cl cells (Sheinin $et$ $al.,$ 1980).

The disaggregation of the facultative heterochromatin requires $de$ $novo$ protein synthesis as indicated by the fact that although cycloheximide and isoleucine starvation rapidly inhibit DNA replication through restriction of protein synthesis, they do not set in motion these events leading to chromatin decondensation (Sheinin $et$ $al.,$ 1980). Similar conclusions are deriving from analogous experiments performed with human lymphocytes prodded into cell cycle by treatment with a mitogen (Setterfield and Kaplan, 1980). These various studies are compatible with the scheme presented in Fig. 12, which attempts to

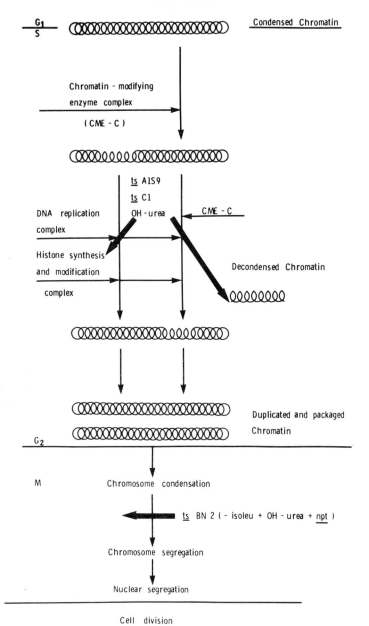

**Fig. 12.** Model for cell cycle progression as reflected in structural modification associated with chromatin replication and segregation. Large arrows denote interruptions in metabolic pathways affected by temperature inactivation of *ts* gene products and by treatment with hydroxyurea (OH-urea) or isoleucine starvation (− isoleu).

relate chromatin structure to the known biochemical events of chromatin replication.

The model suggests that initiation of chromatin replication embodies an event mediated by what is designated as a chromatin-modifying enzyme complex. This [which may be a family of specific, chromatin-bound proteases or histone-modifying enzymes (cf. Elgin and Weintraub, 1975)] is likely to be made late in $G_1$ in association with derepression of chromatin replication, but begins to act at the $G_1$/S traverse, producing localized chromatin decondensation to accommodate the DNA replication complex, the incoming newly-made histones, HMG proteins, and the other chromosomal constituents. In the normal course of events modification of histones would lead to localized recondensation of the chromatin. This cycle would be repeated throughout the entire length of the chromatin fibers, such that by the end of $G_2$, chromatin replication would be completed and the fully recondensed chromatin would be available for the further modeling of mitosis recognized as chromosome condensation, movement, and segregation to the newly forming daughter cells.

The evidence obtained with temperature-inactivated $ts$ A1S9 and $ts$ Cl cells, and with WT-4 cells treated with antimetabolites of DNA synthesis, suggests that the process of decondensation of the facultative heterochromatin is distinct from DNA replication, from histone synthesis, and from the modification reactions of DNA packaging and chromatin modeling. Indeed the decondensation event is only revealed when the biosynthetic events of chromatin replication are aborted (Fig. 12).

It is intriguing that the process of decondensation stops short of involving the constitutive heterochromatin associated with the nuclear membrane and the nucleolus, suggesting independent mobilization mechanisms for site-specific chromatin replication. This is especially interesting since it is already known that different blocks of chromatin are programmed to replicate during early, mid, and late S phase (Back, 1976; Prescott, 1976b; Sheinin *et al.*, 1978b). The various observations indicate that the chromatin modeling processes which move cells into and through mitosis may be triggered during S as a result of normal DNA replication, nucleosome packaging, and chromatin fiber stacking.

This concept is in accord with the observations made with ML-$ts$ 2 cells (Section V,C,4). Temperature-inactivation of the affected gene product appears to be executed in mid-S phase, but is manifest in a terminal phenotype of early anaphase already programmed for progression into telophase and/or cytokinesis.

In this context certain observations made with $ts$ BN2 cells (Nishimoto *et al.*, 1978) become clear. If $ts$ BN2 cells are brought into early DNA synthesis arrest at the *pt* by isoleucine starvation for 72 hours followed by hydroxyurea treatment for 20 hours, and are then incubated at the *npt* the following effects on chromatin organization are observed (see Fig. 12). Chromosome condensation and disassembly of the nuclear membrane, characteristic of prophase, begins within 2 hours and cells with apparently prematurely condensed chromosomes accumulate to reach a plateau about 6 hours later. At $\simeq$ 4–5 hours *pts* a population of cells with micronuclei begins to appear. With prolonged incubation at the *npt* cells with prematurely condensed chromosomes are replaced by cells with micronuclei, reminiscent of those present in cells subjected to procedures known to prevent formation of, or to cause depolymerization of the tubulin of the mitotic spindle. They also resemble the micronucleated cells produced by temperature-inactivation of ML-*ts* 39 cells which arrest in metaphase (see Section V,C,5).

If *ts* BN2 cells are allowed to recover from the combined isoleucine–hydroxyurea treatment for 7 hours (i.e., into late S) prior to temperature upshift, the cells move apparently normally through S, through $G_2$, and through mitosis (Nishimoto *et al.*, 1978). These results make interesting comparison with those obtained with *ts* BN2 cells arrested in $G_1$ by isoleucine starvation, and then upshifted to the *npt* early and very late after release. The former do not move into DNA synthesis; neither do they progress into aberrant mitosis. However, cells upshifted after the cells have passed mid-S phase show by far the greatest accumulation of abnormal mitotic cells over the subsequent 12-hour period of incubation at the *npt*.

The studies with *ts* A1S9, *ts* Cl, and *ts* BN2 cells suggest a complex and multiply coordinated network of events which appear to tie initiation of chromatin replication to chromosome segregation prior to cell division (see Figs. 11 and 12). Temperature-inactivation of DNA replication at different stages of evolution of this process gives rise to two classes of terminal phenotype as evidenced by chromatin morphology. That of disaggregation of the facultative heterochromatin is perhaps associated with an early S phase expression which sets in motion the unfettering of chromatin in preparation for replication. That which manifests itself as progression into mitosis may be associated with a mid- or late-S signal.

In the context it is of interest to discover that BALB/c-3T3 *ts* 2 cells, which arrest in a $G_1$ state at the *npt*, do not exhibit either of these phenomena. The condensed heterochromatin remains apparently unaltered even after 48 hours at the *npt* (Sheinin *et al.*, 1980). This parameter may, therefore, provide yet another means of distinguishing cells which are *dna*[ts] from cells which are *ts* in a function of other cell cycle stages.

*c. Chick Erythrocyte Reactivation in the Study of ts Mammalian Mutants.* The mature chick erythrocyte carries its chromatin in a highly condensed configuration which is inactive in DNA replication. Upon fusion with mammalian cells in S phase, a heterokaryon is formed in which the avian erythrocyte nucleus is reactivated to enter DNA synthesis (cf. Harris, 1974). This process may perhaps be considered as a model of $G_1/S$ traverse in the general sense that it unrolls sequentially as major decondensation of the highly compacted chromatin, followed by formation of the nucleolus, transcription, and eventually, DNA synthesis. Very careful studies suggest that only cells actively engaged in DNA synthesis are able to effect reactivation. However, the precise function(s) or protein(s) involved have not been identified.

Mature chick erythrocytes have been fused with a number of *ts* mammalian cells to assess the cell cycle arrest point upon temperature-inactivation. As noted later in Table VII, of those cells tested only one has yielded heterokaryons which when incubated at the *npt* exhibits reactivated avian nuclei—the *ts* 111 hamster cell which is considered to be temperature-sensitive in cytokinesis.

It is not surprising that the *cs* 4 D3 cell is inactive in this test since it probably arrests in a noncycling $G_0$-state. *ts* AF8 and *ts* CH-K12 cells become temperature-inactivated to an arrest point in $G_1$ prior to that at which derepression of chromatin replication occurs. It is therefore not unexpected to find that such cells are unable to effect analagous derepression in chick erythrocyte nuclei in heterokaryons incubated at the *npt* for some 10 or more hours after expression of the *ts* defect.

Fusion experiments with *ts* A1S9 and *ts* Cl cells were performed with the expectation that temperature-inactivated cells which arrest in S phase might still be capable of reactivating avian erythrocyte nuclei. This did not happen for reasons that are not clear. However, they may have to do with the temporal relationships between temperature inactivation of the *ts* gene product involved, attendant block to essential metabolic pathways, and the reactivation process *per se.*

These problems are exemplified by the experiments performed with *ts* 422E cells. Bramwell (1977) demonstrated that reactivation of chick erythrocyte nuclei in heterokaryons did not occur at 39°C, unless these were first incubated at 33 hours for that interval required for appearance of nucleoli, the organelle which encodes information for ribosomal RNA (rRNA) synthesis (Busch *et al.,* 1978), the *ts* function in *ts* 422E cells (Section V,E,2). These various studies indicate that the use of reactivation of chick erythrocyte nuclei in assessing cell cycle arrest points may be somewhat premature since it does not distinguish between cells

**Table V** Temperature-Sensitive Mammalian Cells with Identified Gene Products

| Mutant designation | Cell of origin | ts Protein | Reference |
|---|---|---|---|
| ts 422E | BHK-21/13 | 60 S Ribosomal subunit protein | Toniolo et al. (1973) |
| AARS$^{ts}$ | CHO, CHL-RJK-O | Aminoacyl tRNA synthetase | See Table VI |
| ts AUX B1 | CHO | Tetrahydrofolate polyglutamyl-transfering enzyme | McBurney and Whitmore (1974b) |
| ts 14 | CHL-V79 | 60 S Ribosomal subunit protein | Haralson and Roufa (1975) |
| 526, 536 | CHL-V79 (A3) | Hypoxanthine-guanine phosphoribosyltransferase | Fenwick and Caskey (1975) |
| ts AMA$^{R}$-1 | CHO | RNA polymerase II | Ingles (1978) |
| ts Cl-B59 | Mouse FM3A | Thymidine kinase | Nakano et al. (1978) |

which arrest in $G_1$ at the *npt,* those which arrest in S, and those which exhibit a random block.

## E. Mutant Cells Temperature-Sensitive in Specific Proteins

Tables V and VI list those *ts* mammalian cells in which it has been possible to identify a specific *ts* gene product. In most instances the energies of the investigators have been directed toward a total biochemical–genetic characterization of the particular *ts* mutants. Their potential for analysis of cell cycle progression has yet to be fully realized.

### 1. ts Aminoacyl-tRNA Synthetases (AARS[ts])

The first isolation of a mammalian cell with a *ts* aminoacyl-tRNA synthetase was made by Thompson *et al.* (1973). These have now been collected in quite large numbers as indicated in Table VI. Genetic and biochemical studies suggest that each cell (designated as AARS[ts]) carries a single, point mutation in a specific aminoacyl-tRNA synthetase. The most extensive analyses have been preformed with *ts* Hl cells which are *ts* in a leucyl-tRNA synthetase. It has been shown by *in vivo* and *in vitro* experiments that the enzyme itself is temperature sensitive (Thompson *et al.,* 1973; Hampel *et al.,* 1978). Revertants in a single-step mutation

**Table VI   Mammalian Cells Temperature-Sensitive in an Aminoacyl tRNA Synthetase**

| AA[a] tRNA synthetase affected | Mutant designation | Cell of origin | Reference |
|---|---|---|---|
| Alanyl | ML-*ts* 3 | Mouse, L5178Y | Sato (1975) |
| Arginyl | AARS[ts] (Arg-1)[b] | CHO | Adair *et al.* (1978) |
| Asparaginyl | AARS[ts] | CHL | Thompson *et al.* (1975, 1977); Adair *et al.* (1978) |
| | RJK-474[c] RJK-472 | CHL-RJK-O | Wasmuth and Caskey (1976) |
| Glutaminyl | AARS[ts] (Gln-1,-4) | CHO | Adair *et al.* (1978) |
| Histidyl | AARS[ts] (His-1) | CHO | Adair *et al.* (1978); Ashman (1978) |
| Leucyl | *ts* H1 | CHO | Thompson *et al.* (1973, 1975, 1977, 1978) |
| | *ts* 025 | | McBurney and Whitmore (1974a) |
| | AARS[ts] | | Haars *et al.* (1976); Ashman (1978) |
| Methionyl | AARS[ts] | CHO | Adair *et al.* (1978) |

[a] Aminoacyl.

[b] Aminoacyl-tRNA synthetase.

[c] RJK-474 is probably exemplary of five other mutants designated 471, 473, 475, and 476.

**Table VII    DNA Synthesis in Chick Erythrocyte Nuclei in Heterokaryons with** *ts*
**Mammalian Cells**

| | Reactivation of chick erythrocyte nuclei | | |
| --- | --- | --- | --- |
| *ts* Cells | *pt* | *npt* | Reference |
| *cs* 4 D3 | + | − | Tsutsui *et al.* (1978) |
| CH-K12 | + | − | Dubbs and Kit (1976) |
| *ts* AF8 | + | − | Tsutsui *et al.* (1978) |
| *ts* AlS9 | + | − | D. R. Dubbs, R. Sheinin, and S. Kit (1979, unpublished) |
| *ts* Cl | + | − | D. R. Dubbs, R. Sheinin, and S. Kit (1979, unpublished) |
| *ts* 111 | + | + | Tsutsui *et al.* (1978) |
| *ts* 422E | + | − | Bramwell (1977) |

have been isolated (Molnar and Rauth, 1979; Molnar *et al.*, 1979). These are characterized by near wild-type growth properties and near normal leucyl-tRNA synthetase activity. Wasmuth and Caskey (1976) have unequivocally demonstrated the temperature sensitivity of an asparaginyl-tRNA synthetase in the RJK-474 isolate of CHL-RJK 0 cells. Three complementation groups affecting the leucyl-, histidyl- and valyl-tRNA synthetases have been identified by Ashman (1978). Adair *et al.* (1978) have also established separate complementation groups for the analagous enzymes utilizing methionine, glutamine, arginine, and asparagine.

Direct use of the AARS$^{ts}$ mutants in the study of cell cycle progression may prove to be difficult if, as is the case with *ts* Hl cells (Thompson *et al.*, 1973; R. Sheinin, unpublished), such cells die and disintegrate within a very short period (less than one generation equivalent) at the *npt*. Extreme temperature sensitivity has also been reported for eight other AARS$^{ts}$ mutants (Ashman, 1978). It is possible, however, to effect phenotypic reversion of AARS$^{ts}$ mutants at the *npt* by adding the affected amino acid to the growth medium (Sato, 1975; Wasmuth and Caskey, 1976; Molnar and Rauth, 1979). Indeed, even unrelated amino acids serve, partially or completely, to protect complexes of aminoacyl-tRNA synthetases at the *npt*. One may therefore be able to manipulate the AARS$^{ts}$ phenotype, in time and extent, to assess the impact on cell cycle progression of a specific aminoacyl-tRNA synthetase, or of protein synthesis generally (Stanners *et al.*, 1978; Lofgren and Thompson, 1979).

A rather unique approach to studying the role of protein synthesis in growth control has been adopted using growth control revertants of *ts*

Hl cells, designated as GRC$^+$ mutants (Pollard and Stanners, 1979; Stanners *et al.*, 1979). These have apparently normal growth control properties but retain their *ts* phenotype. Such cells should be useful in the study of the relationship between capacity for amino acid activation, the stringent control of growth in normal cells, and the loss of such control in transformed cells.

## 2. Mutant Cells Temperature-Sensitive in Ribosomal Proteins (ts 422E and ts 14)

Two mutant mammalian cells have now been isolated which appear to be *ts* in a protein of the ribosomal protein complex, *ts* 422E derived from BHK-21/13 cells (Meiss and Basilico, 1972), and *ts* 14 obtained from the HT-1 clone of V79 Chinese hamster lung cells (Roufa and Reed, 1975).

Temperature inactivation of the formation of 28S rRNA begins in *ts* 422E cells within 2–3 hours after upshift to 39.5°C, and reaches a plateau at 16 hours. Thereafter progression through cytokinesis is blocked, even though DNA and protein synthesis proceed apparently at normal rates for at least 4 days (Meiss and Basilico, 1972; Toniolo *et al.*, 1973; Basilico *et al.*, 1974; Grummt *et al.*, 1979). Synthesis of 18S rRNA and the associated 40 S ribosomal subunit are unaffected under conditions in which formation of 28 S rRNA and the 60 S ribosomal subunit are precluded. The 40 S ribosomal subunit made at the *npt* was shown to be functional in *in vitro* experiments using normal 60 S ribosomal subunits to reconstitute active polysomes.

Subsequent studies revealed a *ts* mechanism for the nuclear processing of rRNA precursors which results in the accumulation of the 32 S rRNA precursor within the nucleolus, to about 10% the level found in wild-type BHK cells (Toniolo and Basilico, 1976). The *ts* 422E defect did not interfere with preceding processing events, since the 45 S rRNA precursor was formed normally at 39.5°C. The *ts* block was apparent in the transient accumulation of the 45 S rRNA (Ouellette *et al.*, 1976) and of a 35 S rRNA precursor, not usually detected in wild type cells (Toniolo and Basilico, 1976). The evidence suggests that abnormal processing may result from the formation of a faulty preribosomal, 32 S rRNA-containing particle. It was suggested that this remains unprocessed, nucleolus-associated, and nonfunctional even after shiftdown to the permissive temperature, because of a *ts* ribosomal protein. Such could also account for the apparent inhibition of the methylation of rRNA and its 45 S precursor molecules observed by Ouellette *et al.* (1976) in temperature-inactivated *ts* 422E cells.

*ts* 14 cells (initially designated 111-53) were shown to undergo extensive temperature inactivation of protein synthesis within 6–7 hours upon upshift to 39°C (Roufa and Reed, 1975), approximately 6 hours before DNA and RNA synthesis were blocked. It was subsequently demonstrated that the *ts* 14 mutation affects the synthesis of a protein of the 60 S ribosomal subunit (Haralson and Roufa, 1975; Roufa and Haralson, 1975). This is indicated by the fact that the 60 S ribosomal particles isolated from temperature-inactivated *ts* 14 cells, unlike their control and wild-type counterparts, do not survive the *in vitro* manipulations of purification. Direct electrophoretic analysis of the proteins of such 60 S particles has not uncovered any differences between the ribosomal proteins of control wild-type or temperature-inactivated cells (Boersma *et al.*, 1979), indicating that the *ts* 14 mutation does not alter the charge of the affected protein.

Roufa and his colleagues have begun to use this mutant to study cell cycle progression. It was initially demonstrated that cells temperature-inactivated for 8 hours resumed DNA synthesis within 1 hour after downshift to 33°C by a process absolutely dependent upon *de novo* protein synthesis (Roufa and Reed, 1975; Roufa and Haralson, 1975). Further experiments were performed using *ts* 14 cells brought to early S phase arrest at 33°C by double thymidine block followed by 18 hours of treatment with 2 m*M* hydroxyurea. When released at 39°C, such *ts* 14 cells carried out semiconservative DNA replication for the remainder of the ongoing DNA-synthetic phase ($\simeq$ 6 hours), with polydeoxyribonucleotide chain formation occurring in the discontinuous mode, under conditions in which protein synthesis was being continuously inactivated such that by 2 hours *pts* a plateau was reached at about 5% of the control rate. Chromatin protein synthesis was clearly affected, as indicated by the fact that the chromatin-bound DNA newly formed at the *npt* was more sensitive to nuclease digestion than was the preformed DNA (Roufa, 1978). Histone synthesis during the first 6 hours after release at 39°C proceeded at one-half the rate effected at the *pt,* whereas the formation of other nuclear proteins was more severely affected.

These observations are in accord with the model presented in Fig. 11 which indicates that S phase chromatin replication proceeds along the separate but coupled metabolic pathways of DNA replication and chromosomal protein replication. The studies with *ts* 14 cells suggest that once normal semiconservative DNA synthesis is initiated, it can proceed to completion even though protein formation, including the synthesis of chromatin structural proteins, is severely restricted. Such a conclusion may reflect the use of excess thymidine and hydroxyurea to bring cells into early S phase arrest, and the concomitant accumulation of the necessary proteins and precursors of DNA synthesis. On the other hand, it may be a valid outcome of the regulation of $G_1/S$ traverse recently demonstrated in Chinese hamster ovary cells brought into metaphase arrest by double thymidine block, followed by colcemid treatment (Moats-Staats *et al.,* 1980). Such cells exhibit a single restriction point for S phase transcription about 4 hours into subsequent $G_1$ traverse, followed within 1 hour by a restriction point for S phase translation 1–3 hours prior to the $G_1$/S interface.

## F. Unbalanced Growth and Enhanced Protein Turnover in *ts* Mammalian Cells

A careful reading of much of the literature describing preliminary characterization of *ts* mammalian cells reveals that temperature-inactivation of the affected gene product gives rise to unbalanced growth. This may be manifest in disruption of coordinate regulation of macromolecular duplication during the aborted cell cycle. It may be evident in the continued increase in the volume of the cell nucleus in temperature-inactivated cells which are unable to divide, but which continue to synthesize protein and RNA. It may be expressed in enhanced turnover of specific cellular proteins.

Such unbalanced growth has been examined in greater detail in the *dna$^{ts}$ ts* A1S9 and *ts* Cl mouse L cells, and in the *ts* 2 mouse fibroblast. It was initially observed that, upon upshift to the *npt,* these cells increased their volume (Thompson *et al.,* 1970, 1971; Setterfield *et al.,* 1978; Sheinin *et al.,* 1980; Sparkuhl and Sheinin, 1980). This protoplasmic growth continued for several days and was seen in the nuclear and cytoplasmic compartments (Setterfield *et al.,* 1978; Sheinin *et al.,* 1980). However, the rate of accumulation of protein and RNA at the *npt* was not commensurate with the rate of synthesis, which remained at control levels for very long periods of time. It was therefore postulated that temperature-inactivation of cell division may provoke uncoupling of protein synthesis from DNA replication and induce protein degradation (cf. Goldberg and St. John, 1976). Specific aspects of such uncoupling are discussed in Section V,D,2.

Sparkuhl and Sheinin (1980) have observed that perhaps the earliest detectable manifestation of temperature inactivation of *ts* A1S9 cells is a marked increase in the rate of turnover of cytoplasmic proteins, which contrasts greatly with the prolonged stability of preformed chromosomal constituents (Sheinin *et al.,* 1978b; Sheinin and Lewis, 1980). Such enhanced turnover is energy dependent and requires *de novo* protein synthesis. It exceeds that which proceeds in wild-type and control cultures, and which by contrast is unaffected by the presence of cycloheximide. Enhanced turnover at the *npt* reduces the $T_{\frac{1}{2}}$ of *ts* A1S9 protein four-fold. Its activation exhibits the same temperature dependence as does temperature-inactivation of DNA synthesis and cell division in *ts* A1S9 cells.

Chloroquine and ammonia, inhibitors of lysosomal proteases, prevent activation of enhanced turnover, suggesting that lysosomes may be the sites of the temperature-induced protein degradation. Because no increase in acid or neutral protease activity was detected in temperature-inactivated *ts* A1S9 cells, it was suggested that enhanced turnover may result from increased transport of faulty gene product to the lysosomes, rather than to an increase in lysosomal enzyme activity per se. This protein translocation and degradation is thought to require energy, an intact microtubule system, and active protein synthesis (cf. Amenta *et al.*, 1977; Segal and Doyle, 1978).

It will be of interest to discover whether enhanced protein turnover is a general property of *ts* mammalian cells undergoing temperature inactivation, or whether it is specific for cells affected in particular cell cycle functions. Studies with temperature-inactivated CHO-*ts* Hl cells and with CHL-*ts* 14 cells (Gunn, 1978) are difficult to interpret. By virtue of their *ts* defects these cell rapidly lose their capacity for protein synthesis and become nonviable. Under such conditions they do not show enhanced protein turnover. Further study of this phenomenon in *ts* cells is indicated to obtain information on protein turnover *per se* and about the mechanisms by which coordinate macromolecule synthesis is regulated during cell cycle progression.


## VI. SUMMARY, CONCLUSIONS, AND PERSPECTIVES

The temporal maps of cell cycle progression developed here are preliminary and will undoubtedly be modified as our knowledge about the *ts* mutants and the biochemical pathways of interest expands and deepens. Such physiological mapping is not simply an interesting exercise in fitting together the pieces of the jigsaw puzzle already in hand. It provides insights into future directions to be taken in biochemical studies, in morphological and physiological analyses, and in the generation and selection of mutant cells *ts* in a specific phase or reaction of the cell duplication cycle.

A major requirement for the pursuit of any aspect of biochemical genetics is to have in hand organisms, preferably in large number, which are mutant in the various genes of interest. The several major efforts aimed at selecting specific cell cycle mutants have not been entirely successful. They have, however, yielded a reasonable number of $G_1$ mutants, and some mutant cells with defects in specific biochemical pathways.

The present imbalance in the spectrum of available *ts* mutants reflects our general dearth of knowledge about the biochemical events of cell cycle progression, the absence of rapid and effective tools for genetic analysis of mammalian cells, and a lack of appreciation that secondary expression of any given *ts* mutation may have serious consequences for cell survival. It is therefore important to recognize the significance of the various experiments which emphasize the fact that incubation of *ts* mammalian cells at the *npt* leads to unbalanced growth, to enhanced turnover of proteins (Section V,E), to uncoupling of normally coordinated processes (Section V,D) and to cell death. Thus the usual selection procedures have undoubtedly resulted in the loss of a large number of *ts* mutant cells; perhaps primarily in S, $G_2$, and M, cell cycle phases in which tightly regulated synthetic processes are mandatory. The relative plethora of $G_1$ *ts* mutants may be due to the fact that these arrest at the *npt* in a state which precludes entry into unbalanced growth, normally triggered by the signal for entry into S phase. Such cells would readily survive the commonly used selection regimen of high-temperature incubation for two or three generation period equivalents.

A number of clues for selection of specific cell cycle mutants are already available. Thus, mutant cells *ts* for movement from $G_0$ into proliferation and for early functions of $G_1$ should remain fully viable at the *npt*, should not enter pathways to unbalanced growth at the *npt*, and should survive suiciding treatments which depend upon proliferation at the *npt*. The *ts* AF8 cell provides another approach to the problem—that of selection of mutants of specific reactions of cell cycle progression. If, as the accumulating evidence indicates, the *ts* AF8 gene encodes information for RNA polymerase II activity of $G_1$, then its specific sensitivity to α-amanitin provides a tool for selection of many more such mutants. More isolations using antimetabolites and specific inhibitors of a variety of enzymes are indicated. Our studies on chromatin replication and structural reorganization of the chromatin in *dna^{ts}* cells and in cells which are *ts* in a $G_1$ function provide tools both for selection and screening of S phase mutants (see Section V,D,2).

The work discussed here clearly establishes the usefulness of *ts* mammalian cells in the study of the biochemical and physiological events of cell cycle progression. The *ts* AF8 cell, the *ts* 422E cell, and the *ts* 14 cell have opened the door to examining the precise roles played by transcription, ribosome synthesis, and protein synthesis in the movement of cells from quiescence to active growth. The *ts* CH-K12 cell may hold a key to understanding the signal for derepression of chromatin replication late in $G_1$. Cells like *ts* K/34C, *cs* 3 D4, and the GRO*$^{cs}$* mutants will undoubtedly shed light on the part played by plasma membrane molecules in triggering a variety of cell cycle events.

Experiments with *dna$^{ts}$* cells are providing systems for *in vitro* dissection of chromatin replication (Humbert and Sheinin, 1978) and for studying the mechanisms by which coupling of synthesis of chromatin DNA and protein is effected. The late $G_1$ mutants have focused on the quite separate but coordinated reactions of DNA synthesis, chromatin protein synthesis, and modification, and the biochemical cycle of chromatin ultrastructural organization, set in motion as a result of derepression of chromatin replication. The *ts* mutants that arrest in mitosis and cytokinesis may yield systems, elusive for so long, for the study of the biochemical reactions of chromosome movement and segregation.

Perhaps most intriguing are those *ts* mutants that exhibit a primary terminal phenotype of late $G_1$, the $G_1$/S traverse, or of S, and that show secondary arrest in mitosis and cytokinesis. These herald elucidation of the several regulatory processes which ensure that the end product of cell cycle progression initiated late in $G_1$ should indeed be the formation of exactly duplicated progeny at cell division.

The enormous potential of biochemical genetics for helping to unravel the sequential processes of proliferation has barely been tapped. Its realization clearly depends upon the identification of the biochemical gene product of the mutants already in hand, as well as the isolation of many more mutants *ts* in proteins already known to function in specific reactions of cellular replication.

## ACKNOWLEDGMENTS

The author's work described herein was financially supported by the Medical Research Council of Canada and the National Cancer Institute of Canada. I am grateful to many for providing preprints and unpublished material, to my scientific colleagues for their critical comments on the manuscript, and to my secretarial co-workers for their cheerful perseverance.

## REFERENCES

Adair, G. M., Thompson, L. H., and Lindl, P. A. (1978). *Somat. Cell Genet.* **4,** 27–44.
Aitken, D. A., and Ferguson-Smith, M. A. (1978). *Cytogenet. Cell Genet.* **22,** 490–492.
Amenta, J. S., Sargus, M. H., and Baccino, F. M. (1977). *Biochem J.* **168,** 223–227.
Ashihara, T., Chang, S. D., and Baserga, R. (1978a). *J. Cell Physiol.* **96,** 15–22.
Ashihara, T., Traganos, F., Baserga, R., and Darzynkiewicz, Z. (1978b). *Cancer Res.* **38,** 2514–2518.

Ashman, C. R. (1978). *Somat. Cell Genet.* **4**, 299–311.
Aubin, J. E., Weber, K., et al. (1979). *Exptl. Cell Res.* **124**, 93–109.
Back, F. (1976). *Int. Rev. Cytol.* **45**, 25–64.
Baserga, R. (1978). *J. Cell Physiol.* **95**, 377–382.
Basilico, C. (1977). *Adv. Cancer Res.* **24**, 223–266.
Basilico, C. (1978). *J. Cell Physiol.* **95**, 367–372.
Basilico, C., Burstin, S. J., Toniolo, D., and Meiss, H. K. (1974). *In* "Viral Transformation and Endogenous Viruses" (A. S. Kaplan, ed.) pp. 75–95. Academic Press, New York.
Beadle, G. W. (1945). *Chem. Rev.* **37**, 15–96.
Beorsma, D., McGill, S., Mollenkamp, J., and Roufa, D. J. (1979). *J. Biol. Chem.* **254**, 559–567.
Berger, N. A., Kaichi, A. S., Steward, P. G., Klevecz, R. R., Forrest, G. L., and Gross, S. D. (1978). *Exp. Cell Res.* **117**, 127–135.
Bonatti, S., Cancedda, R., and Blobel, G. (1979). *J. Cell Biol.* **80**, 219–224.
Bootsma, D., and Ruddle, F. H. (1978). *Cytogenet. Cell Genet.* **22**, 74–91.
Borun, T. W. (1975). *In* "Cell Cycle Differentiation" (J. Reinert and H. Holzer, eds.), Vol. 7. pp. 249–290. Springer-Verlag, Berlin and New York.
Bradbury, E. M. (1979). *Differentiation* **13**, 37–40.
Bramwell, M. E. (1977). *Exp. Cell Res.* **110**, 277–282.
Burstin, S. J., and Basilico, C. (1975). *Proc. Natl. Acad. Sci. U.S.A.* **72**, 2540–2544.
Burstin, S. J., Meiss, H. K., and Basilico, C. (1974). *J. Cell Physiol.* **84**, 397–407.
Busch, H., Ballal, N. R., Rao, M. R. S., Choi, Y. C., and Rothblum, L. I. (1978). *In* "The Cell Nucleus" (H. Busch, ed.), Vol. 5, Pt. B, pp. 415–468. Academic Press, New York.
Chan, A. C., and Walker, I. G. (1975). *Biochem. Biophys. Acta* **395**, 422–432.
Chang, H. L., and Baserga, R. (1977). *J. Cell Physiol.* **92**, 333–343.
Cleaver, J. E. (1972). *In* "Molecular and Cellular Repair Process" (R. F. Beers, Jr., R. M. Herriott, and R. Tilgham, eds.), pp. 195–211. John Hopkins Univ. Press, Baltimore, Maryland.
Crane, M. St. J., and Thomas, D. B. (1976). *Nature (London)* **261**, 205–208.
Cremisi, C. (1979). *Microbiological Reviews* **43**, 297–319.
Crissman, H. A., Mullaney, P. F., and Steinkamp, J. A. (1975). *Methods Cell Biol.* **9**, 179–246.
Dardick, R., Setterfield, G., and Sheinin, R. (1978a). *Brit. J. Hematol.* **39**, 483–490.
Dardick, I., Sheinin, R., and Setterfield, G. (1978b). *Am. J. Exp. Pathol.* **93**, 849–852.
Denney, R. M., Borgaonkor, D., and Ruddle, F. H. (1978). *Cytogenet. Cell Genet.* **22**, 493–497.
Di Liegro, I., Cestelli, A., Ciaccio, M., and Cognetti, G. (1978). *Dev. Biol.* **67**, 266–273.
Dirksen, E. R., Prescott, D. M., and Fox, C. F., eds. (1979). *ICN-UCLA Symp. Mol. Cell. Biol.,* Vol. 13. *Cell Reproduction.*
Dubbs, D. R., and Kit, S. (1976). *Somat. Cell Genet.* **2**, 11–19.
Dustin, P. (1978). "Microtubules." Springer-Verlag, Berlin and New York.
Edgar, R. S., and Lielausis, I. (1964). *Genetics* **49**, 649–662.
Elgin, S. C. R., and Weintraub, H. (1975). *Annu. Rev. Biochem.* **44**, 725–774.
Farber, R. A., and Liskay, R. M. (1974). *Cytogenet. Cell Genet.* **13**, 384–396.
Fenwick, R. G. Jr., and Caskey, C. T. (1975). *Cell* **5**, 115–122.
Ferguson-Smith, M. A., and Westerweld, A. (1978). *Cytogenet. Cell Genet.* **22**, 111–123.
Floros, J., Ashihara, T., and Baserga, R. (1978a). *Cell Biol. Int. Rep.* **2**, 259–269.
Floros, J., Chang, H., and Baserga, R. (1978b). *Science* **201**, 651–653.
Gabiani, G., ed. (1979). "Methods and Achievements in Experimental Pathology," Vol. 8, Cell Biology. S. Karger, Basel, Switzerland.

Georgiev, G. P., Nedospasov, S. A., and Bakayev, V. V. (1978). *In* "The Cell Nucleus (H. Busch, ed.), Vol. VI, pp. 4–34. Academic Press, New York.

Giles, R. E., and Ruddle, F. H. (1976). *Genetics* **83** Suppl. S 26.

Giles, R. E., Shimizu, N., Nichols, E., Lawrence, J., and Ruddle, F. H. (1977). *J. Cell Biol.* **75**, 387A.

Goldberg, A. L., and St. John, A. C. (1976). *Annu. Rev. Biochem.* **45**, 747–803.

Goodwin, G. H., Walker, J. M., and Johns, E. W. (1978). *In* "The Cell Nucleus (H. Busch, ed.), Vol. VI, Pt. C, pp. 182–221. Academic Press, New York.

Graham, F. L. (1977). *Adv. Cancer Res.* **25**, 1–51.

Green, H., and Todaro, G. (1967). *Annu. Rev. Microbiol.* **21**, 573–600.

Grummt, F., Grummt, I., and Mayer, E. (1979). *Eur. J. Biochem.* **97**, 37–42.

Gunn, S. M. (1978). *Exp. Cell Res.* **117**, 448–451.

Gupta, R., Srinivasan, P. R., and Siminovitch, L. (1980). *Somat. Cell Genet.* **6**, 151–169.

Gurley, L. R., Tobey, R. A., Walters, R. A., Hildebrand, C. E., Hohmann, P. G., D'Anna, J. A., Barham, S. S., and Deaven, L. L. (1978). *In* "Cell Cycle Regulation" (J. L. Jeter, I. L. Cameron, G. M. Padilla, and A. M. Zimmerman, eds.), pp. 37–59. Academic Press, New York.

Guttman, S. A., and Sheinin, R. (1979). *Exp. Cell Res.* **123**, 191–205.

Haars, L. S., Hampel, A., and Thompson, L. H. (1976). *Biochim. Biophys. Acta* **454**, 493–503.

Hampel, A. E., Ritter, P. O., and Enger, M. D. (1978). *Nature (London)* **276**, 844–845.

Hanawalt, P. C., Cooper, P. K., Ganeson, A. K., and Smith, C. S. (1979). *Annu. Rev. Biochem.* **48**, 783–836.

Hand, R., Eilen, E., and Basilico, C. (1980). *J. Cell Physiol.* (submitted).

Hand, R., and Kasupski, G. J. (1978). *J. Gen. Virol.* **39**, 437–448.

Haralson, M. A., and Roufa, D. J. (1975). *J. Biol. Chem.* **250**, 8618–8623.

Harris, H. (1974). "Nucleus and Cytoplasm," 3rd ed., pp. 109–141. Oxford Clarendon Press, Oxford.

Hartwell, L. H. (1978). *J. Cell Biol.* **77**, 627–637.

Hatzfield, J., and Buttin, G. (1975). *Cell* **5**, 123–129.

Hayaishi, O., and Ueda, K. (1977). *Annu. Rev. Biochem.* **46**, 95–116.

Howard, S., and Pelc, S. R. (1953). *Heredity* **6**, Suppl. 261–273.

Humbert, J., and Sheinin, R. (1978). *Can. J. Biochem.* **56**, 444–452.

Ingles, C. J. (1978). *Proc. Natl. Acad. Sci. U.S.A.* **75**, 405–409.

Jha, K. K., and Ozer, H. L. (1977). *Genetics* **86** Suppl. S32–S33.

Juliano, R. L., and Ling, V. (1976). *Biochem. Biophys. Acta* **455**, 152–162.

Juliano, R. L., Graves, J., and Ling, V. (1976). *J. Supramol. Struct.* **4**, 521–526.

Kane, A., Basilico, R., and Baserga, R. (1976). *Exp. Cell Res.* **99**, 165–173.

Kedes, L. (1979). *Annu. Rev. Biochem.* **48**, 837–870.

Kit, S., and Jorgensen, G. N. (1976). *J. Cell Physiol.* **88**, 57–64.

Kirschner, M. A. (1978). *Int. Rev. Cytol.* **54**, 1–71.

Lajtha, L. G. (1963). *J. Cell Physiol.* **62**, Suppl. 1, 143–145.

Landy-Otsuka, F., and Scheffler, I. E. (1978). *Proc. Natl. Acad. Sci. U.S.A.* **75**, 5001–5005.

Lewis, W. H., Srinivasan, P. R., Stokoe, N., and Siminovitch, L. (1980). *Somat. Cell Genet.* **6**, 333–348.

Ling, V. (1977). *J. Cell Physiol.* **91**, 209–223.

Ling, V., and Thompson, L. H. (1974). *J. Cell Physiol.* **83**, 103–116.

Lipsich, L. A., Lucas, J. J., and Kates, J. R. (1979). *J. Cell Physiol.* **98**, 637–642.

Liskay, R. M. (1974). *J. Cell Physiol.* **84**, 49–55.

Liskay, R. M. (1978). *Exp. Cell Res.* **114**, 69–77.

Liskay, R. M., and Meiss, H. K. (1977). *Somat. Cell Genet.* **3**, 343–347.

Liskay, R. M., and Prescott, D. M. (1978). *Proc. Natl. Acad. Sci. U.S.A.* **75,** 2823–2877.
Littlefield, J. W. (1977). *In* "Molecular Biology of the Mammalian Genetic Apparatus" (P. Ts'o, ed.), Vol. 2, pp. 181–190. Elsevier North-Holland Biomed. Press, Amsterdam.
Lofgren, D. J., and Thompson, L. H. (1979). *J. Cell Physiol.* **99,** 303–312.
Loomis, W. F., Jr., Wahrmann, J. P., and Luzzati, D. (1973). *Proc. Natl. Acad. Sci. U.S.A.* **70,** 425–429.
McBurney, M., and Whitmore, G. F. (1974a). *J. Cell Physiol.* **83,** 69–74.
McBurney, M., and Whitmore, G. F. (1974b). *Cell,* **2,** 183–188.
Macdonald, H. R., and Miller, R. G. (1970). *Biophys. J.* **10,** 834–842.
McDougall, J. K., Kucherlapati, R., and Ruddle, F. H. (1973). *Nature (London) New Biol.* **245,** 172–175.
Marin, G., and Labella, T. (1977). *J. Cell Physiol.* **90,** 71–78.
Martin, R. F., Radford, I., and Pardee, M. (1977). *Biochem. Biophys. Res. Commun.* **74,** 9–15.
Marunouchi, T. (1979). *Japanese J. of Experimental Medicine* **48,** 155–166.
Marunouchi, T., and Nakano, N. M. (1980). *Cell Structure and Function* **5,** 53–66.
Meiss, H. K., and Basilico, C. (1972). *Nature (London) New Biol.* **239,** 66–68.
Meiss, H. K., Talavera, A., and Nishimoto, T. (1978). *Somat. Cell Genet.* **4,** 125–130.
Melero, J. A. (1979). *J. Cell Physiol.* **98,** 17–30.
Melero, J. A., and Fincham, V. (1978). *J. Cell Physiol.* **95,** 295–306.
Melero, J. A., and Smith, A. E. (1978). *Nature (London)* **272,** 725–727.
Melli, M., Spinelli, G., and Arnold, E. (1977). *Cell* **12,** 167–174.
Miller, O. J., Sanger, R., and Siniscalco, M. (1978). *Cytogenet. Cell Genet.* **22,** 124–128.
Ming, P-M, L., Chang, H. L., and Baserga, R. (1976). *Proc. Natl. Acad. Sci. U.S.A.* **73,** 2052–2055.
Ming, P.-M. L., Lange, B., and Kit, S. (1979). *Cell Biol. Int. Rep.* **3,** 169–178.
Moats-Staats, B. M., Mollenkamp, J. W., and Roufa, D. J. (1980). *J. Biol. Chem.,* submitted.
Molnar, S. J., and Rauth, A. M. (1979). *J. Cell Physiol.* **98,** 315–326.
Molnar, S. J., Thompson, L. H., Lofgren, D. J., and Rauth, A. M. (1979). *J. Cell Physiol.* **98,** 327–339.
Mura, C., Craig, S. W., and Huang, P. C. (1978). *Exp. Cell Res.* **114,** 301–305.
Nadeau, P., Oliver, D. R., and Chalkley, R. (1978). *Biochemistry* **17,** 4885–4893.
Naha, P. M. (1969). *Nature (London)* **223,** 1380–1381.
Naha, P. M. (1970). *Nature (London)* **228,** 166–168.
Naha, P. M. (1973a). *Nature (London) New Biol.* **245,** 266–268.
Naha, P. M. (1973b). *Exp. Cell Res.* **80,** 467–473.
Naha, P. M. (1979). *J. Cell Sci.* **35,** 53–58.
Naha, P. M., Meyer, A. L., and Hewitt, K. (1975). *Nature (London)* **258,** 49–53.
Nakano, M. M., Sekiguchi, T., and Yamada, M. (1978). *Somat. Cell Genet.* **4,** 169–178.
Naylor, S. L., Shows, T. B., and Klebe, R. J. (1979). *Somat. Cell Genet.* **5,** 11–22.
Newrock, K. M., Alfageme, C. R., Nardi, R. V., and Cohen, L. H. (1977). *Cold Spring Harbor Symp. Quant. Biol.* **42,** 421–431.
Nishimoto, T., and Basilico, C. (1978). *Somat. Cell Genet.* **4,** 323–340.
Nishimoto, T., Raskas, H. J., and Basilico, C. (1975). *Proc. Natl. Acad. Sci. U.S.A.* **72,** 328–332.
Nishimoto, T., Okubo, C. K., and Raskas, H. J. (1977). *Virology* **78,** 1–10.
Nishimoto, T., Eilen, E., and Basilico, C. (1978). *Cell* **15,** 475–483.
Ohlsson-Wilhelm, B. M., Freed, J. J., and Perry, R. P. (1976). *J. Cell Physiol.* **89,** 77–88.
Ouellette, A. J., Bandman, E., and Kumar, A. (1976). *Nature (London)* **262,** 619–621.
Owerbach, D., Doyle, D., and Shows, T. B. (1979). *Somat. Cell Genet.* **5,** 281–302.
Pardee, A. B., Dubraw, R., Hamlin, J. L., and Kletzien, R. F. (1978). *Annu. Rev. Biochem.* **47,** 715–750.

Parodi, A. J., and Leloir, L. F. (1979). *Biochim. Biophys. Acta* **559**, 1–38.
Perez Bercoff, R., ed. (1979). "The Molecular Biology of Picornaviruses." Plenum, New York.
Perry, R. P., and Kelley, D. E. (1973). *J. Mol. Biol.* **79**, 681–696.
Pochron, S. R., and Baserga, R. (1979). *J. Biol. Chem.* **254**, 6352–6356.
Pollard, J. W., and Stanners, C. P. (1979). *J. Cell Physiol.* **98**, 571–586.
Prescott, D. M. (1976a). "Reproduction of Eukaryotic Cells." Academic Press, New York.
Prescott, D. M. (1976b). *Adv. Genet.* **18**, 99–177.
Pringle, J. R. (1978). *J. Cell Physiol.* **95**, 393–406.
Reiber, M., and Bacalao, J. (1974a). *Exp. Cell Res.* **85**, 334–339.
Reiber, M., and Bacalao, J. (1974b). *Cancer Res.* **34**, 3083–3088.
Reich, E., Rifkin, D., and Shaw, E., eds. (1975). *In* "Proteases and Biological Control." Cold Spring Harbor Lab., Cold Spring Harbor, New York.
Roscoe, D. H., Read, H., and Robinson, H. (1973a). *J. Cell Physiol.* **82**, 325–332.
Roscoe, D. H., Robinson, H., and Carbonell, A. W. (1973b). *J. Cell Physiol.* **82**, 333–338.
Rossini, M., and Baserga, R. (1978). *Biochemistry* **17**, 858–863.
Rossini, M., Weinmann, R., and Baserga, S. (1979a). *PNAS*, USA **76**, 4441–4445.
Rossini, M., Floros, J., Weinmann, R., and Baserga, R. (1979b). *Conf. on Cell Proliferation* **6**, 393–402.
Rossini, M., Baserga, S., Huang, C. H., Ingle, C. J., and Baserga, R. (1980). *J. Cell Physiol.* (in press).
Roufa, D. J. (1978). *Cell* **13**, 129–138.
Roufa, D. J., and Haralson, M. A. (1975). *In* "DNA Synthesis and Its Regulation" (M. Goulian, F. C. Hanawalt, and C. F. Fox, eds.), pp. 702–712. Benjamin, Menlo Park, California.
Roufa, D. J., and Reed, S. J. (1975). *Genetics* **80**, 549–566.
Roufa, D. J., McGill, S. M., and Mollenkamp, J. W. (1979). *Somat. Cell Genet.* **5**, 97–115.
Ruddle, F. H., and Creagan, R. P. (1975). *Annu. Rev. Genet.* **9**, 407–486.
Ruddle, F. H., Scangos, G., and Klobutscher, L. A. (1979). *In* "Eukaryotic Gene Regulation. ICN UCLA Symp. Cell. Molec. Biology. Ed. In Press.
Sato, K. (1975). *Nature (London)* **257**, 813–815.
Sato, K., and Hama-Inaba, H. (1978). *Exp. Cell Res.* **114**, 484–486.
Sato, K., and Shiomi, T. (1974). *Exp. Cell Res.* **88**, 295–302.
Savard, P., Poirier, G., and Sheinin, R. (1980). *Canad. J. Biochem* (submitted).
Scheffler, I. E., and Buttin, G. (1973). *J. Cell Physiol.* **81**, 199–216.
Schwartz, H. E., Moser, G. C., Holmes, S., and Meiss, H. K. (1979). *Somat. Cell Genet.* **5**, 217–224.
Segal, H. L., and Doyle, D. J., eds. (1978). "Protein Turnover and Lysosome Function." Academic Press, New York.
Setterfield, G., and Kaplan, J. G. (1980). In preparation.
Setterfield, G., Sheinin, R., Dardick, I., Kiss, G., and Dubsky, M. (1978). *J. Cell Biol.* **77**, 246–253.
Sheinin, R. (1967). *In* "The Molecular Biology of Viruses" (J. S. Colter and W. Parachych, eds.), pp. 627–643, Academic Press, New York.
Sheinin, R. (1976a). *Cell* **7**, 49–57.
Sheinin, R. (1976b). *J. Virol.* **17**, 692–704.
Sheinin, R., and Guttman, S. (1977). *Biochim. Biophys. Acta* **479**, 105–118.
Sheinin, R., and Lewis, P. N. (1980). *Somat. Cell Genet.* **6**, 227–241.
Sheinin, R., Darragh, P., and Dubsky, M. (1977). *Can. J. Biochem.* **55**, 543–547.
Sheinin, R., Darragh, P., and Dubsky, M. (1978a). *J. Biol. Chem.* **253**, 922–926.
Sheinin, R., Humbert, J., and Pearlman, R. E. (1978b). *Annu. Rev. Biochem.* **47**, 331–370.

Sheinin, R., Dardick, I., and Doane, F. W. (1980). *Exp. Cell Res.* (submitted).

Sheinin, R., Setterfield, G., Dardick, I., Naismith, L., Kiss, G., Dubsky, M. (1980). *Canad. J. Biochem.* (in press).

Shiomi, T., and Sato, K. (1976). *Exp. Cell Res.* **100**, 297-302.

Shiomi, T., and Sato, K. (1978). *Cell Struct. Function* **3**, 95-102.

Shows, T. B., and McAlpine, P. J. (1978). *Cytogenet. Cell Genet.* **22**, 129-131.

Shows, T. B., Scrafford-Wolff, L., Brown, J. A., and Meisler, M. (1978). *Cytogenet. Cell Genet.* **22**, 219-222.

Simchen, G. (1978). *Annu. Rev. Genet.* **12**, 161-191.

Siminovitch, L., Thompson, L. H., Mankovitz, R., Baker, R. M., Wright, J. A., Till, J. E., and Whitmore, G. F. (1973). *Can. Cancer Res. Conf.* **9**, 59-75.

Slater, M. L., and Ozer, H. L. (1976). *Cell* **7**, 289-295.

Smith, B. J., and Wigglesworth, N. M. (1972). *J. Cell Physiol.* **80**, 253-259.

Smith, B. J., and Wigglesworth, N. M. (1973). *J. Cell Physiol.* **82**, 339-347.

Smith, B. J., and Wigglesworth, N. M. (1974). *J. Cell Physiol.* **84**, 127-133.

Smith, J. A., and Martin, L. (1973). *Proc. Natl. Acad. Sci. U.S.A.* **70**, 1263-1267.

Soderhall, S., and Lindahl, T. (1976). *FEBS Lett.* **67**, 1-8.

Sparkuhl, J. S., and Sheinin, R. (1980). *J. Cell Physiol* (in press).

Stahl, H., and Gallwitz, D. (1977). *Eur. J. Biochem.* **72**, 385-392.

Stanners, C. P., Wightman, T. M., and Harkins, J. L. (1978). *J. Cell Physiol.* **95**, 125-138.

Stanners, C. P., Becker, H., Harkins, J., and Pollard, J. F. (1979). *J. Cell. Physiol.* **100**, 127-138.

Stein, G., Park, W., Thrall, C., Mans, R., and Stein, J. (1975). *Nature (London)* **257**, 764-767.

Stern, H., and Hotta, Y. (1977). *Philos. Trans. R. Soc. London. Ser. B.* **277**, 277-294.

Talavera, A., and Basilico, C. (1977). *J. Cell Physiol.* **92**, 425-436.

Talavera, A., and Basilico, C. (1978). *J. Cell Physiol.* **97**, 429-439.

Tallman, G., Akers, J. E., Burlingham, B. T., and Reeck, G. R. (1977). *Biochem. Biophys. Res. Commun.* **79**, 815-822.

Tarnowka, M. A., Baglioni, C., and Basilico, C. (1978). *Cell* **15**, 163-171.

Tenner, A. J., and Scheffler, I. E. (1979). *J. Cell Physiol.* **98**, 251-265.

Tenner, A., Zieg, J., and Scheffler, I. E. (1977). *J. Cell Physiol.* **90**, 145-160.

Terasima, T., and Tolmach, L. J. (1963). *Exp. Cell Res.* **30**, 344-362.

Thompson, L. H., and Lindl, P. A. (1976). *Somat. Cell Genet.* **2**, 387-400.

Thompson, L. H., Mankovitz, R., Baker, R. M., Till, J. E., Siminovitch, L., and Whitmore, G. F. (1970). *Proc. Natl. Acad. Sci. U.S.A.* **66**, 377-384.

Thompson, L. H., Mankovitz, R., Baker, R. M., Wright, J. A., Till, J. E., Siminovitch, L., and Whitmore, G. F. (1971). *J. Cell Physiol.* **78**, 431-440.

Thompson, L. H., Harkins, J. L., and Stanners, C. P. (1973). *Proc. Natl. Acad. Sci. U.S.A.* **70**, 3094-3098.

Thompson, L. H., Stanners, C. P., and Siminovitch, L. (1975). *Somat. Cell. Genet.* **1**, 187-208.

Thompson, L. H., Lofgren, D. J., and Adair, G. M. (1977). *Cell* **11**, 157-168.

Thompson, L. H., Lofgren, D. J., and Adair, G. M. (1978). *Somat. Cell Genet.* **4**, 423-435.

Tobey, R. A., and Ley, K. D. (1971). *Cancer Res.* **31**, 46-51.

Toneguzzo, F., and Ghosh, H. (1978). *Proc. Natl. Acad. Sci. U.S.A.* **75**, 715-719.

Toniolo, D., and Basilico, C. (1974). *Nature (London)* **248**, 411-413.

Toniolo, D., and Basilico, C. (1976). *Biochim. Biophys. Acta* **425**, 409-418.

Toniolo, D., Meiss, H. K., and Basilico, C. (1973). *Proc. Natl. Acad. Sci. U.S.A.* **70**, 1273-1277.

Tsutsui, Y., Chang, S. D., and Baserga, R. (1978). *Exp. Cell Res.* **113**, 359-367.

Tucker, R. W., Pardee, A. B., and Fujiwara, K. (1979). *Cell* **17,** 527–535.

Walters, R. A., Tobey, R. A., and Hildebrand, C. E. (1976). *Biochem. Biophys. Res. Commun.* **69,** 212–217.

Wang, R. J. (1974). *Nature (London)* **248,** 76–78.

Wang, R. J. (1976). *Cell* **8,** 257–261.

Wang, R. J., and Yin, L. (1976). *Exp. Cell Res.* **101,** 331–336.

Wang, R. J., and Sheridan, W. F. (1974). *Exp. Cell Res.* **84,** 357–362.

Wasmuth, J. J., and Caskey, C. T. (1976). *Cell* **9,** 655–622.

Wickner, S. H. (1978). *Annu. Rev. Biochem.* **47,** 1163–1191.

Wigler, M., Pellicer, A., Silverstein, S., and Axel, R. (1979). *Proc. Natl. Acad. Sci. U.S.A.* **76,** 1373–1376.

Wijnen, L. M. M., Monteba-van Heuvel, M., Pearson, P. L., and Khan, P. M. (1978). *Cytogenet. Cell Genet.* **22,** 232–238.

Wissinger, W., and Wang, R. J. (1978). *Exp. Cell Res.* **112,** 89–94.

Wittes, R. E., and Ozer, H. L. (1973). *Exp. Cell Res.* **80,** 127–136.

Woodland, H. R., and Adamson, E. D. (1977). *Dev. Biol.* **57,** 118–135.

Yanagi, K., Talavera, S., Nishimoto, T., and Rush, M. G. (1978). *J. Virol.* **25,** 42–50.

Yen, A., and Pardee, A. B. (1978). *Exp. Cell Res.* **114,** 389–395.

Zlatanova, J., and Swetly, P. (1978). *Nature (London)* **276,** 276–277.

$\mathbb{5}$

# Initiation of DNA Synthesis in S Phase Mammalian Cells

### ROGER HAND

> So, naturalists observe, a flea
> Hath smaller fleas that on him prey;
> And these have smaller still to bite 'em;
> And so proceed ad infinitum.
>
> Jonathan Swift

## I. INTRODUCTION

The lineal descendants of Swift's naturalists are today's biologists. While they, for the most part, studied whole organisms, we, for the most part, sequence genomes. Despite the differences in experimental systems, there are many similarities in our observations. An example is the subject of this chapter, the initiation of DNA replication. In the mammalian cell, there are several levels at which this takes place. A cell initiates DNA synthesis when it leaves G1 and enters S phase. But throughout S phase, the cell initiates synthesis on large sections of chromatin within chromosomes. The beginning of synthesis on these subchromosomal sections then permits initiation on individual replication units that make

**167**

NUCLEAR–CYTOPLASMIC INTERACTIONS
IN THE CELL CYCLE

up these sections. Finally, initiation on these units permits the initiation of synthesis of small replication intermediates—Okazaki pieces. While these different levels are easily distinguished conceptually, such distinction has not always been made in the literature and this has led to some confusion. The purpose of this chapter is to discuss the, levels of initiation of DNA replication in mammalian cells in the hope of developing a reasonable model for the regulation of this process.

## II. STRUCTURAL AND FUNCTIONAL FEATURES OF MAMMALIAN GENOMES

An ordered series of biochemical events in $G_1$ leads up to the beginning of S phase. These have been reviewed recently (Pardee *et al.*, 1978) and are covered in other chapters in this volume. Suffice it to say here that cytoplasmic factors exert a positive control over the initiation of synthesis at the beginning of S phase (Graham *et al.*, 1966; Johnson and Harris, 1969; Rao and Johnson, 1970). These factors have been identified as proteins or as low molecular weight heat-stable substances (Grummt, 1978). There is most probably an orderly cascade mechanism leading to the synthesis of the precursors and enzymes required for DNA replication. What determines exactly when and where the first nucleotide will be placed on the template is unknown, but certainly a prerequisite for synthesis is the buildup to a critical level of nucleotide triphosphates (Skoog *et al.*, 1973; Walters *et al.*, 1973) and adequate amounts of DNA polymerase alpha (Chang *et al.*, 1973; Spadari and Weissbach, 1974). Other enzymes and cofactors directly involved in replication have not been defined.

Genome DNA in mammalian cells is complexed with histones in a highly ordered structure: two molecules of each of four histones, H2A, H2B, H3, and H4 form a nucleus around which roughly 140 nucleotide pairs are coiled. This structure, the nucleosome, is linked to neighboring nucleosomes by a linker of about 60 nucleotide pairs of DNA. The linker is complexed with histone H1. The 200 nucleotide pairs with the attached histones make up the basic repeating structural unit of chromatin (Kornberg, 1977; Felsenfeld, 1978). The nucleosomes are coiled then into higher-order structures so the interphase chromatin has a regular architecture in which the DNA is more highly compacted. These have been called solenoids (Finch and Klug, 1976) or supersuperhelices (Worcel and Benyajati, 1977). Further supercoiling to higher-order structures in interphase chromatin may occur (Finch and Klug, 1976; Sedat and Mannelidis, 1977) to which RNA and non-histone chromosomal proteins contribute (Benyajati and Worcel, 1976). An alternative model in which the higher-order structures are formed by loops of nucleosomes has also been proposed (Marsden and Laemmli, 1979). Although many of the studies on the higher order of chromatin structure apply to metaphase chromosomes (Bak *et al.,* 1977; Marsden and Laemmli, 1979), topological constraints and compaction are also present in interphase chromatin (Cook and Brazell, 1975) and undoubtedly some degree of high-order coiling or looping is present. A representation of the models for interphase chromatin structure is shown in Fig. 1. In these models unique structural aspects of the DNA–protein interaction occur at intervals of 200 nucleotide pairs, and also at much longer intervals determined by the higher-order coiling or looping of the nucleosomes. Therefore, multiple classes of repeating structures are present in chromatin.

The question then arises whether these structural features are involved in the initiation of DNA replication at any or all of the levels discussed above. There are several possible choices for the structural features that define an initiation point of any sort. First, a particular

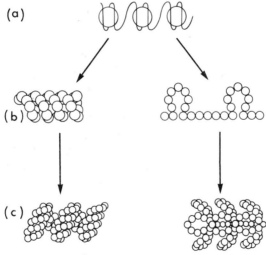

**Fig. 1.**  Models for interphase chromatin structure. (a) The lowest order of chromatin structure. The circles represent the core of histones around which the DNA double helix, represented by the solid line, is coiled. (b) A higher-order structure in which the nucleosomes are compacted into a coil or solenoid on the left (Finch and Klug, 1976) or compacted into loops on the right (Marsden and Laemmli, 1979). The DNA helix, which coils around the cores, has been omitted for clarity. (c) Further coiling (left) or looping (right) yields a higher degree of compaction. The DNA helix again is omitted for clarity.

sequence of nucleotides could define such a point. In several mammalian viruses for which sequence data are available, palindromes are
found in the region of the origin of replication (Subramanian *et al.*,
1977; Dhar *et al.*, 1978; Soeda *et al.*, 1978). These sequences could form
cruciform structures as a result of intrastrand nucleotide pairing and as
such could serve as recognition signals for the attachment of proteins
involved in the replication of the DNA of these viruses. Similar structures might serve the same purpose in mammalian DNA (Tamm *et al.*,
1979). On the other hand, some unique aspect of the protein–DNA
interaction in chromatin could also serve as the recognition site for ini-
tiator proteins. Finally, it is possible that a combination of nucleotide
sequences and DNA–protein interactions serve to define an initiation
point.

It has been known for many years that the replication of mammalian
DNA during S phase is a highly ordered process. Subsections of chromosomes replicate at defined times in S. This is shown by the pattern of
metaphase chromosomes pulse-labeled during the preceding S phase,
using either a radioactive tracer (Taylor, 1960) or a fluorescent probe
(Latt, 1975) to identify replicating DNA. The DNA replicated at a particular time during the S phase is likely to be replicated at the same time
in a subsequent S phase (Mueller and Kajiwara, 1966; Adegoke and
Taylor, 1977). Also, different subclasses of DNA, such as satellites, are
replicated at defined times during S (May and Bello, 1974; Tapiero *et
al.*, 1974). DNA with a high proportion of adenine and thymine bases
tends to be late replicating (Bostock and Prescott, 1971; Comings, 1972).
By electron microscope autoradiography following [³H]thymidine
pulse-labeling, definite patterns of nuclear DNA synthesis are seen in S
phase: early S phase cells have grains localized over nucleoplasm in areas
containing euchromatin, while late S phase cells have grains peripherally
located at the membrane or over the nucleolus in areas containing
heterochromatin (Williams and Ockey, 1970; Erlandson and De Harven,
1971). More recently it has been shown that the synthesis of late replicating DNA depends on the structural integrity and the completion of
synthesis of early replicating DNA (Hamlin and Pardee, 1978). Also if a
late S phase cell is fused to a $G_1$ cell, the nucleus of the $G_1$ cell is induced
into synthesis, but with the pattern of early S phase DNA synthesis
(Yanishevsky and Prescott, 1978). This evidence argues very strongly for
a stringent program for DNA synthesis with definite classes of DNA
being replicated at specific points in the S phase and, further, that late
replication events are dependent upon early replication events, at least
in the normal cell.

## III. FUNCTIONAL SUBUNITS OF MAMMALIAN GENOMES

To this point, we have looked at the genome as a whole from a structural and functional point of view. Is there evidence for functional subgenomic units that regulate the highly ordered pattern of replication? The answer is yes—certainly at three different levels of genome organization and perhaps at more. The evidence for the subgenomic organization has been reviewed recently (Hand, 1978) and I will discuss it here briefly, while citing some more recent data as well. The three levels are illustrated in Fig. 2.

There is evidence that relatively large segments of the genome—up to 1 mm in length—are replicated as units and that initiation of synthesis on these sections is controlled by single initiation events. Low doses of ultraviolet and X-irradiation inhibit the formation of new DNA chains without affecting the rate of polymerization on nascent chains (Painter and Young, 1975, 1976; Povirk, 1977). The target size of the inactivated regions is about $10^9$ daltons of DNA (note that the mammalian genome contains some $10^{12}$ daltons of DNA). The length of DNA inactivated therefore is about 1 mm long. Metaphase chromosomes labeled with a tracer for DNA synthesis for a 1-hour period during the preceding S phase display bands of labeled DNA that are visible in the light microscope (Latt, 1975; Stubblefield, 1975). Taking into account the compaction ratio of the DNA in the metaphase chromosome, these bands contain 0.5 to 1 mm of DNA. Thus DNA up to 1 mm in length is replicated coordinately and the initiation of synthesis on these stretches of the genome can be inactivated by a single hit from radiation.

There is abundant evidence for coordinate initiation of synthesis on the 15 or more replication units within these 1-mm segments. The replication unit is a structure with a central origin from which two replication forks proceed bidirectionally (Huberman and Riggs, 1968). Its size is variable, but most reported measurements fall within a range of 4–400 $\mu$m with averages of about 50 $\mu$m (Edenberg and Huberman, 1975; Hand, 1978). The very fact that these units can be visualized as tandem arrays by fiber autoradiography after [3H]thymidine pulses of less than 1 hour indicates that neighboring units initiate synthesis with a degree of synchrony. There is also statistical support for this (Hand, 1975). In addition, hydrodynamic methods have shown that simultaneously operating replication units are located within 50 $\mu$m of each other (Planck and Mueller, 1977). A careful analysis of the timing of initiation events on adjacent units has indicated that a unit will initiate synthesis within 5 minutes of its neighbor up to 40% of the time and within 30 minutes of

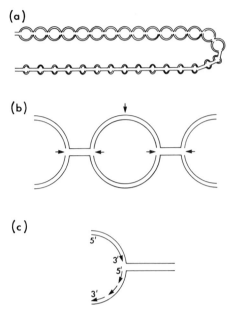

**Fig. 2.** The functional levels of genome organization. (a) Large sections of chromatin containing 15 or more replication units initiate synthesis synchronously. The upper row of replication bubbles represents a group of units that began synthesis at the same time, the lower row represents a group of neighboring units that began synthesis slightly later. (b) Initiation on individual replication units. A complete unit flanked by two half-units is shown. The vertical arrow represents the point of initiation on the middle unit; the horizontal arrows show the direction of movement of the replication forks. (c) Initiation of Okazaki piece synthesis. A half-replication unit is shown. The arrows and primed numbers indicate the direction of elongation of the daughter chains. The nascent strand in the lower part of the diagram is elongated discontinuously with the overall direction $3' \to 5'$, while its individual Okazaki pieces are synthesized in the conventional $5' \to 3'$ direction.

its neighbor up to 90% of the time (Jasny and Tamm, 1979). Since there are about 15 distinct initiation events in a short time along one stretch of the genome, and all can be eliminated by a single hit by a physical agent such as radiation, it is clear that the events governing initiation on these individual units must be distinct from the events governing the initiation of synthesis on the group as a whole.

In nonmammalian eukaryotes, there are additional classes of replication units. These are much smaller than the standard replication units. They have been visualized by electron microscopy as small replication bubbles—structures that have all the characteristics of replicating DNA. They are present in embryonic cells from insects (Blumenthal *et al.,* 1973; Zakian, 1976) and sea urchins (Baldari *et al.,* 1978). There is also evidence for their presence in mammalian cells (Burks and Stambrook, 1978). These clearly do not represent one end of the distribution of sizes of standard replication units; in fact a bimodal distribution of replication bubbles is seen in sea urchin embryonic cells (Kurek *et al.*, 1979). We cannot be certain whether these small replication bubbles in invertebrates represent distinct classes of replication units, since it has been suggested that they may be formed by the reannealing of single-stranded regions of the templates in standard replication units (Baldari *et al.*, 1978). Further studies are required before we can conclude definitely that a distinct class of small units is present in mammalian cells.

Finally, the smallest molecules that replicate as units are the Okazaki pieces. These are less than 1 $\mu$m long. Recent work suggests that only one of the two nascent strands is synthesized discontinuously as Okazaki pieces, that on which the overall direction of chain elongation is 3' to 5'. On the other strand, on which overall elongation is 5' to 3', synthesis is continuous (Perlman and Huberman, 1977; Hunter *et al.*, 1977).

There is a degree of flexibility in the organizational pattern of DNA, and this is found at all levels of control of replication.

The order of replication of the large chromosomal subsections is not fixed. There is cell-to-cell and tissue-to-tissue variation in the regions of the chromosome replicated in late S (Willard and Latt, 1976; Willard, 1977). The pattern of replication of chromosomes in hybrid cells differs from that observed in the parental lines (Farber and Davidson, 1978). These studies indicate that these subchromosomal sections are autonomous foci of replication, with only some degree of regulation between them (Willard, 1977).

Within these foci, there is also some flexibility in the control of synthesis. The replication units that make up these foci vary in size over two orders of magnitude. The origins of the units are located at nonrandom intervals along the chromosome (Cohen *et al.*, 1978). In nonmammalian eukaryotes, the size of replication units changes at different stages of development of the cells (Callan, 1972; Blumenthal *et al.*, 1973; Van't Hof, 1975; Van't Hof and Bjerknes, 1977). This phenomenon has not been demonstrated in mammalian cells, but there is evidence that experimental manipulation can change the size of replication units. In simian virus 40-transformed cells, the interval between active initiation sites is different than in their nontransformed counterparts. By fiber autoradiography, the intervals are shorter in transformed cells grown in optimal conditions, and longer in transformed cells grown in suboptimal conditions, compared to untransformed cells (Martin and Oppenheim, 1977; Oppenheim and Martin, 1978). When a hydrodynamic method based on bromodeoxyuridine-substitution and photolysis is used to measure replication unit size, it is larger in SV40-transformed mouse and human cells than in untransformed controls (Kapp *et al.*, 1979). The exact growth conditions of the cells (cell density, time in conditioned medium, percentage of cycling cells) are not given in this hydrodynamic study, but the results are not necessarily inconsistent with those obtained by fiber autoradiography, and the main point is that there is a change in the interval between active origins. When DNA synthesis resumes following the release of a fluorodeoxyuridine block, there is activation of new unit origins so that the interval between functional sites is reduced (Taylor and Hozier, 1976; Taylor, 1977). These studies indicate that there are many more potential origins of synthesis with the subchromosomal sections than are normally used to complete replication.

Flexibility probably exists at the level of initiation of Okazaki pieces. Available evidence suggests that the sites of initiation of Okazaki pieces are not determined by base sequence (Reichard *et al.*, 1974; Tseng and Goulian, 1975; Anderson *et al.*, 1977). Therefore, any site along the template might serve for initiation, and this implies that such sites are not fixed. If this is the case, then a particular section of DNA might serve for continuous synthesis of DNA if it is template for the leading strand, or for discontinuous synthesis if it is template for the lagging strand.

## IV.  A MODEL FOR THE INITIATION OF DNA REPLICATION

The picture therefore emerges of a regulated yet flexible program for DNA synthesis in the mammalian cell. Based on this evidence, I would like to present a model for the initiation of DNA synthesis during S phase.

The critical set of initiation events is that controlling the beginning of synthesis on the large subchromosomal sections of chromatin. These will be referred to as primary initiation events. There are probably about 4000 such events during an S phase, based on estimates of 15 replication units per segment. This number corresponds approximately to the number of subbands defined by staining on extended prometaphase chromosomes (Comings, 1978). Although no functional role has yet been assigned to these subbands, the structural features that determine staining may also play a role in functional organization of the genome. In view of the evidence that there is stringent control within the large replicating segments, but not between them (Willard, 1977), any one of several might begin the S phase, with the determining factor being accessibility of the origins to the proteins and other factors required for the primary initiation events. Accessibility could be determined by proximity of a particular chromatin site (the origin) to a particular nuclear structure, such as some unique site of the nuclear matrix or nuclear membrane. Since the weight of evidence is against the involvement of the membrane in the initiation of replication in mammalian cells (Edenberg and Huberman, 1975; Sheinin *et al.*, 1978), a unique site or set of sites on the matrix seems more likely. Suppose then that at the time when a critical level of precursors and enzymes for DNA synthesis has been achieved, synthesis initiates at several of these sites throughout the nucleus. The primary initiation events must decompact the chromatin partially at these sites to allow access of replication factors. This decompaction would then have to spread over a localized stretch of chromatin of at least several replication units. This partial decompaction would allow a second set of replication proteins to have access to the chromatin, and these would initiate synthesis at the replication units within the immediate area of the primary initiation site. Initiation events at the level of replication units will be called secondary initiation events. Further decompaction occurs allowing initiation events down to the level of Okazaki pieces. Some aspects of this model are shown in Fig. 3.

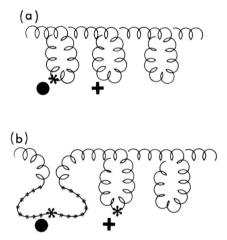

**Fig. 3.** Primary and secondary initiation events in DNA synthesis. The coiled line represents chromatin compacted into a high-order structure. (a) A particular point along the coil (large asterisk) defined in part by its proximity to a unique site in the nuclear matrix (filled circle) serves as a primary initiation site. An adjacent loop also has a primary initiation site defined by its proximity to another unique point on the matrix (cross). (b) Following the initiation on the coil on the left, the chromatin is decompacted, allowing secondary initiation events (small asterisks) on replication units near the primary site. The large asterisk on the middle loop represents a subsequent primary initiation event.

In this sort of model then, a primary initiation event on a large chromosomal subsection then sets off a cascade mechanism governed in part by the decompaction of the localized stretch of chromatin. Subsequent secondary initiation events take place when the chromatin decompaction proceeds and Okazaki piece synthesis begins when the chromatin is decompacted to the level of primary nucleosomes. Decompaction and recompaction of chromatin are intimately involved in the replication of mammalian DNA (Setterfield *et al.*, 1978; Chaly *et al.*, 1979) and the nucleosome has been proposed as the structure controlling Okazaki piece synthesis (Hewish, 1976; Rosenberg, 1976).

What then determines the sites of subsequent primary initiation events? There are at least two possibilities. The replicated chromatin might be dislocated from the points on the matrix that serve as recognition sites for the initiation proteins and be replaced by unreplicated chromatin. More likely, I believe, is the possibility that further initiations depend on availability of initiation factors. As S phase progresses, there are increased amounts of replication factors synthesized, and as the concentration of critical primary initiation factors increases, then more and more primary initiation events take place at numerous loci throughout the nucleus. A mechanism of this sort does away with the requirements for termination events. Each primary initiation event would begin a replication process that would last several hours, since the secondary initiation events are coordinated in an imperfect temporal fashion and this temporal synchrony decreases as the distance between initiation sites increases (Jasny and Tamm, 1979). It is not unreasonable to suggest that those secondary initiation events located further from the primary site occur later. Replication on a chromosomal subsection would then terminate when it meets and fuses with replicating DNA from a neighboring section in a fashion analogous to that postulated for the termination of synthesis on replication units (McFarlane and Callan, 1973; Hand, 1975).

Synthesis at the end of S phase would decrease in this model as the concentration of critical initiation factors decreases. Another mechanism for ending S phase is also possible. Reinitiation of synthesis on DNA already replicated has not been observed in mammalian cells. Perhaps as the replicated chromatin is recompacted following synthesis, it takes on a specific configuration that prevents reinitiation. This specific configuration might be determined by the presence of two sister chromatids over a stretch of chromosome, rather than one. The S phase would then end simply when all the chromatin is in this configuration, and could not begin again until sometime after mitosis when the configuration associated with single chromatids is present. This sort of mechanism would explain why certain factors such as triphosphate precursors remain at high levels in $G_2$ in the absence of DNA synthesis (Walters *et al.*, 1973). The replication factors, such as triphosphates and enzymes, would exert positive control, while conformational changes in the chromatin that occur after replication would exert a negative control.

## V. PREDICTIONS OF THE MODEL

The model presented is consistent with what we know of mammalian DNA replication. It also allows several specific predictions which can be tested experimentally. Although specific base sequences at the different classes of origins are not ruled out, the specificity of origins must be determined in part by chromatin structure. There should be definable origins for the primary initiation events, and they should be associated with the nuclear matrix. Subchromosomal sections whose replication is controlled by primary initiation events should be larger if they are active early in S phase than if they are activated late in S. The size of the late S sections would be limited by the fusion with sections that had begun synthesis much earlier. Secondary initiation events should spread in an ordered fashion from the primary initiation sites. Finally, *in vitro* systems containing all factors required for the initiation of DNA synthesis should be active when $G_1$ chromatin is used as template, but not when the template is $G_2$ chromatin.

How can one go about testing the assumptions and predictions of the model? The most important task is to define functionally and structurally the subchromosomal section governed by the primary initiation events. The use of X- and uv-irradiation as probes for DNA replication will be important here, since these physical agents inhibit primary initiation events by a single hit. Autoradiography of [$^3$H]thymidine-labeled chromatin should also prove useful. As various higher-order structures in chromatin are defined, they should be studied using light microscope autoradiography and the spreading techniques used for electron microscopy in the same manner that replicating DNA segments have been studied by fiber autoradiography. Autoradiography of spread chromatin should be a convenient procedure since the compaction of the DNA should produce relatively dense patterns after short exposure. Direct testing of many of the predictions must await the development of an *in vitro* system that faithfully reproduces the *in vivo* replication process.

## REFERENCES

Adegoke, J. A., and Taylor, J. H. (1977). *Exp. Cell Res.* **104,** 47–54.
Anderson, S., Kaufman, G., and De Pamphilis, M. L. (1977). *Biochemistry* **16,** 4990–4998.
Bak, A. L., Zeuthen, J., and Crick, F. (1977). *Proc. Natl. Acad. Sci. U.S.A.* **74,** 1595–1599.
Baldari, C. T., Amaldi, F., and Buongiorno-Nardelli, M. (1978). *Cell* **15,** 1095–1107.
Benyajati, C., and Worcel, A. (1976). *Cell* **9,** 393–407.
Blumenthal, A. B., Kriegstein, H. J., and Hogness, D. S. (1973). *Cold Spring Harbor Symp. Quant. Biol.* **38,** 205–223.
Bostock, C. J., and Prescott, D. M. (1971). *J. Mol. Biol.* **60,** 151–162.
Burks, D. J., and Stambrook, P. J. (1978). *J. Cell Biol.* **77,** 762–773.
Callan, H. G. (1972). *Proc. R. Soc. London Ser. B* **181,** 19–41.
Chaly, N., Johnstone, M., and Hand, R. (1979). *Clin. Invest. Med.,* **2,** 141–152.
Chang, L. M. S., Brown, M., and Bollum, F. J. (1973). *J. Mol. Biol.* **74,** 1–8.
Cohen, J. E., Jasny, B. R., and Tamm, I. (1978). *J. Mol. Biol.* **128,** 219–245.
Comings, D. E. (1972). *Exp. Cell Res.* **71,** 106–112.
Comings, D. E. (1978). *Annu. Rev. Genet.* **12,** 25–46.
Cook, P. R., and Brazell, I. A. (1975). *J. Cell Sci.* **19,** 261–279.
Dhar, R., Lai, C. J., and Khoury, G. (1978). *Cell* **13,** 345–358.
Edenberg, H. J., and Huberman, J. A. (1975). *Annu. Rev. Genet.* **9,** 245–284.
Erlandson, R. A., and DeHarven, E. (1971). *J. Cell Sci.* **8,** 353–397.
Farber, R. A., and Davidson, R. L. (1978). *Proc. Natl. Acad. Sci. U.S.A.* **75,** 1470–1474.
Felsenfeld, G. (1978). *Nature (London)* **271,** 115–122.
Finch, J. T., and Klug, A. (1976). *Proc. Natl. Acad. Sci. U.S.A.* **73,** 1897–1901.
Graham, C. F., Arms, K., and Gurdon, J. B. (1966). *Dev. Biol.* **14,** 349–381.
Grummt, F. (1978). *Proc. Natl. Acad. Sci. U.S.A.* **75,** 371–375.
Hamlin, J. L., and Pardee, A. B. (1978). *In Vitro* **14,** 119–127.
Hand, R. (1975). *J. Cell Biol.* **64,** 89–97.
Hand, R. (1978). *Cell* **15,** 317–325.
Hewish, D. R. (1976). *Nucl. Acids Res.* **3,** 69–78.
Huberman, J. A., and Riggs, A. D. (1968). *J. Mol. Biol.* **32,** 327–341.
Hunter, T., Francke, B., and Bacheler, L. (1977). *Cell* **12,** 1021–1028.
Jasny, B. R., and Tamm, I. (1979). *J. Cell Biol.* **81,** 692–697.
Johnson, R. T., and Harris, H. (1969). *J. Cell Sci.* **5,** 625–649.
Kapp, L. N., Park, S. D., and Cleaver, J. E. (1979). *Exp. Cell Res.,* in press.
Kornberg, R. (1977). *Annu. Rev. Biochem.* **46,** 931–954.
Kurek, M. P., Billig, D., and Stambrook, P. (1979). *J. Cell Biol.* **81,** 698–703.
Latt, S. A. (1975). *Somat. Cell Genet.* **2,** 293–321.
McFarlane, P. W., and Callan, H. G. (1973). *J. Cell Sci.* **13,** 821–839.
Marsden, M. P. F., and Laemmli, U. K. (1979). *Cell* **17,** 849–858.
Martin, R. G., and Oppenheim, A. (1977). *Cell* **11,** 859–869.
May, M. S., and Bello, L. J. (1974). *Exp. Cell Res.* **83,** 79–86.
Mueller, G. C., and Kajiwara, K. (1966). *Biochem. Biophys. Acta* **114,** 108–115.
Oppenheim, A., and Martin, R. G. (1978). *J. Virol.* **25,** 450–452.
Painter, R. B., and Young, B. R. (1975). *Radiat. Res.* **64,** 648–656.
Painter, R. B., and Young, B. R. (1976). *Biochim. Biophys. Acta* **418,** 146–153.
Pardee, A. B., Dubrow, R., Hamlin, J. L., and Kletzien, R. F.(1978). *Annu. Rev. Biochem.* **47,** 715–750.
Perlman, D., and Huberman, J. A. (1977). *Cell* **12,** 1029–1043.
Planck, S. R., and Mueller, G. C. (1977). *Biochemistry* **16,** 1808–1813.

Povirk, L. F. (1977). *J. Mol. Biol.* **114,** 141–151.
Rao, P. N., and Johnson, R. T. (1970). *Nature (London)* **225,** 159–164.
Reichard, P., Eliasson, R., and Soderman, G. (1974). *Proc. Natl. Acad. Sci. U.S.A.* **71,** 4901–4905.
Rosenberg, B. H. (1976). *Biochem. Biophys. Res. Commun.* **72,** 1384–1391.
Sedat, J., and Manuelidis, L. (1977). *Cold Spring Harbor Symp. Quant. Biol.* **42,** 331–350.
Setterfield, G., Sheinin, R., Dardick, I., Kiss, G., and Dubsky, M. (1978). *J. Cell Biol.* **77,** 246–263.
Sheinin, R., Humbert, J., and Perlman, R. C. (1978). *Annu. Rev. Biochem.* **47,** 277–316.
Skoog, L., Nordenskjold, B. A., and Bjursell, G. (1973). *Eur. J. Biochem.* **33,** 428–432.
Soeda, E., Kimura, G., and Miura, K.-C. (1978). *Proc. Natl. Acad. Sci. U.S.A.* **75,** 162–166.
Spadari, S., and Weissbach, A. (1974). *J. Mol. Biol.* **86,** 11–20.
Stubblefield, E. (1975). *Chromosoma* **53,** 209–221.
Subramanian, K. N., Dhar, R., and Weissman, S. M. (1977). *J. Biol. Chem.* **252,** 355–367.
Tamm, I., Jasny, B. R., and Cohen, J. E. (1979). *Alfred Benzon Symp.* **13,** in press.
Tapiero, H., Shaool, D., Monier, M. P., and Harel, J. (1974). *Exp. Cell Res.* **89,** 39–46.
Taylor, J. H. (1960). *J. Biophys. Biochem. Cytol.* **7,** 455–464.
Taylor, J. H. (1977). *Chromosoma* **62,** 291–300.
Taylor, J. H., and Hozier, J. C. (1976). *Chromosoma* **57,** 341–350.
Tseng, B. Y., and Goulian, M. (1975). *J. Mol. Biol.* **99,** 339–347.
Van't Hof, J. (1975). *Exp. Cell Res.* **93,** 95–104.
Van't Hof, J., and Bjerknes, C. A. (1977). *Chromosoma* **64,** 287–294.
Walters, R. A., Tobey, R. A., and Ratliff, R. L. (1973). *Biochim. Biophys. Acta* **319,** 336–347.
Willard, H. F. (1977). *Chromosoma* **61,** 61–73.
Willard, H. F., and Latt, S. A. (1976). *Am. J. Hum. Genet.* **28,** 213–227.
Williams, C. A., and Ockey, C. H. (1970). *Exp. Cell Res.* **63,** 365–372.
Worcel, A., and Benyajati, C. (1977). *Cell* **12,** 83–100.
Yanishevsky, R. M., and Prescott, D. M. (1978). *Proc. Natl. Acad. Sci. U.S.A.* **75,** 3307–3311.
Zakian, V. A. (1976). *J. Mol. Biol.* **108,** 305–332.

# 6

# Tissue- and Species-Specific
# Nuclear Antigens and the Cell Cycle

ROBERT C. BRIGGS, WANDA M. KRAJEWSKA,
LUBOMIR S. HNILICA, GLORIA LINCOLN,
JANET STEIN, and GARY STEIN

## I. INTRODUCTION

The diversity in the transcription of cellular DNA is thought to be a mechanism whereby the variety of phenotypically different types of cells originate in an organism. The elucidation of the components involved in regulating the expression of the information encoded within the DNA has been intensively studied in recent years. One natural area of emphasis has centered on the proteins closely associated with the DNA. Some of these chromosomal proteins are known to be involved in chromatin architecture. It appears that the nucleosome core, a flattened sphere of approximately 110 Å in diameter containing two copies of each of the four core histone proteins (H2a, H2b, H3, and H4) and 140 base pairs of DNA coiled around the outside, represents the primary level of genome organization (Dubochet and Noll, 1978). Nucleosome

**181**

NUCLEAR–CYTOPLASMIC INTERACTIONS
IN THE CELL CYCLE

cores are interconnected by a variable segment of linker DNA that has the H1 histone associated with it. The core particles with their DNA linkers form the nucleosome. Additional packaging of nucleosomes may account for higher-order chromatin structure (Rattner and Hamkalo, 1979).

In addition to the basic histone proteins that are of very limited heterogeneity, there is a group of non-histone proteins associated with the nuclear DNA that are integral components of the chromatin and chromosomes. Because of their heterogeneity and suggestions of tissue- and species-specific distributions, the chromosomal non-histone proteins have been considered as possible regulators of transcription (Chiu and Hnilica, 1977; Stein *et al.*, 1974). The interest in this group of proteins has resulted in the development and application of more sensitive analytical methods for their study. The non-histone proteins have now been resolved into hundreds of different polypeptides by two-dimensional PAGE; many of these, especially the major components, appear to be common to many tissues and even between species. The predominant chromosomal protein components are probably not involved in determining specificity of gene transcription simply by their presence or absence, but they are more likely required for the normal functioning of all nuclei independent of cell type. However, a few less universally distributed components have also been detected using these high resolution electrophoretic methods.

Some studies have revealed non-histone proteins that were found only in certain tissues, species, or malignant tumors. In addition, areas within chromatin of defined sequence show variable structural organization (Wu *et al.*, 1979a,b). The DNA associated with transcriptionally active areas is more sensitive to DNA digestion than other areas of the genome (Weintraub and Groudine, 1976; Garel and Axel, 1976). This sensitivity does not depend on actual transcription but rather is determined by the stage of cellular differentiation and whether or not an area of the genome is organized in such a manner that it is capable of being transcriptionally expressed by that cell type in response to an appropriate stimulation. It is thought that some proteins, capable of recognizing specific sequences in the DNA, are either associated with these areas or cause modifications of some other components within these areas, thus facilitating transcriptional specificity and sensitivity to nuclease (Weisbrod and Weintraub, 1979). Since the transcriptional expression of genes is unique for a cell type, the existence of cell type-specific components in chromatin can be anticipated. Indeed, through the use of sensitive and selective immunological methods, chromosomal non-histone proteins have been detected that are specific for species, tissues, cells, and malignancy (Hnilica and Briggs, 1980). In some cases, the intrachromosomal localization has been studied and the results indicate that these specific antigens are not uniformly distributed throughout the chromatin. These findings, considered in context with changes of immunological specificity of chromosomal non-histone protein antigens observed during normal development and with malignant transformation, are consistent with their relation to the process of cellular differentiation. Some of the characteristics of these specific nuclear antigens especially their relationship to the cell cycle will be discussed in this chapter.

## II.  CELL- AND TISSUE-SPECIFIC CHROMATIN ANTIGENS

Antisera to chromatin antigens have been elicited with total chromatin, dehistonized chromatin, or chromatin extracts. Use of dehistonized chromatin has most frequently been reported to produce antisera that recognize components specific for cell types or tissues. In the initial use of this material for immunization, Chytil and Spelsberg (1971) showed that dehistonized chromatin prepared from oviducts of immature chicks stimulated with diethylstilbesterol for 15 days elicited antibodies in rabbits that reacted in complement fixation tests only with dehistonized chromatin prepared from the stimulated chick oviduct. Dehistonized chromatin from other tissues or unstimulated immature oviducts were all nonreactive. However, once chicks were started on daily diethylstilbesterol treatments, the oviduct chromatin showed progressively increasing immunological activity until maximal activity was reached after 15 days (Spelsberg et al., 1973). A fraction of non-histone protein was responsible for the antigenic specificity, and this activity could be transferred by reconstituting this fraction with DNA from other sources (Spelsberg et al., 1972). Use of dehistonized chromatin for the production of antisera that recognize tissue-specific antigens was quickly repeated with rat liver (Chytil et al., 1974; Wakabayashi et al., 1974) and calf thymus (Wakabayashi et al., 1974). More recently this has been extended to include chicken erythroid cells (Hardy et al., 1978) and human neutrophilic granulocytes (Briggs et al., 1978, 1980). In these studies the antigens were specific for normal tissue or cell type and appeared during cellular differentiation. Many of these studies were done in the authors' laboratory where particular emphasis was placed on elucidating the relationship between the non-histone part of the antigen and the DNA component. The antigenic specificity appears to depend on the complexing of the protein(s) with the DNA (NPAg–D complexes). The isolated histone, DNA, or non-histone protein fractions were not immunologically active and only reconstituting the non-histone protein fraction with homologous DNA restored the specific immunological reaction.

## III. SPECIES-SPECIFIC CHROMATIN ANTIGENS

Total chromatin prepared from human and mouse fibroblast tissue culture cell lines were used by Zardi and co-workers to elicit antibodies in rabbits and chickens (Zardi *et al.*, 1973, 1974; Okita and Zardi, 1974; Zardi, 1975). The antisera reacted with species-specific non-histone protein components either free or as a complex with DNA. However, the immunological reaction was stronger when the complex was tested. The antiserum to the human fibroblast chromatin was able to distinguish between the normal cell chromatin and that from the cells virally transformed counterpart. The possibility of cell or tissue specificity was not directly tested in these studies. However, antisera recognizing chromatin antigens showing species specificity, but not cell or tissue specificity, have been described by other investigators. Antisera produced to a non-histone protein extract from rat liver chromatin showed stronger species rather than tissue specificity as determined by measuring ability of the non-histone protein fraction from various sources to compete for antibody in a radioimmunoassay (Cohen *et al.*, 1978). Also, antisera raised to dehistonized chromatin prepared from human normal and transformed cells reacted with other human cell nuclei in immunofluorescence tests, but not with nuclei in cells from other species (Tsutsui *et al.*, 1976, 1977). It appears likely that within the group of non-histone proteins there are components that display different types of distributions, some being common to nuclei in all or most cell types in an organism and others being restricted to certain cell types or tissues. Our laboratory is concerned with chromatin antigens having the latter characteristics. One of these that is currently being studied is a chicken erythroid cell-specific chromatin antigen (Hardy *et al.*, 1978).

Antisera were raised in rabbits to dehistonized chromatin prepared from reticulocytes obtained from chickens made anemic by phenylhydrazine injections (Hardy *et al.*, 1978; Krajewska *et al.*, 1979). The immunological activity of the antisera with chromatin preparations was determined by microcomplement fixation test (Wasserman and Levine, 1961). The antisera reacted only with chicken erythroid chromatin when tested with chromatin obtained from chicken tissues or nonerythroid tissues from other species (Fig. 1). To assure that the antigen(s) (NPAg–D) was of nuclear origin, it was localized in acetone fixed blood

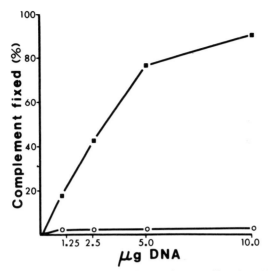

**Fig. 1.** Extent of complement fixation when antiserum (diluted 1:100) to dehistonized chicken reticulocyte chromatin was assayed with reticulocyte chromatin (■), chicken liver (○), rat liver (○), and Novikoff hepatoma (○). Chicken red blood cells were collected by cardiac puncture or decapitation. Reticulocytes were obtained from chickens made anemic by phenylhydrazine injections (10 mg/kg for 5 days). The red blood cell nuclei were isolated as previously described (Hardy *et al.*, 1978) or by nitrogen cavitation (Shelton *et al.*, 1976). Chromatin was prepared (Spelsberg *et al.*, 1971) and adjusted for use in immunological assays. Some chromatin was dehistonized and used for immunization. Nuclei were isolated from liver tissue by the method of Blobel and Potter (1966) and from Novikoff hepatoma cells by hypotonic shock. The method of preparing chromatin was the same for nuclei obtained from all sources.

smears by the peroxidase–antiperoxidase immunocytochemical staining method of Sternberger (1974). The specific immunological reaction was clearly localized over the nucleus of the reticulocyte and the erythrocyte nucleus (Fig. 2). No immunocytochemical reaction was observed with erythroid cells incubated with rabbit serum obtained before immunization (Fig. 2B). No staining reaction was observed in the nuclei of the parenchymal cells in chicken liver or kidney tissue frozen sections, further attesting to the antigen's cell-specific distribution.

To investigate species specificity, nucleated erythrocytes were obtained from a number of organisms and chromatin was prepared. When tested with antiserum to chicken reticulocyte dehistonized chromatin (Fig. 3), only erythrocyte chromatins from closely related birds showed significant cross-reactivity. The relative levels of activity were in agreement with the known taxonomic groupings ·(Welty, 1962). No cross-reacting antigen was detected in nucleated erythrocytes from different

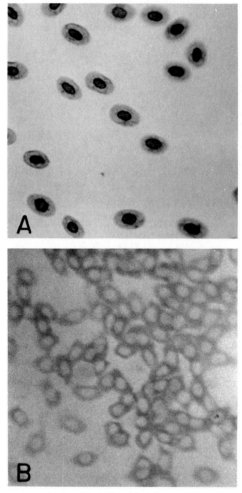

**Fig. 2.** Immunocytochemical localization of chicken erythroid cell antigen(s). Antiserum diluted 1:200 was reacted with acetone fixed smears of chicken reticulocytes (A). Serum (1:200) collected from the rabbit prior to its immunization (B).

classes of vertebrates. The variable extent of cross-reaction with chromatin from the same cell type in closely related species indicates a similar requirement for the antigen in this cell in other species, but that evolutionary changes have occurred.

Although antigen was detected in chicken reticulocyte and erythrocyte chromatin preparations, less activity was observed in the material obtained from the mature erythrocytes (Fig. 4). Interestingly, once

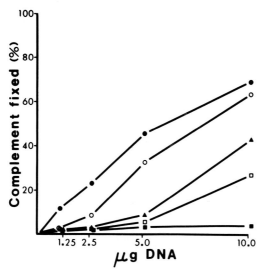

**Fig. 3.** Complement fixation test of erythrocyte chromatins isolated from various species with antiserum (diluted 1:100) to dehistonized chicken reticulocyte chromatin. Chicken (●), turkey (○), quail (▲), goose (□), duck (■), turtle (■), bullfrog (■), and catfish (■).

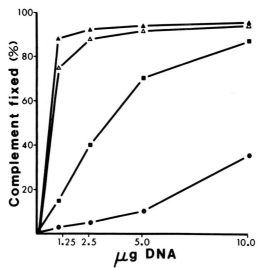

**Fig. 4.** Effects of removing histones on the immunological detection of the chicken erythroid cell-specific chromatin antigen. Antiserum (diluted 1:100) to dehistonized chicken reticulocyte chromatin was reacted in the complement fixation test with total (■) and dehistonized (▲) reticulocyte chromatins and total (●) and dehistonized (△) erythrocyte chromatins. Histones were extracted from chromatin with 2.5 $M$ NaCl, 5 $M$ urea, 0.1 $M$ sodium phosphate buffer pH 6.0, and 0.1 m$M$ PMSF. Non-histone proteins and DNA were collected by centrifugation (100,000 $g$, 36 hours).

chromatins from both sources had been dehistonized the same amount of increased antigen activity was detected (Fig. 4). It was also possible to elevate the immunological activity of the erythroid cell chromatin by exposing it to polyanions prior to testing. Since both treatments are known to remove the histone proteins it appeared likely that they were responsible for lowering antigen detection. Chromatin fractionation and reconstitution experiments were done to elucidate moré clearly the histone effects.

Once chromatin was fractionated into histones, non-histones, and DNA, none of the isolated fractions contained any immunological activity. When the non-histones were complexed with DNA, the immunological activity was restored to a level comparable to that of dehistonized chromatin (Krajewska *et al.*, 1980). This could have been anticipated since the immunogen (dehistonized chromatin) is a complex of non-histone proteins with DNA. The inhibitory effect of histones on the immunological activity was then demonstrated by adding the histone fraction to dehistonized chromatin prior to testing. The immunological activity of reticulocyte dehistonized chromatin was lowered by adding histones from either reticulocytes or erythrocytes, but the level was higher than the level reached by adding reticulocyte or erythrocyte histones to dehistonized erythrocyte chromatin (Fig. 5). From these results it is apparent that chromatin organization, as reflected by immunological detection of a non-histone protein antigen(s) (NPAg–D), is determined at two levels. One level is caused by histones and results in nonspecific reduction of immunological activity and another by non-histones, which is responsible for differences in level of antigen detection between chromatin from an earlier stage cell and that from the mature erythrocyte. The structural/organizational implications of these results led to a consideration for the intrachromosomal localization of the antigen(s).

One of the methods used for examining the interdependence between the protein and DNA components in the NPAg–D complex was to digest the DNA in chromatin with micrococcal nuclease and test the reaction products for immunological activity. When the same amount of reticulocyte chromatin was digested with increasing amounts of nuclease, the immunological activity was destroyed only at the highest level of enzyme (Fig. 6), thereby demonstrating the requirement for intact high molecular weight DNA. The results of these experiments also show that antigen activity is not affected by low levels of nuclease treatment. This was an important observation in regard to studying chromatin structure since these low levels of nuclease cause sufficient strand breakage to allow the release of soluble material that can be separated into mononucleosomes, oligonucleosomes, and high molecular weight frac-

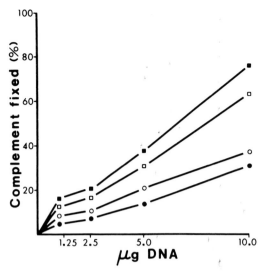

**Fig. 5.** The effects of reconstituted histone on the immunological detection of the chicken erythroid cell-specific chromatin antigen. Dehistonized chicken reticulocyte chromatin reconstituted with reticulocyte (■) and erythrocyte (□) histones. Dehistonized erythrocyte chromatin reconstituted with reticulocyte (○) and erythrocyte (●) histones. Histone and dehistonized chromatin preparations were obtained as described in Fig. 4. After reconstitution, chromatins were reacted in complement fixation test with antiserum (diluted 1:100) raised to dehistonized chicken reticulocyte chromatin.

tion on BioGel A-50 m columns (Campbell *et al.*, 1979). When these materials were tested for immunological activity, more than 90% was associated with the high molecular weight material and almost no activity with the mononucleosomes. The DNA associated with the antigenically active fraction was apparently protected to some extent from digestion. The accumulation of antigen in this nuclease resistant high molecular weight fraction suggests that it may be part of a specialized nuclear structure. Sanders (1978) reported experimental evidence for the presence in chromatin of specialized nucleosomes which may be involved in maintaining the condensed state of chromatin and thereby determining the overall morphological features of the cell nucleus. The chromosome scaffold is another nuclear structure first isolated from metaphase chromosomes by Adolph *et al.*, 1977a) which reportedly also persists in interphase chromatin (Adolph *et al.*, 1977b). Finally, a nuclear fraction called the matrix is also thought to be responsible for the overall nuclear shape (Berezney and Coffey, 1977). Each of these structural materials to some extent preferentially survives the DNase digestion and various salt extractions of chromatin. The relationships of the cell- and tissue-

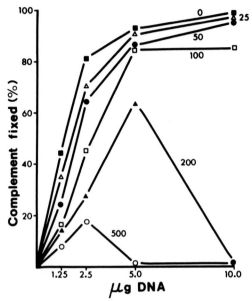

**Fig. 6.** Sensitivity of chicken erythroid cell-specific chromatin antigen to nuclease digestion. Antiserum (diluted 1:100) to dehistonized chicken reticulocyte chromatin reacted in the complement fixation test with dehistonized chromatin treated with various levels of micrococcal nuclease. Digestion of 200 μg chromatin measured as DNA (37°C, 10 minutes) in 40 mM NaCl, 10 mM Tris-HCl pH 7.5, 1.0 mM CaCl₂ at the levels of nuclease (units) indicated.

specific nuclear antigens to the structural components of chromatin are of considerable interest and will be further discussed later in this chapter.

## IV.   CHROMATIN ANTIGENS IN MALIGNANT GROWTH

Both nucleolar and total chromatin materials (total chromatin, dehistonized chromatin, and chromatin extracts) when used as immunogens have been reported to elicit antisera that recognize antigenic changes unique to malignant growth.

### A.   Nucleolus

In the case of nucleolar activity, autoantibodies were previously discovered that specifically recognize human cancer cells (Priori *et al.,* 1971; McBride *et al.,* 1972; Ioachim *et al.,* 1976). Recently, Busch and his associates raised antisera to isolated HeLa cell nucleoli or nuclear extracts that gave strong nucleolar immunofluorescence staining with nearly all of 62 human malignant tumor specimens examined (Busch *et al.*, 1979; Davis *et al.*, 1979). Although antigen characterization is not yet available, it does appear to be protein in nature. Previous studies by this group using nucleoli from the transplantable rat Novikoff hepatoma ascites cells led to the detection of liver-specific, tumor-specific, and tumor- and fetal-specific nucleolar antigenic proteins (Busch *et al.*, 1974; Davis *et al.*, 1978). One of the nucleolar antigens, NoAG-1, was purified to homogeneity and partially characterized (Marashi *et al.*, 1979). It has a molecular weight of 60,000 and a pI of 5.1. This antigen was detected only in Novikoff hepatoma nucleoli and not in normal rat liver.

## B. Chromatin

Although the use of total chromatin from normal fibroblasts was reported to result in the production of antisera that detected primarily species-specific antigens, the antigenic activity associated with the chromatin prepared from transformed fibroblasts was not the same (see Section III). The antisera reacted with total chromatin and to a lesser extent with chromosomal proteins free from DNA. The proteins were probably non-histones (Zardi, 1975).

The most definitive work in the area of isolating and characterizing chromatin antigens specific for malignancy has come from studies more directed at the chromosomal non-histone proteins. These studies require the use of various extracts of chromatin known to contain non-histone proteins or dehistonized chromatin as the immunogen preparation. Using the former approach an antigen was detected in 0.6 $M$ NaCl extracts of Novikoff hepatoma chromatin (NAg-1) which was also present in Walker carcinosarcoma 256, and in 18-day fetal rat liver chromatins (Yeoman *et al.*, 1976a). This NAg-1 was not detected in normal or regenerating rat liver, rat heart, or kidney chromatins. The antigen was isolated and found to be a 26,000-dalton glycoprotein. Its presence in fetal and malignant cells suggests a possible relationship to the production of oncofetal gene products. Since the 18-hour regenerating rat liver did not contain this antigen, the NAg-1 is not a growth-specific antigen. It should be noted that the antigen is reactive when free from DNA and that in positively reacting tissue, some antigen is also detected in the cytoplasm. More recently Yeoman *et al.* (1978) have isolated and partially characterized five additional nuclear antigens from 0.6 $M$ NaCl extracts of Novikoff hepatoma chromatin. All of the antigens are acidic proteins with MW ranging from 12,000 to 129,000. Three of the antigens are glycoproteins. Of the six antigens in this extract, three are tumor-associated and three are present in fetal tissue while the NAg-1 is considered an oncofetal type antigen (Yeoman *et al.*, 1978).

Another method of obtaining solubilized non-histone proteins for use as an immunogen resulted in the production of antiserum that showed a specific reaction for a mouse osteogenic sarcoma (Kono *et al.*, 1977). A non-histone protein fraction was responsible for the immunological reaction and cross-reacting antigen(s) was not found in Ehrlich ascites tumor, normal mouse liver, or calf thymus. This immunologically active non-histone protein fraction is probably quite heterogeneous and the antigenic components were not further identified (Kono *et al.*, 1977).

Although immunization with chromatin extracts has been successful in eliciting antisera that detect non-histone protein components specific for malignancy or malignant and fetal tissue, other investigators prefer to employ protein and DNA complexed together for immunization. The immuno-potentiating effect of DNA on the proteins as well as the ability of this material to elicit antisera that recognize antigen differences between normal and malignant tissues was first realized about two decades ago (Messineo, 1961). However, this early, and some later work (Carlo *et al.*, 1970) was done with poorly defined materials and it was difficult to attribute the antigenic activity to any fraction of chromatin other than to suspect a DNA and protein composition (Carlo *et al.*, 1970). More recently a method for immunization was introduced by Chytil and Spelsberg (1971), which involved the preparation of chromatin from isolated nuclei which was then dehistonized in high salt and urea solution at pH 6.0 (Spelsberg *et al.*, 1971). Once the histones were removed, the complexes of non-histone protein and DNA (NPAg–D complexes) were used to immunize rabbits. The rat Novikoff hepatoma has been the most thoroughly studied system using these methods. Antisera raised to Novikoff dehistonized chromatin react in microcomplement fixation test with chromatin from Novikoff, but not from normal rat liver and *vice versa* (Wakabayashi and Hnilica, 1973; Wakabayashi *et al.*, 1974; Chiu *et al.*, 1976). The antisera to Novikoff reacted equally well with chromatin from Walker carcinosarcoma 256, but chromatin from slow growing and well differentiated Morris hepatomas showed lower levels of cross-reactivity (Chiu *et al.*, 1974). Some cross-reaction was also observed with chromatin from human tissues (Chiu *et al.*, 1977). These results suggest that the activity specific for malignancy may in some cases cross species boundaries. However, antisera raised to dehistonized chromatin prepared from human cancer tissues did not react with Novikoff material or that from human cancers originating in other human tissues (Chiu *et al.*, 1977). The human cancer antisera including those to HeLa S3 cells did not react with chromatins derived from normal human tissues (Chiu *et al.*, 1977; Briggs *et al.*, 1979).

In order to determine the effects of chemical carcinogens on the immunological specificity of chromatin, Fisher rats were fed a diet containing the hepatocarcinogen *N,N*-dimethyl-*p*-(tolylazo) aniline (3′MDAB). At 7-day intervals the immunological activity of the liver chromatin was evaluated. Within 2 weeks of feeding the immunological activity began to change to that of Novikoff hepatoma chromatin or that of dye-produced tumors (Chiu *et al.*, 1975). Very similar results were also obtained in rats fed dimethylhydrazine, which produces malignant tumors of the large bowel (Chiu *et al.*, 1979).

The isolation of these specific antigens has been a high priority, but the characteristics of the NPAg–D complexes have complicated the process. Because the immunogen preparation is a complex of protein and DNA, the immunological activity is dependent upon both components. Fractionation schemes must begin with the total non-histone proteins as they are nearly all present in the immunogen preparation (dehistonized chromatin) and, once fractionated, they must be reconstituted to high molecular weight DNA in order to locate and recover the immunological activity. Fortunately, the antigenicity of most non-histone proteins is adversely affected by urea and only some appear to be active after the dehistonization of chromatin. The antigenic proteins, in general, have poor solubility characteristics and various denaturants are frequently used during fractionation. These reagents must be removed prior to performing the immunological assay. The microcomplement fixation test is the assay of choice since in most cases the antisera produced are nonprecipitating. In spite of these problems, three non-histone polypeptides with molecular weights of 45,000–60,000 daltons have been found to represent protein component of the Novikoff hepatoma-specific NPAg–D complexes (Fujitani *et al.*, 1978). Recently the three peptides have been isolated and partially characterized (Zimmer *et al.*, 1980). Each component is immunologically active and PAS staining of the electrophoretically separated proteins indicates that they are glycosylated. The same fraction isolated from normal rat liver was not immunologically active with the Novikoff antiserum. The methodology developed for the isolation of the Novikoff antigen proteins is now being applied to the other systems in the hope of elucidating the biological significance of this unusually specific group of chromosomal proteins (those involved in the NPAg–D complexes).

During the study of the Novikoff-specific chromatin antigens, the results of some experiments indicated that the NPAg–D antigens might be growth associated. Chromatins isolated from livers of rats 24 and 48 hours after partial hepatectomy reacted strongly with antibodies to Novikoff hepatoma dehistonized chromatin (Chiu *et al.*, 1976). Chromatins isolated at 6, 12, and 72 hours did not react or only marginally so. Additional experiments revealed that the activity to regenerating rat liver could be absorbed out of the anti-Novikoff serum using 24- or 48-hour regenerating rat liver chromatin, but that the antiserum was still active with Novikoff hepatoma chromatin in a complement fixation test. These findings revealed heterogeneity in the NPAg–D antigens, some specific for malignancy, and others probably growth associated and showing cell cycle dependence. The question of growth association of the NPAg–D antigens was reexamined in an *in vitro* system where the induction of cell cycle synchrony was more easily achieved. The results of those studies are discussed in Section V.

## V. THE CELL CYCLE

Information concerning the growth-associated properties or characteristics of nucleolar or chromatin antigens during the cell cycle has been introduced into the previous sections when that data were available. Because of its relative importance, one investigation was specifically oriented to study the cell cycle dependence of the NPAg–D antigens (Briggs *et al.*, 1979).

HeLa cells were maintained in suspension cultures and synchronized by double thymidine block. The S phase cells were harvested 3 hours after release from the second block and $G_2$ cells collected 8 hours after release. Chromatin was prepared from these cells (Stein and Borum, 1972) and dehistonized for use as an immunogen (Chytil and Spelsberg, 1971). Antisera obtained to chromatin prepared from the S and $G_2$ phase synchronized cells reacted specifically with chromatin from HeLa S 3 cells (Fig. 7). Chromatins from other normal and malignant human and rat tissues were not immunologically active with either antiserum. In order to test whether the antisera reacted preferentially with chromatin from one phase of the cycle, five batches of three chromatins were prepared for testing, one from each of $G_1$, S, and $G_2$ synchronized HeLa cells. The $G_1$ phase cells were obtained 2 hours after selective detachment of mitotic cells. Cells were monitored by determination of mitotic index and autoradiographic analysis of $^3$H-TdR incorporation. The level of synchrony in $G_2$ cells was between 70% and 80%, whereas that of $G_1$ and S phase cells was greater than 95%. In the testing of the first batch of chromatins from the three phases of the cycle, both the anti-S and anti-$G_2$ sera reacted with a slight preference for the S phase chromatin (Fig.

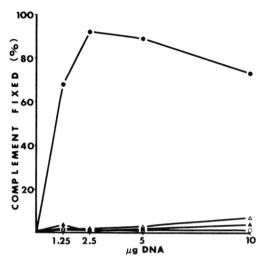

**Fig. 7.** Complement fixation of chromatins isolated from a number of sources when reacted with antiserum (1:200) to dehistonized chromatin prepared from S phase synchronized HeLa cells. Chromatins isolated from HeLa (●), WI-38 (○), normal human lung (○), human lung cancer (○), normal rat liver (▲), and Novikoff hepatoma (△) cells. From Briggs *et al.* (1979), reprinted with permission.

8). Both antisera showed the same maximal levels of activity with all chromatins in the different batches. However, within a batch of chromatins any one of the three cycle phases could show slightly stronger activity. The preference for $G_1$, S, or $G_2$ derived materials appeared as a slight shift in the complement fixation curve (Fig. 8). Within each batch, the higher immunological activity with chromatin from one phase of the cycle was recognized to the same extent by both anti-S and anti-$G_2$ sera. The failure to consistently observe one cycle phase chromatin preparation displaying the highest activity within batches of three preparations led us to conclude that there was no change in the NPAg–D antigens during the cell cycle. Both antisera showed equivalent immunological activity when tested with chromatin from HeLa cells collected in logarithmic growth, synchronized in $G_1$, S, or $G_2$, or in mitosis. The identity of antigens in chromatin prepared from cells synchronized in the different phases of the cycle was also demonstrated by absorption. Chromatin prepared from $G_1$ or $G_2$ phase synchronized cells was able to absorb out all of the immunological activity in the anti-$G_2$ serum against chromatins isolated from each of the three different phases (Fig. 8). In addition, consistent with the lack of cell cycle specificity, the antisera gave a uniform immunocytochemical staining reaction with all cells in a

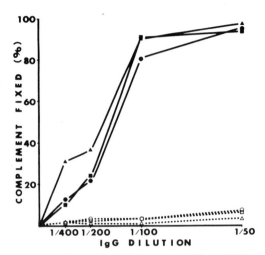

**Fig. 8.** Complement fixation activity of chromatins (10 μg DNA) prepared from $G_1$ (●), S (▲), and $G_2$ (■) phase synchronized HeLa cells reacted with varying dilutions of IgG from antiserum to dehistonized chromatin isolated from synchronized $G_2$ phase HeLa cells. IgG was prepared by ammonium sulfate precipitation and DEAE-cellulose chromatography. The protein concentration of the undiluted IgG fraction was 880 μg/ml. Open symbols and broken lines represent the reaction of the same chromatins after absorbing the IgG fraction with chromatin isolated from synchronized $G_2$ phase HeLa cells. From Briggs *et al.* (1979), reprinted with permission.

population of logarithmically growing HeLa cells (Fig. 9), where autoradiographic analysis of $^3$H-TdR incorporation from a 30-minute exposure showed about 38% labeled nuclei in the cultures. A reaction with only a fraction of these cells would have indicated restriction of antigen to only parts of the cell cycle. Evidently the NPAg–D antigens unique to the HeLa cell chromatin are not simply growth related, as would be indicated by cell cycle dependence. However, it should be noted that the detection of some tumor antigens has been reported to be cell cycle dependent (Cikes and Friberg, 1971; Burk and Drewinko, 1976; Burk *et al.*, 1976; Williams *et al.*, 1979).

The present findings are consistent with earlier studies that showed very little qualitative change in non-histone proteins of synchronously dividing HeLa cells when analyzed electrophoretically (Bhorjee and Pederson, 1972; Karn *et al.*, 1974). The total quantity of non-histone protein changed markedly, but only as a whole (Karn *et al.*, 1974). Our immunological results do not conform with the overall quantitative changes, and although consistent with the lack of qualitative electrophoretic alterations, the strict cell type specificity of the antigens suggests that the two analytical tools are probably not directed to the

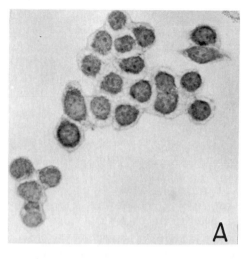

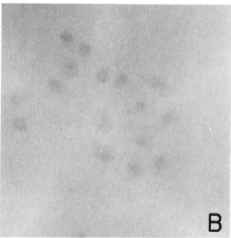

**Fig. 9.** Immunocytochemical staining of antiserum (1:200) to dehistonized chromatin isolated from $G_2$ phase synchronized cells when reacted with HeLa cells grown on glass slides (A). Immunocytochemical staining of logarithmically growing HeLa cells with serum (1:200) collected from the rabbit prior to immunization (B). From Briggs *et al.* (1979), reprinted with permission.

same components. The major polypeptides detected in gels are known to be common constituents of many tissues and cell types (Elgin and Bonner, 1970; Shaw and Huang, 1970; MacGillivray and Rickwood, 1974; Yeoman *et al.*, 1976b; Takami and Busch, 1979). The HeLa-specific chromatin antigens could be minor unique components not detected in gels or be some of the major polypeptide components after having undergone a cell type-specific modification or organizational change that confers antigenic uniqueness. In either case, their antigenic specificity and constant amount in chromatin throughout the cell cycle is consistent with their being fundamental cellular characteristics determined by the state of differentiation and not just related to growth.

Additional work with the HeLa-specific NPAg–D′ antigen(s) has revealed the antigenic fraction to be within the chromosomal scaffold material; in limited nuclease digestions of HeLa chromatin the antigen was found in high molecular weight materials and not with mononucleosomes (Campbell *et al.*, 1979). The presence of such highly specific antigens in apparently organizational components of chromatin suggests unique functional roles for these complexes.

In another approach to discovering possible cell cycle roles for the antigenic non-histone protein–DNA complexes, Baserga and co-workers (Tsutsui *et al.*, 1977, 1978) have employed cell fusion. In one set of experiments an antiserum to human non-histone protein–DNA complexes (dehistonized chromatin) isolated from human placenta was used to detect species-specific chromatin antigens in the nucleus of human-hamster cell hybrids (Tsutsui *et al.*, 1977). Interestingly, activity was directed to an antigen that could be detected in the hybrid cell only in the $G_0$ state. Once the cells were induced to proliferate, antigen detection declined. Other experiments involved detection of hamster-specific antigenic non-histone protein–DNA complexes after fusion of temperature-sensitive hamster cell lines ($G_1$ arrest) with chick erythrocytes (Tsutsui *et al.*, 1978). While DNA synthesis was induced after fusion at the permissive temperature, the chick erythrocytes were not reactivated at the nonpermissive temperature unless the hamster cells were in S phase. Apparently S phase-specific factors were required for the complete reactivation of the erythrocyte nucleus. The phase-specific factors were probably not the non-histone protein antigens, since they were taken up by the erythrocyte nuclei even at the nonpermissive temperature when the erythrocytes were not activated.

## VI. CONCLUSIONS

The results discussed in this chapter demonstrate the feasibility of studying chromatin structure and function using highly selective, sensitive immunological methods. This approach allows for the detection of specific components and facilitates their isolation. In addition, the specificity of the immunological reaction for complex structures will provide a tool for elucidating specific interactions between some of the many components in chromatin. The discovery of these interactions, at all levels of organization, is essential for understanding chromatin function. However, the results discussed also indicate that nuclear antigens display a variety of specificities (species, cell type, malignancy) and production of useful antisera, when employing chromatin or chromatin fractions, can not usually be predicted. With regard to the cell cycle, immune sera were used to detect an uncharacterized antigen common to human $G_0$ cells. Such an antigen could be involved in regulating cell cycle transition. On the other hand, the NPAg–D antigens (complexes of non-histone protein and DNA) did not vary during the cell cycle and therefore are probably not involved in the dynamic interactions required for cellular replication. The results with the NPAg–D suggest that antisera to more easily extracted chromosomal proteins may be more valuable for detecting nuclear components involved in cell cycle related nuclear–cytoplasmic interactions.

## ACKNOWLEDGMENT

We gratefully acknowledge the excellent technical assistance of Beverly Bell and Judy Briggs. This work was supported by USPHS Grant CA-18389.

## REFERENCES

Adolph, K. W., Cheng, S. M., Paulson, J. R., and Laemmli, U. K. (1977a). *Proc. Natl. Acad. Sci. U.S.A.* **74,** 4937–4941.
Adolph, K. W., Cheng, S. M., and Laemmli, U. K. (1977b). *Cell* **12,** 805–816.
Berezney, R., and Coffey, D. S. (1977). *J. Cell Biol.* **73,** 616–637.
Bhorjee, J. S., and Pederson, T. (1972). *Proc. Natl. Acad. Sci. U.S.A.* **69,** 3345–3349.
Blobel, G., and Potter, V. R. (1966). *Science* **154,** 1662–1665.
Briggs, R. C., Chiu, J. F., Hnilica, L. S., Chytil, R., Rogers, L. W., and Page, D. L. (1978). *Cell Differ.* **7,** 313–323.
Briggs, R. C., Campbell, A., Chiu, J. F., Hnilica, L. S., Lincoln, G., Stein, J., Stein, G. (1979). *Cancer Res.* **39,** 3683–3688.
Briggs, R. C., Forbes, J. T., Hnilica, L. S., Montiel, M. M., and Thor, D. E. (1980). *J. Immunol.* **124,** 243–249.
Burk, K. H., and Drewinko, B. (1976). *Cancer Res.* **36,** 3535–3538.
Burk, K., Drewinko, B., Lichtiger, B., and Trujillo, J. M. (1976). *Cancer Res.* **36,** 1278–1283.
Busch, H., Gyorkey, F., Busch, R. K., and Davis, F. M. (1979). *Cancer Res.* **39,** 3024–3030.
Busch, R. K., Daskal, I., Spohn, W. H., Kellermayer, M., and Busch, H. (1974). *Cancer Res.* **34,** 2362–2367.
Campbell, A. M., Briggs, R. C., Bird, R. E., and Hnilica, L. S. (1979). *Nucleic Acids Res.* **6,** 205–218.

Carlo, D. J., Bigley, N. J., and Van Winkle, Q. (1970). *Immunology* **19,** 879–889.

Chiu, J. D., and Hnilica, L. S. (1977). *In* "Chromatin and Chromosome Structure" (H. J. Li and R. A. Eckhardt, eds.), pp. 193–254. Academic Press, New York.

Chiu, J. D., Craddock, C., Morris, H. P., and Hnilica, L. S. (1974). *FEBS Lett.* **42,** 94–97.

Chiu, J. F., Hunt, M., and Hnilica, L. S. (1975). *Cancer Res.* **34,** 916–919.

Chiu, J. F., Chytil, F., and Hnilica, L. S. (1976). *In* "Onco-Developmental Gene Expression" (W. H. Eishman and S. Sell, eds.), pp. 271–280. Academic Press, New York.

Chiu, J. F., Hnilica, L. S., Chytil, F., Orrahood, J. T., and Rogers, L. W. (1977). *J. Natl. Cancer Inst.* **59,** 151–153.

Chiu, J. F., Decha-Umphai, W., Markert, C., and Little, B. W. (1979). *J. Natl. Cancer Inst.* **63,** 313–317.

Chytil, F., and Spelsberg, T. C. (1971). *Nature (London) New Biol.* **233,** 215–218.

Chytil, F., Glasser, W. R., and Spelsberg, T. C. (1974). *Dev. Biol.* **37,** 295–305.

Cikes, M., and Friberg, S. (1971). *Proc. Natl. Acad. Sci. U.S.A.* **68,** 566–569.

Cohen, M. E., Kleinsmith, L. J., and Midgley, A. R. (1978). *Methods Cell Biol.* **18,** 143–149.

Davis, F. M., Busch, R. K., Yeoman, L. C., and Busch, H. (1978). *Cancer Res.* **38,** 1906–1915.

Davis, R. M., Gyorkey, F., Busch, R. K., and Busch, H. (1979). *Proc. Natl. Acad. Sci. U.S.A.* **76,** 892–896.

Dubochet, J., and Noll, M. (1978). *Science* **202,** 280–285.

Elgin, S. C. R., and Bonner, J. (1970). *Biochemistry* **9,** 4440–4447.

Fujitani, H., Chiu, J. F., and Hnilica, L. S. (1978). *Proc. Natl. Acad. Sci. U.S.A.* **75,** 1943–1946.

Garel, A., and Axel, R. (1976). *Proc. Natl. Acad. Sci. U.S.A.* **73,** 3966–2970.

Hardy, K., Chiu, J. F., Beyer, A. L., and Hnilica, L. S. (1978). *J. Biol. Chem.* **253,** 5825–5831.

Hnilica, L. S., and Briggs, R. C. (1980). *In* "Cancer Markers: Developmental and Diagnostic Significance" (S. Sell, ed.), pp. 463–483. Humana Press, Clifton, New Jersey.

Ioachim, H. L., Dorsett, D. H., and Paluch, E. (1976). *Cancer* **38,** 2296–2309.

Karn, J., Johnson, E. M., Vidali, G., and Allfrey, V. G. (1974). *J. Biol. Chem.* **249,** 667–677.

Kono, N., Shima, I., and Ohta, G. (1977). *J. Biochem. (Tokyo)* **81,** 1549–1555.

Krajewska, W. M., Briggs, R. C., and Hnilica, L. S. (1979). *Biochemistry* **18,** 5720–5725.

Krajewska, W. M., Briggs, R. C., Chiu, J. F., and Hnilica, L. S. (1980). *Biochemistry* (in press).

McBride, C. M., Bowen, J. J., and Dmochowski, L. L. (1972). *Surg. Forum* **23,** 92–93.

MacGillivray, A. J., and Rickwood, D. (1974). *Eur. J. Biochem.* **41,** 181–190.

Marashi, F., Davis, F. M., Busch, R. K., Savage, H., and Busch, H. (1979). *Cancer Res.* **39,** 59–66.

Messineo, L. (1961). *Nature (London)* **190,** 1122–1123.

Okita, K., and Zardi, L. (1974). *Exp. Cell Res.* **86,** 59–62.

Priori, E. S., Seman, G., Dmochowski, L., Gallager, H. S., and Anderson, D. E. (1971). *Cancer* **28,** 1462–1471.

Rattner, J. B., and Hamkalo, B. A. (1979). *J. Cell Biol.* **81,** 453–457.

Sanders, M. M. (1978). *J. Cell Biol.* **79,** 97–109.

Shaw, L. M. V., and Huang, R. C. C. (1970). *Biochemistry* **9,** 4530–4542.

Shelton, K. R., Cobbs, C. S., Povlishock, J. T., and Burkat, R. K. (1976). *Arch. Biochem. Biophys.* **174,** 177–186.

Spelsberg, T. C., Hnilica, L. S., and Ansevin, A. T. (1971). *Biochim. Biophys. Acta* **228,** 550–562.

Spelsberg, T. C., Steggles, A. W., Chytil, F., and O'Malley, B. W. (1972). *J. Biol. Chem.* **247,** 1368–1374.

Spelsberg, T. C., Mitchell, W. M., Chytil, F., Wilson, E. M., and O'Malley, B. W. (1973). *Biochim. Biophys. Acta* **312**, 765–778.
Stein, G. S., and Borun, T. W. (1972). *J. Cell Biol.* **52**, 292–307.
Stein, G. S., Spelsberg, T. C., and Kleinsmith, L. V. (1974). *Science* **183**, 817–824.
Sternberger, L. A. (1974). "Immunocytochemistry." Prentice-Hall, Englewood Cliffs, New Jersey.
Takami, H., and Busch, H. (1979). *Cancer Res.* **39**, 507–518.
Tsutsui, Y., Suzuki, I., and Iwai, K. (1976). *Exp. Cell Res.* **101**, 202–206.
Tsutsui, Y., Chang, H. L., and Baserga, R. (1977). *Cell Biol. Int. Rep.* **1**, 301–308.
Tsutsui, Y., Chang, S. D., and Baserga, R. (1978). *Exp. Cell Res.* **113**, 359–367.
Wakabayashi, K., and Hnilica, L. S. (1973). *Nature (London) New Biol.* **252**, 153–155.
Wakabayashi, K., Wang, S., and Hnilica, L. S. (1974). *Biochemistry* **13**, 1027–1032.
Wasserman, E., and Levine, L. (1961). *J. Immunol.* **87**, 290–295.
Weintraub, H., and Groudine, M. (1976). *Science* **193**, 848–854.
Weisbrod, S., and Weintraub, H. (1979). *Proc. Natl. Acad. Sci. U.S.A.* **76**, 630–634.
Welty, J. C. (1962). *In* "Life of Birds," pp. 15–25, Saunders, Philadelphia, Pennsylvania.
Williams, J. L., Pickering, J. W., and Wolcott, M. (1979). *J. Immunol.* **122**, 1121–1125.
Wu, C., Bingham, P. M., Livak, K. V., Holmgren, R., and Elgin, S. C. R. (1979a). *Cell* **16**, 797–806.
Wu, C., Wong, Y. C., and Elgin, S. C. R. (1979b). *Cell* **16**, 807–814.
Yeoman, L. C., Jordan, J. J., Busch, R. K., Taylor, C. W., Savage, H. E., and Busch, H. (1976a). *Proc. Natl. Acad. Sci. U.S.A.* **73**, 3258–3262.
Yeoman, L. C., Seeber, S., Taylor, C. W., Fernbach, D. J., Falletta, J. M., Jordan, J. J., and Busch, H. (1976b). *Exp. Cell Res.* **100**, 47–55.
Yeoman, L. C., Woolf, L. M., Taylor, C. W., and Busch, H. (1978). *In* "Biological Markers of Biological Neoplasia: Basic and Applied Aspects" (R. W. Ruddon, ed.), pp. 409–418. Elsevier, North Holland, Amsterdam.
Zardi, L. (1975). *Eur. J. Biochem.* **55**, 231–238.
Zardi, L., Lin, J. D., and Baserga, R. (1973). *Nature (London) New Biol.* **245**, 211–213.
Zardi, L., Lin, J. C., Petersen, R. O., and Baserga, R. (1974). *In* "Control of Proliferation in Animal Cells" (B. Clarkson and R. Baserga, eds.), Vol. I, pp. 729–741. Cold Spring Harbor Lab., Cold Spring Harbor, New York.
Zimmer, M. S., Briggs, R. C., and Hnilica, L. S. (1980), submitted.

# 7

# Tubulin Synthesis during the Cell Cycle

R. C. BIRD, S. ZIMMERMAN, and A. M. ZIMMERMAN

## I. INTRODUCTION

Since the identification of microtubules almost three decades ago, a wealth of scientific research has revealed the ubiquitous nature of microtubules, in living forms. Recent research has shown that microtubules play a major role in cell morphology, chromosome movement, cilia and flagella motility, as well as transport and secretion (cf. reviews, 75,30,20). However, the specific functions that microtubules perform in these various roles are not precisely known. A better understanding of microtubule functions may be gained from studies on microtubule synthesis. Such studies must be focused on tubulin synthesis and its regulation since microtubules are formed from the polymerization of tubulin protein. Knowledge of how tubulin synthesis changes during the cell cycle and early development may shed light on the nature of the elements that

NUCLEAR–CYTOPLASMIC INTERACTIONS
IN THE CELL CYCLE

control essential cell activities. The purpose of this chapter is to review selected research concerning the regulation of tubulin synthesis in the cell cycle and during early development. In addition, we shall report on our studies of tubulin synthesis in a model cilia regenerating system and in synchronously dividing cells (6–8).

## II.  CHEMICAL COMPOSITION AND SYNTHESIS OF TUBULIN

Tubulin, the major protein of microtubules, exists as a heterodimer ($\alpha$ and $\beta$) of 110,000 MW (82,13,22,85,21) which is highly conserved among widely divergent species (46,47). The $\alpha$- and $\beta$-tubulin subunits can be separated by sodium dodecyl sulfate (SDS) polyacrylamide gel electrophoresis; difference in migration of the $\alpha$- and $\beta$-tubulin has been shown to be due to differences in amino acid composition (13). Development of a system in which microtubules could be polymerized *in vitro* without addition of nucleating sites (81) has led to the purification of mammalian brain tubulin by several rounds of temperature-dependent polymerization and depolymerization (71,10). The dependence of tubulin polymerization upon the presence of associated proteins that copurify with tubulin prepared by this method has been demonstrated (19,55,39,80,73,74,86,58,15). Microtubule associated proteins (MAPS) (73) are composed of two groups of related proteins: the high molecular weight protein (HMW) (55) and the tau protein (80). Their presence has suggested a possible regulatory mechanism for microtubule assembly. Several models have been proposed for the assembly of microtubules *in vitro* (40,37,36,56). The many functions attributed to tubulin seem to be controlled by a finely regulated steady state between polymerized and depolymerized states. This implies that size of the soluble pool of tubulin subunits is an important factor in considering the polymerization of microtubules.

Size of the soluble pool of tubulin is dependent upon the amount of tubulin polymerized into microtubules, the synthetic rate of nascent tubulin production, and the rate of tubulin degradation. This makes the study of tubulin pool regulation particularly complex. In addition, the size of the soluble pool of tubulin varies considerably between different organisms and cell types. Sea urchin embryos contain a relatively large tubulin pool which can support several rounds of deciliation and cilia regeneration without the need for synthesizing additional tubulin (3,76). Both the mitotic apparatus and the cilia have been shown to form from the preexisting pool of tubulin subunits (83,84,50,3,11,63,26,5). In the biflagellate alga, *Chlamydomonas*, there are suggestions that the size of the tubulin pool or the amount of another limiting flagella component is small. Following deflagellation of *Chlamydomonas,* in the presence of cycloheximide, both flagella are regenerated to only about one-fourth of their original length (67). Further, amputation of only one flagellum results in partial resorption of the other flagellum and then partial regeneration of both, the total length of regenerated flagellum being greater than when both are removed. This demonstrates a reutilization of flagellar proteins (18). The situation in the ameboflagellate, *Naegleria,* is different. Upon stimulation to differentiate, amebas synthesize and selectively utilize most of the tubulin that is polymerized into new flagella despite the presence of cellular tubulin (43,27,28). This implies that flagellar and cytoplasmic tubulins are separate; the study also points out the possibility that soluble tubulin may be divided into at least two separate pools, each of which can be selectively utilized by the cells. In support of this contention the tetraflagellate alga, *Polytomella agilis,* has been shown to have four isoelectrically different $\alpha$-tubulins (51). The $\alpha_1$ has been shown to be the major cytoskeletal $\alpha$-tubulin while $\alpha_3$ has been shown to be the major flagellar $\alpha$-tubulin. *In vitro* translation of poly(A)$^+$mRNA shows no trace of $\alpha_3$, suggesting that $\alpha_3$ may be a posttranslationally modified product of $\alpha_1$ (52). Preliminary results in *Chlamydomonas* suggest a similar modification may be occurring. The differences imposed by modification or perhaps differences in separate tubulin genes may allow cells to distinguish between separate pools of tubulin.

The evidence that at least some of the amono acid residues are identical in $\alpha$- and $\beta$-tubulin (46,47) has suggested that, though $\alpha$- and $\beta$-tubulin genes are related and probably arose through gene duplication, they are products of separate genes. The presence of both $\alpha$- and $\beta$-tubulin in *in vitro* translation assays of mRNA precludes posttranslational modification unless this occurs *in vitro* (38,68,23). Separation of poly(A)$^+$RNA from embryonic chick brain by native gel electrophoresis (14,16) has resulted in two RNA fractions, one of which translates $\alpha$ and the other of which translates $\beta$-tubulin in an *in vitro* assay system. Further analysis has shown that the apparent difference in mRNA mobility is due to secondary structure since $\alpha$- and $\beta$-tubulin mRNAs migrate together on denaturing gels containing methyl mercury. The molecular weights of tubulin mRNAs range from 620,000 to 650,000 daltons. In two cases (14,49) the presence of two major bands corresponding to $\alpha$-tubulin in *in vitro* translation products using homologous or heterologous cell-free systems has been reported. Unless a posttranslational modification system is operating *in vitro,* and this seems unlikely as heterologous systems derived from rabbit reticulocyte and wheat germ were employed, the results suggest that at least two separate mRNAs for α-tubulin exist. This implies that tubulin is coded for by at least three separate genes. Data relating the number of different tubulin genes and their multiplicity will require the production of cDNA clones for α- and β-tubulin. Preliminary reports of such clones have been reported (1,2,72). Subtle differences in tubulin genes may reflect the different functions of microtubules and allow the cell to selectively differentiate between separate pools of tubulin maintained for different functions.

## III. TUBULIN SYNTHESIS DURING DEVELOPMENT AND THE CELL CYCLE

Our understanding of tubulin synthesis comes primarily from studies on model systems used to simplify the complex events occurring during development and the cell cycle. Sea urchin oocytes and zygotes, *Naegleria gruberi*, mammalian cell lines in culture, *Chlamydomonas*, and *Tetrahymena* have been developed as model systems for the investigation of tubulin synthesis.

### A.   Tubulin Synthesis in Sea Urchins

It has been reported that sea urchin embryos, in spite of their demonstrably large tubulin pool, synthesize tubulin during the first division cycle (53) and that polysomes isolated from five times deciliated embryos bind more [³H]colchicine than polysomes from control embryos (35). These early studies lack specificity since they rely on assays of vinblastine and ammonium sulfate precipitable radioactivity or drug specificity. Nevertheless, this work demonstrates the applicability of sea urchin development to the study of tubulin synthesis. The tubulin pool remains relatively constant in size (0.4% of total protein) during early sea urchin development (63,64,60). Tubulin synthesis is also detectable throughout early development (63,62,34) and incorporation of labeled amino acids into tubulin is unaffected by actinomycin D or centrifugal enucleation of embryos (61). These data suggest that the large pool of tubulin available within sea urchin embryos is constantly being turned over and that the synthesis may be directed by stored maternal mRNAs.

In order to study the dependence of tubulin synthesis on new transcription, Merlino *et al.* (54) have used *in vitro* translation of poly(A)⁺RNA to investigate tubulin synthesis. After noting a 2 to 3-fold enhancement of [³⁵S]methionine incorporation into tubulin in five times deciliated sea urchin embryos, they measured a concomitant 2 to 3-fold increase in tubulin synthesized *in vitro* in a cell free translation system directed by poly(A)$^+$RNA from five times deciliated embryos. Induction of tubulin synthesis required a prior depletion of the cytoplasmic tubulin pool and was dependent upon the activation of a newly translatable message. This message appeared to be the result of increased gene transcription since the induction of tubulin synthesis was actinomycin D sensitive. Although induction of tubulin synthesis was inhibited, synthesis of tubulin was maintained at the levels found in nondeciliated embryos. Double isotope labeling of RNA demonstrated that the fraction of poly(A)$^+$RNA containing most of the tubulin mRNA activity also corresponded to the RNA fraction with the highest specific activity. This suggests an enhanced specific activity for tubulin mRNA in deciliated embryos implying new synthesis occurs as a result of increased transcription of tubulin genes.

Long-lived cytoplasmic mRNA (possibly of maternal origin) and the demonstrably inducible transcription of new tubulin message account for new tubulin synthesis in sea urchin embryos. Since there is no appreciable change in tubulin pool size during early development, changes in utilization and degradation probably account for the synthesis of extra tubulin. In the case of the deciliated embryos, induction of tubulin synthesis occurred only after substantial depletion of the soluble pool and did not rise appreciably until after first or second deciliation. This implies that the size of the soluble pool of tubulin plays a role in the regulation of tubulin synthesis in sea urchin embryos.

## B.  Tubulin Synthesis in *Naegleria*

The unicellular eukaryote, *Naegleria gruberi,* can change from an ameboid form to a flagellate form when stimulated by a change in environment such as transfer from a nutritional medium to starvation buffer (25) or when shifted to a lower temperature while being maintained in growth medium (17). Amebas undergo a rapid and relatively synchronous differentiation to flagellates which occurs optimally within about 1 hour, with the period of morphological change requiring less than 30 minutes. This differentiation has been used as a model developmental system and has been analyzed with respect to tubulin synthetic activity (43). Uniformly [$^{35}$S]methionine-labeled amebas were stimulated to transform and chased with cold substrate. Subsequently, tubulin was isolated from the newly formed flagellar outer doublets. The specific activity of tubulin isolated from the methionine-chased cultures contained only 30% as much radioactivity as the unchased cultures. Thus approximately 70% of the tubulin in flagellar outer doublets is synthesized *de novo*. These results support the radioimmune assay study in which antibodies specific for *Naegleria* outer doublet tubulin were prepared. Data from the immune assay indicate that most of the flagellar tubulin antigen appeared during transformation (42).

In order to determine at which level tubulin synthesis was being regulated, Lai et al. (44) assayed tubulin mRNA by translating total RNA from differentiating cells *in vitro*. Tubulin was the major protein synthesized, amounting to 5% of the total mRNA activity; at least 92% of the tubulin mRNA appeared during differentiation. Tubulin mRNA first appeared 20 minutes postinitiation, and peaked at 60 minutes postinitiation, the same time as flagella appeared. Actinomycin D blocked the appearance of new tubulin mRNA as well as differentiation (29). These results suggest that during *Naegleria* differentiation the peak of tubulin synthesis is a direct reflection of the amount of tubulin mRNA present and may be transcription-dependent. Since tubulin mRNA appears prior to formation of the flagella it is unlikely that fluctuations in pool size play an important role in the regulation of tubulin synthesis in *Naegleria*.

## C.   Tubulin Synthesis in Mammalian Cells

Mammalian cells have also been used to study tubulin synthesis. During mouse oocyte growth the absolute rate of tubulin synthesis increases about 50% as the oocyte grows from 40 to 85 $\mu$m and tubulin accounts for about 1.8% of total protein synthesis (70). The total amount of protein synthesis devoted to tubulin declines slightly during this time. Tubulin is a major protein synthesized during mouse oocyte growth which is similar to early sea urchin development.

Several lines of Chinese hamster cells have been synchronized, both by mechanical selection and metabolic inhibitors, and used for assaying [$^3$H]colchicine-binding activity (41). Such studies have demonstrated an increase in colchicine binding during $G_2$. Fluctuations in the measurable amount of tubulin present suggest that degradation of tubulin as well as synthesis play important roles in the regulation of the size of the tubulin pool.

The morphological pattern of microtubules in rat kangaroo cells (strain PtKl) during parts of the cell cycle has been investigated by indirect immunofluorescent staining with monospecific antibodies raised against bovine brain tubulin (12). The elaborate array of fine fluorescent filaments normally observed in interphase cells becomes diminished during prophase and was accompanied by rounding as the cells entered mitosis. The mitotic apparatus initially formed as one or two bright fluorescent spots near the nucleus with filaments radiating out from each spot. The stained kinetochores of the condensing chromosomes were also visible. By metaphase the mitotic apparatus was brightly fluorescent with the dark bodies of the chromosomes near the center of the spindle. Cytokinesis occurred following the movement of the chromosomes to the poles. A cytoplasmic bridge filled with microtubules joined the two daughter cells. The bridge fluoresced brightly except for a midbody. During early $G_1$ phase the bridge became thinner and the cytoplasmic microtubules reformed. Disappearance of the cytoplasmic microtubules during mitosis and their reformation afterward suggests that reutilization of existing tubulin can occur.

The regulation of tubulin synthesis has been investigated in mammalian cells (strains 3T6 and CHO) by measuring the rate of tubulin synthesis and tubulin mRNA activity (4). Microtubule depolymerizing agents such as colchicine and nocodazole lead to a rapid inhibition of tubulin synthesis. Tubulin mRNA activity is also greatly reduced in cells treated with these drugs. These data have been used to develop a model for the regulation of tubulin synthesis in which elevated levels of soluble tubulin inhibit new tubulin mRNA synthesis. Since tubulin synthesis drops off quickly following drug treatment the model also proposes a short half-life for tubulin mRNA. In support of this, the half-life of tubulin mRNA in actinomycin D treated cells has been measured at about 2 hours. In addition, depolymerization of microtubules without an increase in soluble tubulin, mediated by vinblastine-induced precipitation of tubulin as paracrystals, does not inhibit tubulin synthesis and, in fact, enhances it. The model implies that control of tubulin synthesis occurs in such a way as to maintain a constant pool of soluble tubulin despite the amount of polymerized tubulin present (4). This concept is compatible with the measured increase in tubulin synthesis during $G_2$. The increase in tubulin synthesis is followed by a rapid decline as the cells approach mitosis (41) and the depolymerization of cytoplasmic microtubules occurs (12). In order for the pool size to remain constant, tubulin synthesis must decrease as the cells enter mitosis since, at this time, the cytoplasmic microtubules are depolymerizing and swelling the soluble pool.

## D.  Tubulin Synthesis in *Chlamydomonas*

Induction of tubulin synthesis has been demonstrated in deflagellated *Chlamydomonas reinhardii* along with the induction of other flagellar proteins (31). This system has been used to dissect the regulation of tubulin synthesis (78). Following deflagellation of *Chlamydomonas* cells during the gametic phase of their life cycle, tubulin mRNA activity was assayed by translating fractions of polyribosomes in an *in vitro* translation system derived from wheat germ. Nondeflagellated cells show no tubulin mRNA activity. Flagella first appear about 15–20 minutes following deflagellation; tubulin mRNA activity also appears at this time and continues to rise until it peaks about 60 minutes and then declines. Tubulin represents greater than 14% of the translation products at its peak. The large cytoplasmic tubulin pool in *Chlamydomonas* does not seem to fluctuate following deflagellation. It has been suggested that since *de novo* synthesis of flagellar tubulin occurs in the presence of a large cellular tubulin pool, the flagellar tubulin may differ from cellular tubulin as in *Naegleria*.

Further investigation of tubulin synthesis has resulted in a detailed outline of tubulin synthesis during the vegetative cell cycle and sexual life cycle of *Chlamydomonas* (79). Cells were labeled with $[^{35}S]H_2SO_4$ radioactivity *in vivo* and the whole cell proteins were subjected to electrophoresis and autoradiography. Tubulin synthesis was not detected in gametic and early zygotic cells unless they were deflagellated. About 90 minutes after zygote formation, cells lose the ability to induce tubulin synthesis and to regenerate their flagella. Little tubulin synthesis is noted in vegetatively growing cells that have been light-dark synchronized until late in the light period 1.5–2 hours prior to cytokinesis. Tubulin synthesis can be induced by deflagellation of vegetative cells at any time during the cell cycle except near cytokinesis when tubulin is normally induced. Deflagellation near the time of cytokinesis results in little induction of tubulin synthesis.

In both sexual and vegetative cycles tubulin synthesis is not inducible (above normal levels) prior to, or during, periods of flagella resorption. In vegetative cells induction of tubulin synthesis normally occurs just prior to cytokinesis. The induction of tubulin synthesis in these cells precedes the cellular requirements concerned with the formation of the mitotic spindle, cleavage furrow, and new flagellar apparatus in each of the 4–8 daughter cells (79). Induction of tubulin synthesis during the vegetative cell cycle has been shown to be the result of increased mRNA production and is likely the result of increased transcription (2).

## IV. TUBULIN SYNTHESIS IN *TETRAHYMENA*

### A. Cilia Regeneration and Division Synchrony

*Tetrahymena pyriformis* GL has been used in our laboratory as a model system to investigate tubulin synthesis during cilia regeneration and during the cell cycle. *Tetrahymena* provide an ideal system in which to study this problem since they can be induced to undergo cilia regeneration or synchronous division by simple manipulations of their environment.

The relatively synchronous regeneration of cilia occurs following a brief treatment with calcium (66). Cilia regeneration takes about 90–100 minutes during which time the cells regain motility and cilia attain the predeciliation length of 6 $\mu$m. In *Tetrahymena* cilia regeneration is an inducible system in which the cells undertake the assembly of a new set of organelles and any obligatory synthesis of new protein.

*Tetrahymena* are induced to undergo a series of synchronous divisions by employing the one heat shock per generation technique described by Zeuthen (87). The cells are given seven heat shocks (34°C) separated by 157-minute periods (the duration of a normal generation) at 28°C. In order to avoid any heat shock-induced changes in the cell cycle, the first free running division or the time between the first and second synchronous divisions was studied. Since the cells divide prior to each heat shock they remain about the same size during synchronization, which avoids the problems inherent in other methods of division synchrony (69). The use of this technique also allows the cell cycle events that precede division to be synchronized, the best example of which is S phase (9). The advantage of inducing synchrony, by the technique of one heat shock per generation, is that a high degree of synchrony is obtained as well as induction of synchronous biochemical events, without the necessity of using drugs or inhibitors. The synchronous cell cycle of *Tetrahymena* presents a simple model system in which to investigate the cell cycle related regulation of tubulin synthesis.

## B.  Tubulin Synthesis during Regeneration

Log growth *Tetrahymena* can be deciliated by a brief calcium pulse treatment (66,7). Following resuspension in non-nutrient recovery medium the cells remain motionless for about 50 minutes (Fig. 1). During this period any residual cilia that are not removed from the cells by the calcium deciliation treatment are resorbed followed by the regeneration of a new complement of cilia. Sporadic movement of the cells is observed as the new cilia begin to move. Greater than 90% of the cell population is motile by 80 minutes after the cells are resuspended in recovery medium and by 90 minutes the cells recover their normal rate of movement. Measurement of cilia reveal that they reach their full length (6.0–6.2 $\mu$m) about 100 minutes postdeciliation. In experiments where low concentrations (1 and 10 $\mu$g/ml) of cycloheximide were added to deciliated cells, regeneration was completely inhibited. Actinomycin D

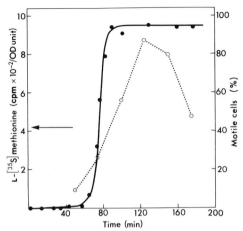

**Fig. 1.** Specific activity of tubulin during cilia regeneration. Log growth *Tetrahymena* were deciliated and allowed to regenerate their cilia in buffered recovery medium. Aliquots of cells were labeled for 20 minutes with L-[35S]methionine, and acetone extracted proteins were solubilized in Laemmli buffer (45) and subjected to polyacrylamide gel electrophoresis. Gels were sectioned and radioactivity was determined. The dotted line indicates the specific activity of tubulin from regenerating cells. The arrow indicates the specific activity of tubulin in untreated log growth cells. The solid line indicates the kinetics of motility recovery. From Bird and Zimmerman (7).

treatment (25 μg/ml) retarded the cilia regeneration kinetics and limited the total number of cells able to regenerate their cilia.

Sucrose density gradient sedimentation profiles of polyribosomes isolated from log growth and cilia regenerating cells (30, 60, and 90 minutes postdeciliation) were studied in order to provide information on the rate of protein synthetic activity in these cells. Profiles from log growth cells show 83–87% of the extractable ribosomes are engaged as polysomes, suggesting a high rate of protein synthetic activity. In the cilia regenerating cells, by comparison, the polyribosomes comprise a very small fraction of the extractable ribosomes immediately after deciliation and up to 30 minutes postdeciliation. As the new cilia emerge and motility begins the fraction of polyribosomes increases relative to the total amount of ribosomal material. The percentage of ribosomes engaged as polysomes increases continuously as cilia regeneration proceeds (Fig. 2).

*Tetrahymena* were assayed for tubulin synthetic activity during log growth and cilia regeneration with short pulses of L-[35S]methionine. Proteins from whole cells were solubilized and subjected to polyacrylamide gel electrophoresis according to the method of Laemmli (45) as modified by Bird and Zimmerman (7). Coomassie blue stained proteins from log growth cells and from cells at various stages of cilia regeneration were compared (Fig. 3). There are marked qualitative changes in the electrophoretic pattern of proteins particularly when comparing log growth and cilia regenerating cell proteins of molecular weight greater than 68,000. The α- and β-tubulin bands were identified by comigration with purified *Tetrahymena* cilia tubulin. This is important to note since we have found that α-tubulin from cilia of *Tetrahymena pyriformis* GL does not comigrate with α-tubulins from other sources (notably sea urchin and neurotubules), while β-tubulin from *Tetrahymena* does comigrate with other β-tubulins. This has also been shown for other strains of *Tetrahymena* (23). The β-tubulin band is obscured by a close but slightly higher molecular weight band. There is little detectable change in the intensity of the Coomassie stained tubulin bands derived from whole cells during cilia regeneration. However, an autoradiogram of the same

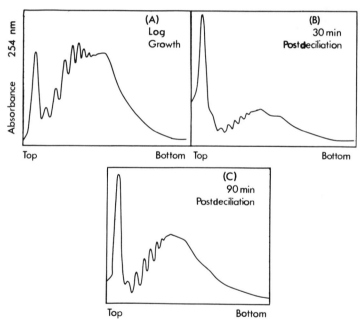

**Fig. 2.** Polysome profiles from log growth and deciliated *Tetrahymena*. Polysomes extracted from log growth cells and from cells at progressive times after deciliation were subjected to sucrose density gradient centrifugation. Sedimentation was from left to right. (A) Sedimentation profile of polysomes extracted from log growth cells. Sedimentation profiles of polysomes extracted from postdeciliation cells at (B) 30 minutes postdeciliation and (C) 90 minutes postdeciliation. Up to 30 minutes postdeciliation, prior to any sign of motility, there was only a small amount of polysomal material recovered. As the cells regain motility the number of recoverable polysomes increases. From Bird and Zimmerman (7).

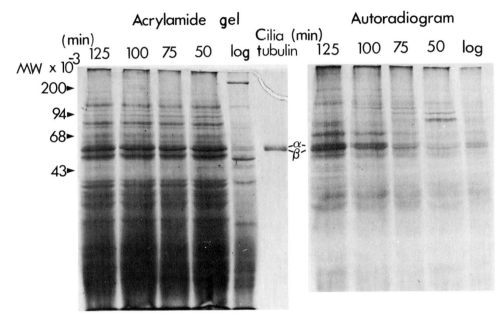

**Fig. 3.** Polyacrylamide gel electrophoresis of whole cell proteins derived from log growth *Tetrahymena* and from cells at progressive times after deciliation. Equal numbers of cells were labeled with L-[$^{35}$S]methionine for 20 minutes and the proteins were acetone-extracted and subjected to polyacrylamide gel electrophoresis on 10% gels followed by autoradiography. The horizontal times indicate the end of the radioactive labeling periods. Purified $\alpha$ and $\beta$ *Tetrahymena* cilia tubulin were also subjected to electrophoresis and are labeled. The vertical numbers indicate the position of molecular weight markers. In the Coomassie blue stained gel there is little detectable change in the total amount of tubulin present during cilia regeneration. The autoradiogram of the same gel shows the prominent pair of tubulin bands increasing in intensity with time as cilia regeneration proceeds. Maximum incorporation occurs at approximately 125 minutes postdeciliation. From Bird and Zimmerman (7).

gel shows the prominent pair of tubulin bands increasing in intensity as a function of cilia regeneration (Fig. 3). Maximum radioactive incorporation of methionine occurs at approximately 125 minutes postdeciliation and is much greater than the activity in similarly labeled log growth cells. In the samples collected at 50 minutes postdeciliation there is a transient increase in the activity of several higher molecular weight proteins of approximately 87,000 and 77,000; the identities of these proteins are unknown.

Polyacrylamide gels identical to those subjected to autoradiography were sliced into 2-mm sections; each section was emulsified and the radioactivity was determined (Fig. 4). The specific activity of tubulin increased continuously to a peak which occurred at approximately 125 minutes postdeciliation; at its peak tubulin synthesis accounted for 7–10% of the incorporated label. Synthetic activity decreased after this time. If the change in tubulin-specific activity is compared to the kinetics of cilia regeneration, it is obvious that the peak in tubulin activity occurs approximately 35 minutes after the cells become fully motile. Cilia regeneration is virtually complete when tubulin synthesis reaches a maximum (Fig. 1).

The results demonstrate an unequivocal induction of tubulin synthesis. As the cells recover from deciliation, the rate of protein synthesis rises. Simultaneously, the rate of tubulin synthesis increases proportionately much greater than general protein synthesis. The peak in tubulin synthetic activity occurs approximately 45 minutes after greater than 90% of the cells become motile (125 minutes postdeciliation). This is about 25 minutes after the cilia reach full length. It seems clear that the existing pools of tubulin available within the cell are utilized to resynthesize cilia, at least during the first hour of regeneration. Since the data demonstrate a disproportionately high induction of tubulin synthesis (7–10% of the incorporated label at its peak) after the cilia reach full length, it is clear that the nascent tubulin replenishes the soluble tubulin

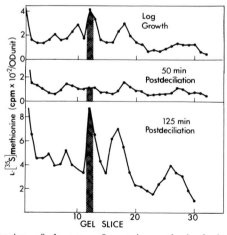

**Fig. 4.** Quantification of the rate of protein synthesis during cilia regeneration. Polyacrylamide gels, as described in Fig. 3, were sliced in 2-mm sections; the radioactivity of each section was determined. The samples shown are from log growth cells and from cells at representative times after deciliation. Migration was from left to right. The shaded area indicates the activity in tubulin. The specific activity of tubulin increased continuously until it peaked at about 125 minutes postdeciliation well above log growth levels. From Bird and Zimmerman (7).

pool depleted during cilia regeneration. This conclusion is supported by the work of others (32,65,77,57).

Nutritionally deprived *Tetrahymena* have also been used to study tubulin synthesis. Guttman and Gorovsky (32) used *Tetrahymena* that were *nutritionally deprived* for 18–20 hours to study tubulin synthesis following deciliation. They found that the rate of tubulin synthesis peaked approximately 100–120 minutes postdeciliation and then declined; their studies suggest that the induction of tubulin synthesis accompanies pool depletion. At its peak tubulin synthesis accounted for 7–8% of total protein synthesis. The authors, however, were unable to detect an increase in the rate of tubulin synthesis in deciliated growing cells. They concluded that high background levels of protein synthetic activity masked any induction that might have occurred. Nelsen (57) detected little difference in the specific activity of cilia tubulin from deciliated and nondeciliated, nutritionally deprived *Tetrahymena*. This suggests that preexisting tubulin is mainly used in the formation of new cilia following deciliation. An induction of at least β-tubulin mRNA has been shown in starved deciliated *Tetrahymena* (48) although α- and β-tubulin have been shown to be coded for by separate messenger molecules (59).

## C.    Tubulin Synthesis during the Cell Cycle

*Tetrahymena* were synchronized by the one heat shock per generation technique described by Zeuthen (87). Log growth cultures were submitted to seven heat shocks (34°C) for 30 minutes duration each, separated by growth (28°C) for 157 minutes. Following the end of the last heat shock (designated EH) the division index was assayed by determining the percentage of cells, in a population, which exhibited furrowing on both sides. The first synchronous division occurred at about 80 minutes after EH while the second synchronous division occurred about 200 minutes after EH (Fig. 5). Almost all the cells divided since cell number virtually doubled following each division. Sedimentation profiles of polyribosomes, taken at three time points, during the synchronous cell cycle showed little change in the relative amount of polysomes related to total ribosomal material. However, profiles became somewhat more heavily laden with higher molecular weight polysomes toward the end of the cell cycle. Fluctuation in the amount of polysomes during the cell cycle has also been determined in *Tetrahymena* synchronized by a different procedure (33).

Tubulin synthesis was assayed during consecutive 20-minute intervals between first and second synchronous divisions in the manner described for cilia regenerating cells. Whole cells were pulse-labeled with

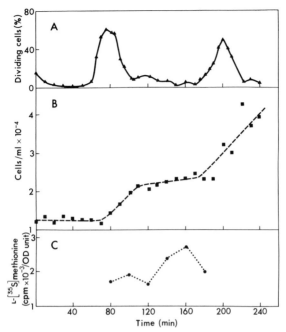

**Fig. 5.** Tubulin synthesis in synchronously dividing *Tetrahymena*. Log growth cultures were synchronized by the one heat shock per generation technique of Zeuthen (87). Cultures were submitted to seven 30-minute heat shocks (34°C) each spaced a normal cell generation time apart (157 minutes at 28°C). Aliquots of cells were labeled for 20-minute periods with L-[35S]methionine and the proteins, acetone extracted, were subjected to electrophoresis and radioactivity was determined as described for deciliated cells. (A) The division schedule of synchronized *Tetrahymena*. The "free running cell cycle" occurs between the first (80 minutes EH) and second (200 minutes EH) induced synchronous divisions. (B) The cell density expressed as cells per milliliter. (C) The specific activity of tubulin during the free running cell cycle. There is little change in the specific activity of tubulin during the first half of the cycle. Prior to the second synchronous division as the cells enter $G_2$, the specific activity of tubulin rises to a peak value. From Bird and Zimmerman (8).

L-[35S]methionine for 20 minutes. Proteins from the labeled cells were extracted, solubilized, and subjected to polyacrylamide gel electrophoresis; the gels were subjected to autoradiography or they were sliced into 2-mm sections and the radioactivity in each section was quantified. The specific activity of tubulin does not change appreciably during the first half of the synchronous cell cycle (Fig. 5). However, at 140 minutes after EH just prior to when the cells enter $G_2$, tubulin synthesis is induced. At its peak (160 minutes after EH) tubulin synthesis is 60% higher in specific activity than it was during the first half of the cell cycle;

as the cells enter division tubulin synthesis declines to preinduction levels.

## V. CONCLUDING REMARKS

Tubulin synthesis and its regulation have been reviewed in several cellular systems. It is evident that the regulation of tubulin synthesis is complex and may be subject to more than one control mechanism. Although the organisms discussed are distinct from one another, comparisons of the control of tubulin synthesis can be made.

The regulatory mechanism operating during *Chlamydomonas* regeneration and *Naegleria* differentiation appears to be exclusive of regulation by pool size. In each of these organisms induction of tubulin synthesis occurs without prior depletion of the soluble pool. Induction appears to be the result of some direct trigger mechanism induced by deflagellation or the signal to differentiate. In mammalian cells, sea urchin embryos, and *Tetrahymena* the situation is quite different; in these systems, induction of tubulin synthesis seems to be a response to pool depletion since these systems respond with greater rates of tubulin synthesis when the soluble pool has been depleted.

Tubulin synthesis in *Tetrahymena* is controlled by a pool replacement mechanism as shown by the work of Bird and Zimmerman (6–8). This control is indirect and results from an attempt to maintain a constant soluble tubulin pool size. In the *Tetrahymena* cilia-regeneration system, tubulin synthesis occurs as a response to pool depletion rather than as a consequence of direct activation by deciliation. In division synchronized *Tetrahymena*, tubulin synthesis remains at a relatively constant level during the first half of the "free running cell cycle." As the cells leave S phase and enter $G_2$, tubulin synthesis is increased approximately 60%. Tubulin synthesis returns to preinduction levels as the cells enter division.

The dramatic rise in tubulin synthesis which occurs in the $G_2$ phase may reflect a commitment by the cell to polymerize large numbers of microtubules for the impending division as well as for the development of new oral apparatus. Microtubules play an important role in the formation of the oral apparatus (24). This represents a major commitment of tubulin by the cell which seems to occur just prior to the measured induction of tubulin synthesis during $G_2$. Thus it would seem likely that the increase in tubulin synthesis, at this time, is a response to pool depletion during stomatogenesis.

It may be concluded from the current study that two types of regulatory mechanisms exist for tubulin synthesis in the cellular systems discussed, i.e., a direct trigger mechanism characteristic of *Naeglaeria* and *Chlamydomonas,* and an indirect mechanism found in mammalian cells, sea urchin embryos, and *Tetrahymena.* The data presented do not exclude either regulatory mechanism from any of the model systems nor limit an organism to any one regulatory mechanism, exclusively. In both regulatory mechanisms, the level of regulation appears to be at the mRNA. It is as yet unclear whether this control is transcriptional, post-transcriptional, or the result of unmasking of mRNA. Further discrimination between these regulatory mechanisms and the levels at which they operate may be aided by specific cDNA probes for tubulin mRNA sequences. Such research should lead to a clearer understanding of how tubulin synthesis is regulated.

## ACKNOWLEDGMENT

The authors wish to express their appreciation to Helen Laurence for her technical assistance. The research reported in this chapter was supported by a research grant from National Science and Engineering Research Council of Canada to A. M. Zimmerman. R. C. Bird was a recipient of a National Science and Engineering Research Council of Canada Postgraduate Scholarship.

## REFERENCES

1. Alexandraki, D., and Ruderman, J. V., *J. Cell Biol.* **83,** 342a (1979).
2. Ares, M., and Howell, S. H., *J. Cell Biol.* **83,** 10a (1979).
3. Auclair, W., and Siegel, B. W., *Science* **154,** 913 (1966).
4. Ben-Ze'ev, A., Farmer, S. R., and Penman, S., *Cell* **17,** 319 (1979).
5. Bibring, T., and Baxandall, J., *Dev. Biol.* **55,** 191 (1977).
6. Bird, R. C., and Zimmerman, A. M., *J. Cell Biol.* **83,** 331a (1979).
7. Bird, R. C., and Zimmerman, A. M., *Exp. Cell Res.,* in press (1980).
8. Bird, R. C., and Zimmerman, A. M., Manuscript submitted (1981).
9. Bols, N. C., and Zimmerman, A. M., *Exp. Cell Res.* **108,** 259 (1977).
10. Borisy, G. G., Marcum, J. M., Olmstead, J. B., Murphy, D. B., and Johnson, K. A., *Ann. N.Y. Acad. Sci.* **253,** 107 (1975).
11. Borisy, G. G., and Taylor, E. W., *J. Cell Biol.* **34,** 525 (1967).
12. Brinkley, B. R., Fuller, G. M., and Highfeild, D. P., *in* "Cell Motility" (R. Goldman, T. Pollard, and J. Rosenbaum, eds.), p. 435. Cold Spring Harbor Lab., Cold Spring Harbor, New York 1976.
13. Bryan, J., and Wilson, L., *Proc. Natl. Acad. Sci. U.S.A.* **68,** 2273 (1971).
14. Bryan, R. N., Cutter, G. A., and Hayashi, M., *Nature (London)* **272,** 81 (1978).
15. Cleveland, D. W., Hwo, S.-Y., Kirschner, M. W., *J. Mol. Biol.* **116,** 207 (1977).
16. Cleveland, D. W., Kirschner, M. W., and Cowan, N. J., *Cell* **15,** 1021 (1978).
17. Corff, S., and Yuyama, S., *Exp. Cell Res.* **114,** 175 (1978).

18. Coyne, B., and Rosenbaum, J. L., *J. Cell Biol.* **47**, 777 (1970).
19. Dentler, W. L., Granett, S., and Rosenbaum, J. L. *J. Cell Biol.* **65**, 237 (1975).
20. Dustin, P., "Microtubules." Springer-Verlag, Berlin and New York, 1978.
21. Eipper, B. A., *Proc. Natl. Acad Sci. U.S.A.* **69**, 2283 (1972).
22. Feit, H., Dutton, G. R., Barondes, S. H., Shelanski, M. L., *J. Cell Biol.* **51**, 138 (1971).
23. Fliss, E. R., and Suyama, Y., *J. Protozool.* **26**, 505 (1979).
24. Frankel, J., and Williams, N. E., *in* "Biology of *Tetrahymena*" (A. M. Elliot, ed.), p. 375. Dowden, Hutchinson, Ross, Stroudsburg, Pennsylvania, 1973.
25. Fulton, C. M., and Dingle, A. D., *Dev. Biol.* **15**, 165 (1967).
26. Fulton, C., Kane, R. E., and Stephens, R. E., *J. Cell Biol.* **50**, 762 (1971).
27. Fulton, C. M., and Kowit, J. D., *Ann. N.Y. Acad. Sci.* **253**, 318 (1975).
28. Fulton, C., and Simpson, P. A., *in* "Cell Motility" (R. Goldman, T. Pollard, and J. Rosenbaum, eds.), p. 987. Cold Spring Harbor Lab., Cold Spring Harbor, New York, 1976.
29. Fulton, C., Simpson, P. A., and Lai, E. Y., *in* "Cell Reproduction" (E. R. Dirksen, D. M. Prescott, and C. F. Fox, eds.), p. 337. Academic Press, New York, 1978.
30. Goldman, R., Pollard, T., and Rosenbaum, J. L., "Cell Motility." Cold Spring Harbor Lab., Cold Spring Harbor, New York, 1976.
31. Gorovsky, M. A., Carlson, K., Rosenbaum, J. L., *Anal. Biochem* **35**, 359 (1970).
32. Guttman, S. D., and Gorovsky, M. A., *Cell* **17**, 307 (1979).
33. Hermolin, J., and Zimmerman, A. M., *Cytobios* **3**, 247 (1969).
34. Hynes, R. O., Raff, R. A., and Gross, P. R., *Dev. Biol.* **27**, 150 (1972).
35. Iverson, R. M., *Exp. Cell Res.* **66**, 197 (1971).
36. Johnson, K. A., and Borisy, G. G., *in* "Molecules and Cell Movement" (S. Inoue, and R. E. Stephens, eds.), p. 119. Raven, New York, 1975.
37. Johnson, K. A., and Borisy, G. G., *J. Mol. Biol.* **117**, 1 (1977).
38. Jorgensen, A. O., and Heywood, S. M., *Biochim. Biophys. Acta.* **414**, 321 (1975).
39. Keates, R. A. B., and Hall, R. H., *Nature (London)* **257**, 418 (1975).
40. Kirschner, M. W., Williams, R. C., Weingarten, M., and Gerhart, J. C. *Proc. Natl. Acad. Sci. U.S.A.* **71**, 1159 (1974).
41. Klevecz, R. R., and Forrest, G. L., *Ann. N.Y. Acad. Sci.* **253**, 292 (1975).
42. Kowit, J. D., and Fulton, C. M., *J. Biol. Chem.* **249**, 3638 (1974a).
43. Kowit, J. D., and Fulton, C., *Proc. Natl. Acad. Sci. U.S.A.* **71**, 2877 (1974b).
44. Lai, E. Y., Walsh, C., Wardell, D., and Fulton, C., *Cell* **17**, 867 (1979).
45. Laemmli, U. K., *Nature (London)* **227**, 680 (1970).
46. Luduena, R. F., and Woodward, D. O., *Proc. Natl. Acad. Sci. U.S.A.* **70**, 3594 (1973).
47. Luduena, R. F., and Woodward, D. O., *Ann. N.Y. Acad. Sci.* **253**, 272 (1975).
48. Marcaud, L., and Hayes, D., *Eur. J. Biochem.* **98**, 267 (1979).
49. Marotta, C. A., Strocchi, P., and Gilbert, J. M., *J. Neurochem.* **33**, 231 (1979).
50. Mazia, D. *In* "The Cell" (J. Brachet and A. E. Mirsky, eds.), p. 77. Academic Press, New York, 1961.
51. McKeithan, T. W., and Rosenbaum, J. L., *J. Cell Biol.* **79**, 297a (1978).
52. McKeithan, T. W., Lefebvre, P. A., Silflow, C. D., and Rosenbaum, J. L., *J. Cell Biol.* **83**, 338a (1979).
53. Meeker, G. L., and Iverson, R. M., *Exp. Cell Res.* **64**, 129 (1971).
54. Merlino, G. T., Chamberlain, J. P., and Kleinsmith, L. J., *J. Biol. Chem.* **253**, 7078 (1978).
55. Murphy, D. B., and Borisy, G. G., *Proc. Natl. Acad. Sci. U.S.A.* **72**, 2896 (1975).
56. Murphy, D. B., Johnson, K. A., and Borisy, G. G., *J. Mol. Biol.* **117**, 33 (1977).
57. Nelsen, E. M., *Exp. Cell Res.* **94**, 152 (1975).

58. Penningroth, S. M., Cleveland, D. W., and Kirschner, M. W., *in* "Cell Motility" (R. Goldman, T. Pollard, and J. Rosenbaum, eds.), p. 1233. Cold Spring Harbor Lab., Cold Spring Harbor, New York, 1976.
59. Portier, M.-M., Milet, M., and Hayes, D. H., *Eur. J. Biochem.* **97,** 161 (1979).
60. Raff, R. A., *Am. Zool.* **15,** 661 (1975).
61. Raff, R. A., Brandis, J. W., Green, L. H., Kaumeyer, J. F., and Raff, E. C., *Ann. N.Y. Acad. Sci.* **253,** 304 (1975).
62. Raff, R. A., Colot, H. V., Selvig, S. E., and Gross, P. R., *Nature (London)* **235,** 211 (1972).
63. Raff, R. A., Greenhouse, G., Gross, K. W., and Gross, P. R., *J. Cell Biol.* **50,** 516 (1971).
64. Raff, R. A., and Kaumeyer, J. F., *Dev. Biol.* **32,** 309 (1973).
65. Rannestad, J., *J. Cell Biol.* **63,** 1009 (1974).
66. Rosenbaum, J. L., and Carlson, K., *J. Cell Biol.* **40,** 415 (1969).
67. Rosenbaum, J. L., Moulder, J. E., and Ringo, D. L., *J. Cell Biol.* **41,** 600 (1969).
68. Saborio, J. L., Palmer, E., and Meza, I., *Exp. Cell Res.* **114,** 365 (1978).
69. Scherbaum, O., and Zeuthen, E., *Exp. Cell Res.* **6,** 221 (1954).
70. Schultz, R. M., Letourneau, G. E., Wassarman, P. M., *Dev. Biol.* **73,** 120 (1979).
71. Shelanski, M. L., Gaskin, F., and Cantor, C. R., *Proc. Natl. Acad. Sci. U.S.A.* **70,** 765 (1973).
72. Silflow, C., and Rosenbaum, J., *J. Cell Biol.* **83,** 409a (1979).
73. Sloboda, R. D., Rudolph, S. A., Rosenbaum, J. L., and Greengard, P. *Proc. Natl. Acad. Sci. U.S.A.* **72,** 177 (1975).
74. Sloboda, R. D., Dentler, W. L., and Rosenbaum, J. L., *Biochemistry* **15,** 4497 (1976).
75. Soifer, D., "The Biology of Cytoplasmic Microtubules" *Anal. N.Y. Acad. Sci.* **253** (1975).
76. Stephens, R. E., *Biol. Bull.* **142,** 489 (1972).
77. Tamura, S., *Exp. Cell Res.* **68,** 180 (1971).
78. Weeks, D. P., and Collis, P. S., *Cell* **9,** 15 (1976).
79. Weeks, D. P., and Collis, P. S., *Dev. Biol.* **69,** 400 (1979).
80. Weingarten, M. D., Lockwood, A. H., Hwo, S.-Y., and Kirschner, M. W., *Proc. Natl. Acad. Sci. U.S.A.* **72,** 1858 (1975).
81. Weisenburg, R. C., *Science* **177,** 1104 (1972).
82. Weisenburg, R. C., Borisy, G. G., and Taylor, E. W., *Biochemistry* **7,** 4466 (1968).
83. Went, H. A., *J. Biophys. Biochem. Cytol.* **6,** 447 (1959).
84. Went, H. A., *Ann. N.Y. Acad. Sci.* **90,** 442 (1960).
85. Witman, G. B., Carlson, K., Berliner, J., and Rosenbaum, J. L., *J. Cell Biol.* **54,** 507 (1972).
86. Witman, G. B., Cleveland, D. W., Weingarten, M. D., and Kirschner, M. W., *Proc. Natl. Acad. Sci. U.S.A.* **73,** 4070 (1976).
87. Zeuthen, E., *Exp. Cell Res.* **68,** 49 (1971).

8

# Water–Macromolecular Interactions during the Cell Cycle

### PAULA T. BEALL

## I. INTRODUCTION

Water, as the major constituent of living cells, needs to be placed in proper perspective in any study of the cell cycle and the processes that control division. Although water makes up 70–90% of the mass of a living cell, the great majority of cell biology research has concentrated on interesting macromolecular components such as DNA, RNA, enzymes,

**223**

NUCLEAR–CYTOPLASMIC INTERACTIONS
IN THE CELL CYCLE

membranes, and structural proteins. In the past the cell was considered to be an aqueous solution of biochemicals surrounded by a semipermeable membrane. However, today as our understanding of the complexity of the cooperation between the macromolecular components of living cells increases, it is time to reexamine the role of water in cellular structure and function.

Every ion, metabolite, and macromolecule in a cell is surrounded by water molecules in which it is dissolved or hydrated. They swarm about the slow moving macromolecules like a flight of bees; some crawling slowly over the surface in the hydration shell, and others flying about the surface constantly exchanging places with others on the surface. Every active site of an enzyme is filled with water molecules which must be moved out of the way before a substance can interact with the crucial amino acid sidechain. In each of these reactions the major hydrogen donor is water. The product of the reaction then must diffuse away from the enzyme through the surrounding cytosol, where water–macromolecular interactions determine its solubility. The enzymes, chromatin, structural proteins, and membranes depend a great deal on hydrophobic and hydrophilic interactions with the surrounding water molecules for their shape and stability. For example, microtubules depend on water interactions to stabilize the polymerized form and protein–lipid bilayer membranes exist only because of the hydrophobic nature of the lipid components. Thus, there are very close interrelationships between water and the other molecules in cells.

Modern concepts in cell biology view the living cell as much more than an aqueous solution surrounded by a semipermeable membrane, and there is a definitive place for water in this concept. As the membrane becomes more complex with long trailing molecules on its surface and networks of structural and functional proteins integrated into and underneath its structure, water molecules also become an integral part of the membrane. The physical properties of the cytosol depend on the physical properties of its major component, water. The dynamics of every macromolecular group and the kinetics of enzyme reactions become integrated with a new view of water itself. The field of biochemistry may well have to be reconsidered from this viewpoint.

As one studies those events and processes that make up the cell cycle it will no longer be entirely sufficient to concentrate on only one molecule at a time. Some effort must be made to understand water–macromolecular interactions in dividing cells. This chapter will review the historical importance of water in living systems and some of the early work done on plants and animals which pointed out the important role of water. In recent years variations in the water content and the physical properties of water during the cell cycle have been examined in several cell systems, which will be discussed. Most importantly, the underlying water–macromolecular interactions in dividing cells have been studied and will be discussed for three types of macromolecules. The future of this area of research holds the hope of actively manipulating macromolecular structure to intervene in the processes that involve water in the control of cellular division.

## A. Historical Importance of the Study of Water

Since macromolecular evolution took place in the primitive oceans millions of years ago, it is not surprising that water should play such a ubiquitous and important role in life. Had we evolved in seas of liquid ammonia, that liquid might now be the topic of study, but it is $H_2O$ which was the mother liquor of life on Earth.

The early Greeks recognized the primary role of water by assigning to it one of the three elements to exist as "earth, fire, and water." Historical investigations on the role of water in biological systems were by necessity quantitative instead of qualitative. Some of the earliest scientific experiments may have been to measure the difference in weight between wet and dry plants and animals. Dessication or mummification of the human body also showed the importance of water to man.

In the seventeenth and eighteenth centuries as the science of chemistry was being formulated, the theory of osmotic regulation began to be applied to living cells, especially plants. However, the greatest impetus for the consideration of water came at the turn of the twentieth century when the chemists van't Hoff and Arrhenius laid much of the foundation for dilute solution theory. The biologists de Vries and Pfeffer capitalized on the theories of van't Hoff and Arrhenius. From these ideas came much of electrophysiology as we know it today (Hodgkin and Huxley, 1939). Chemists such as Bull (1943), Sponsler and Bath (1942), and Giese (1957) developed a colloid theory for the structure of protoplasm, which has failed with time. The application of the theory of osmotic regulation to living cells in the 1950s by such scientists as Ponder (1948), Lucké and McCutcheon (1952), and Dick (1959) began to reveal some startling facts about water in cells. The cell did not behave as a perfect osmometer. A special factor called Ponder's R was developed to bring experimentally determined osmotically induced volume changes into agreement with osmotic theory of the cell as an osmometer. R is a measure of how far cells deviate from perfect behavior and becomes a measure of all the factors that combine to produce nonideal behavior of water in the cell. R has been determined for many kinds of cells from erythrocytes to plant cells (Ponder, 1948; Dick, 1959). Values for nonosmotically active cell water have ranged from 20 to 60%. Several theories have been developed to attempt to account for the nonideal behavior of water molecules in biological systems.

The so-called membrane theory of cellular function derives from dilute solution theory and depends on the chemical and physical properties of the cell membrane to produce an assymetric distribution of ions between the cell interior and exterior. This theory was first elucidated by Bernstein (1902) and further developed by Hodgkin and Katz (1949) and Goldman (1943). This theory depends on membrane enzyme pumps to move ions such as $Na^+$ and $K^+$ by an energy-requiring process and would not predict any changes in the physical properties of water during the cell cycle.

An alternative hypothesis in which the structure of water and proteins of the protoplasm play an important role has been proposed by several authors (Ling, 1962; Troshin, 1966; Ernst, 1970) and can be called the adsorption theory. The best development of such a view has been done by Ling (1962, 1979; Ling and Ochsenfeld, 1965) in his Association–Induction Hypothesis in which ions are mostly associated with charged side groups on cellular macromolecules and water has altered physical properties due to interactions with cellular macromolecules which causes exclusion of certain solutes in the cytoplasm and results in the asymmetric distribution of sodium and potassium. This type of theory would predict changes in the physical properties of water during the cell cycle as electrolyte concentrations change and groups of macromolecules undergo cyclic conformational changes. The evidence given in this chapter tends to support the alternative view of the cell as a whole entity in which all components are interrelated through their effect on the structure of water in the cell.

## B. The Study of Water in Plant Cells

Plant cells have a unique relation with water that animal cells do not commonly experience. Because of the plant cell wall which resists uncontrolled cell swelling during exposure to environmental osmotic pressure changes, plant cells experience a turgor pressure due to the hyperosmolarity of intracellular fluids and the consequent uptake of water into the cells. The internal hyperosmolarity is usually caused by an active uptake of salts and the biochemical synthesis of organic products. In nature the turgor pressure allows the plant to stand erect and spread its leaves. Since plants do not have a heart, the fluid movement in plants is controlled by osmotic gradients in which water plays a crucial role. In fact it was first in plants that the nonideal osmotic behavior of biological water was discovered (Gortner, 1932). The majority of work investigating water in plants has been of an agricultural type and little work in the field has dealt with single plant cells. Some examples of this type of work are discussed with references in Section II; however, very little work has been done that relates to water in the plant cell cycle. Perhaps with the new methods of culturing plant cells without cell walls in suspension medium there will be renewed investigations of the plant cell cycle.

## C. The Study of Water in Animal Cells

The amount of data that has accumulated on the water content and physical properties of water in animal cells under different conditions is staggering. Several books have been published on the subject, i.e. "Mammalian Cell Water" (Olmstead, 1966), "Water Relations in Membrane Transport in Plants and Animals" (Jungreis *et al.,* 1977), "Cell Associated Water" (Drost-Hansen and Clegg, 1979), and "Water Transport in Cells and Tissues" (House, 1974). The majority of work presented in such volumes is descriptive of cell volume regulation, ion fluxes and movements of water with ions, the absorption of water as a nutrient, and alterations in measurable properties of water in disease states. Only a few experiments have been done to study changes in water content or physical properties of water during the cell cycle. Those experiments will now be discussed in detail and some mechanisms for these changes will be proposed.

## II. CHANGES IN THE PROPERTIES OF WATER DURING CELL DIVISION

## A. Methods for the Study of the Properties of Water in Biological Systems

Since the 1950s there has been a resurgence of interest in the study of biological water which is due in some part to technological advances that make possible the application of a number of physical techniques to complex biological systems. Interpretation of data from complex systems is difficult, but for the first time some of the most exacting physical instrumentation is being used on cells. On-line minicomputers, fast response data recording, electronic amplifiers, accurate signal averaging, fast fourier transform, and less expensive commercial versions of previously home built instruments, as well as a new willingness among physicists seeking to fund research efforts with biologists, have contributed to new biophysical data on water. Some of these new methods are described below.

The study of the influences of macromolecular surfaces on water in living cells has depended on the development of a technology for studying dynamic molecular motion in a nondestructive manner. Nuclear magnetic resonance spectroscopy (nmr) is such a technique and has been used successfully by many investigators on living tissues (Odeblad *et al.*, 1956; Damadian, 1973a; Cope, 1969; Hazlewood *et al.*, 1969; Bratton *et al.*, 1965; Swift and Fritz, 1969). In pulsed nmr measurements, a cell or tissue sample is placed directly into the nmr glass sample tube and placed between the poles of a strong electromagnet. A pulse of radio frequency energy at the resonance frequency of the nuclei to be studied (hydrogen for water, or phosphorous, nitrogen, oxygen, fluorine, carbon, and deuterium can be used) is transmitted into the sample. The nuclei under study are specifically excited and raised to a higher energy state. After the pulse is turned off, the nuclei give up their energy to their surroundings and relax back to the equilibrium distribution. The excitation energy can be dissipated as vibrational, rotational, and translational energies by interaction with the surrounding molecules. In the case of hydrogen protons of water molecules the time it takes for the system to relax is related to the mean average freedom of motion of the water molecules. The relaxation time characteristic of interactions with the surrounding lattice of molecules is called the spin–lattice relaxation time, $T_1$, and the relaxation time characteristic of the hydrogen–hydrogen spin energy exchange is called the spin–spin relaxation time, $T_2$. Both $T_1$ and $T_2$ can be derived from the original Bloch (1946) equations for nmr and the experimental data. In the case of water the $T_1$ of pure water undergoing rapid motion with a correlation time of $10^{-11}$ seconds is ~3000 msec at 25°C, and the $T_2$ value is ~2700 msec. Relaxation times lower than these would indicate a reduced freedom of motion of water molecules or some artifactual problem. Extensive analysis of water relaxation times in biological samples has shown a reduction of three- to sixfold in $T_1$ and five- to twentyfold in $T_2$. The majority of artifactual problems have been eliminated (Hazlewood, 1973) and the arguments are now at the stage of interpretation of the data. Models for interpretation of nmr data on biological water range from a tiny fraction of water in the cell being tightly associated with the surface of macromolecules and the remainder being like pure water, to a slight pertubation in the overall structure of water in the cell (Zimmerman and Britton, 1957; Bratton *et al.*, 1965; Cooke and Wein, 1973; Hazlewood, 1973, 1979;

Chang *et al.*, 1972). Controversy over interpretation of nmr data remains, but whereas the amount of water with altered physical properties was once thought to be approximately 1–3% of total cellular water, this percentage has increased to 15–30% over the last few years (Hazlewood, 1979).

Differential scanning microcalorimetry has gained acceptance as a technique to study the thermodynamic properties of water in cells. The most consistent finding has been that the heat capacity of water in biological systems is greater than in pure water and that a significant fraction of cell water remains unfrozen at −15°C (Clegg, 1979; Andronikashvili and Mrevlishvili, 1976; Hoeve and Kakivaja, 1976). The anomalous freezing behavior of water in biological systems has revealed that water in muscle may be supercooled to very low temperatures and that ice crystals once initiated grow in different patterns in relaxed and contracted muscle indicating a change in the arrangement of water molecules around proteins (Chambers and Hale, 1932; Miller and Ling, 1970).

Measurements of the dielectric relaxation properties of water can yield information about the instantaneous structure of water around macromolecules. Currently an oscillating microwave field is applied to a sample and the frequency increased until the water molecules can no longer relax between pulses. Schwan and Foster (1977) have found no great differences between cell water and pure water below 20 GHz. However, Fricke and Jacobsen (1939) and Clegg (1979) working at higher frequencies see some water molecules with perturbed motion. This is a new technique which can be applied as well to living cells during the cell cycle. Clegg (1979) has applied it to the developing brine shrimp.

The measurement of viscosity of cytoplasm can be accomplished by a number of methods. The measurement of macroviscosity using the motion of iron filings pulled through the cytoplasm by an external magnet (Crick and Hughes, 1950) reported the modulus of rigidity of chick fibroblast cytoplasm to be of the order of $10^2$ dynes/cm$^2$. Wilson and Heilbrunn (1960) reported the viscosity of the immature egg of *Spisula* as 4.3 cP. Measurements of microviscosity at the level of individual water molecules are more difficult. The technique of electron spin resonance (esr) has been utilized to follow the molecular motion of spin-labeled probe molecules in the aqueous phase of the cell as well as the membranes (Sachs and Latorre, 1974; Belagyi, 1975; Cooke, 1976; Haak *et al.*, 1976). The results indicate that the microviscosity of cytoplasm is between 3 and 10 cP compared with 1 cP for pure water.

Drost-Hansen (1971) has presented evidence that water near surfaces, called vicinal water, displays thermal anomalies in its behavior due to

**Table I    nmr Relaxation Times and Diffusion Coefficient of Water in Various Cells and Tissues**[a]

| Animal and tissue | nmr Frequency (MHz) | $T_1$ (msec) | $T_2$ (msec) | D ($10^{-5}$cm$^2$/second) |
|---|---|---|---|---|
| Pure water 25°C | 30 | 3000 | 2700 | 2.5 |
| Bacteria | | | | |
| E. Coli | 100 | 557 | 30 | |
| Halobacterium | 14 | 40 | 21 | |
| Slime mold | 20 | 131 | 31 | 0.49 |
| Invertebrates | | | | |
| Barnacle muscle | | | | 1.35 |
| Brine shrimp cysts | 30 | 255 | 55 | 0.16 |
| Frog | | | | |
| Egg | 30 | 216 | 36 | 0.67–0.28 |
| Skeletal muscle | 30 | 700 | 48 | 0.46 |
| Liver | 20 | 270 | 46 | 0.42 |
| Chicken | | | | |
| Kidney | 32 | 403 | | |
| White muscle | 32 | 598 | | |
| Lung | 32 | 670 | | |
| Mammals | | | | |
| Lens—rabbit | | | | 0.9 |
| Muscle | | | | |
| Rat | 30 | 720 | 45 | 0.50–1.5 |
| Mouse | 30 | 185 | 49 | |
| Human | 24 | 459 | | |
| Red blood cells—human | 6 | 559 | | 0.2–0.55 |
| Liver | | | | |
| Rat | 60 | 293 | 52 | |
| Mouse | 24 | 350 | 51 | |
| Human | 24 | 383 | | |
| Heart | | | | |
| Rat | 60 | 873 | 46 | 0.36 |
| Mouse | 30 | 650 | 45 | |
| Human | 24 | 873 | | |
| Kidney | | | | |
| Rat | 60 | 685 | 56 | |
| Mouse | 30 | 503 | 47 | |
| Kidney—human | 24 | 685 | | |
| Thyroid gland—human | 24 | 586 | | |
| Ovary—human | 100 | 989 | | |

[a] For extensive tables of data on properties of water, see Hazlewood (1979). For data on isolated cell systems see Table IV.

interactions with that surface. Etzler and Drost-Hansen (1979) have utilized these anomolies to correlate with water structure changes in living plant and animal cells under various stresses. The selectivity for ions, amount of growth, germination rate, incorporation or synthesis rate of metabolites, and cytoplasmic viscosity can be measured as a function of temperature and minimums and maximums in these properties found to correlate with water property anomolies. Some of these specific experiments will be discussed later.

In addition, simple dessication experiments are utilized to determine water contents of cells. Precautions observed in this method require complete evaporation of water at temperatures low enough to prevent the loss of volatile lipids or degradation of the sample. Drying at 105°C or less under a vacuum or freeze drying can be used.

Other methods have been used in the study of water in biological systems including quasi-elastic neutron scattering, infrared and Raman spectroscopy, ion isotope exclusion, diffusion of tritium-labeled water, and microelectrode conductance. Such methods provide a new way of looking at and measuring the properties of water in cells. It is now possible to make good quantitative and qualitative measurements on the physical properties of water in living systems. A sample of some of the properties of water measured for various cells and tissues is given in Table I.

## B. Bacteria and Fungi

The variety among bacteria make it difficult to come to any generalization about the role of water in their cell cycle. Most information must be inferred from the effect of environmental stresses. Bacteria, for example, can be separated into three thermophilic classes: cryophiles that grow well below 5°C, mesophiles that grow best in the vicinity of 37°C, and thermophiles that grow well above 40°C. Damadian (1973b) proposes that the separation of growth around 4°C may be a function of the molecular packing of water molecules preparing to undergo the ice transition. Foter and Rahn (1936) have pointed out that few if any mesophilic bacteria exhibit growth below 4°C even though they remain unfrozen. *Streptococcus lactis* and *Lactobacillus acidophilus* stop growing at 5°C and five strains of *Salmonella* cease growing between 5.5° and 6.1°C. Minkoff and Damadian (1976) further develop a cytotonus hypothesis to explain the regulation of cellular physiology of bacteria through the quantitative control of cell water. The hypothesis utilizes the discovery of actinlike proteins in CBH *E. coli* to couple the bacterial cell wall, cell membrane, contractile proteins, ion selectivity properties, and cell water content for the total regulation of cell growth. nmr Measurements for water protons in *E. coli* show a $T_1$ value of 557 msec and a $T_2$ value of 30 msec (Zaner, 1973), significantly reduced from that of free water. These values increase upon hypotonic swelling and decrease with hypertonic shrinking in a monotonic manner depending on the salt used (Damadian, 1973b). The growth rate of the *E. coli* population decreases on either side of the normal salt concentration. Other studies of bacteria subjected to osmotic stresses indicate that bacteria resist swelling and shrinking better than mammalian cells possibly due to the cell wall. Water contents of various bacterial strains have been measured (see Table I), but systematic analyses of water content and physical properties of water have not been undertaken for bacteria.

Water transport and content as functions of the cell cycle have been little studied in fungi. In general, fungi seem to be readily adaptable to a wide variety of osmotic concentrations due to an active transport system and the ability to osmoregulate with changes in internal carbohydrate synthesis (Slayman, 1977). One outstanding example of a cell cycle variable which correlates with water properties is the intracellular pH change in *Physarum polycephalum* as a function of the cell cycle (Gerson, 1979). Using the fluorescent pH indicator 6,7-dihydoxy-4-methylcoumarin, Gerson and Burton (1977) showed that the intracellular pH of *polycephalum* varied from 5.9 during interphase to 6.9 during mitosis. The rapid increase in pH near mitosis may correspond to an uptake of water prior to cell or nuclear division. An early nmr measurement on interphase *Physarum* by Walter and Hope (1960) found the $T_1$ to be $131 \pm 13$ msec and $T_2$ to be $30 \pm 2$ msec. However, this is an example of a problem with nmr because *Physarum* is grown in a medium containing soluble iron in sufficient quantity to depress nmr values due to paramagnetic impurities. With modification of the growth medium and current nmr techniques the entire cell cycle of *Physarum* could be explored. Variation in intracellular pH may be a good indicator of changes in water content and water properties. The obvious effects of pH changes on growth rate of all types of cells remind one of the importance of the ability of water to disassociate in considerations of pH control mechanisms.

## C.  HeLa Cells

The one system where extensive investigations of the water content and physical properties of water have been done is the synchronized HeLa cell cycle (Beall *et al.*, 1976). HeLa cells from the original line of Gey *et al.*, (1952) were grown as monolayers in Eagle's minimal medium supplemented with 10% fetal calf serum and antibiotics at 37°C in equilibrium with 10% $CO_2$. A population of cells partially synchronized by an excess thymidine block was treated either with colcemid (0.05 mg/ml) or $N_2O$ gas at 80 psi (Rao, 1968) to yield 98% mitotic cells. These cells were allowed to grow into the cycle and sampled at 0, 30 minutes, 4 hours, 8 hours, and 12 hours of a 22-hour cell cycle. A double thymidine block produced a population of S phase cells at 12 hours in the cycle which then grew into the stages of $G_2$. $G_2$ cells were harvested at 18, 19, and 20 hours of the cell cycle with good synchrony. Water content of cell pellets and nmr relaxation times were determined (see Table II). A phase specific pattern of changes in the $T_1$ and $T_2$ relaxation times was found. The highest $T_1$ value (or most mobile water molecules) was found in mitosis ($T_1 = 1000$ msec compared to 3000 msec for pure water). The lowest $T_1$ relaxation times were found in S phase ($T_1 = 534$ msec) indicating the most restricted water of the cell cycle. Changes in $T_1$ could not be contributed entirely to increases in the ratio of water to dry solids. While water content or the ratio of water to dry solids did decrease from early mitosis to S phase along with $T_1$ values, $T_1$ increased independently of the water ratio from 12 to 20 hours of the cell cycle. As cells grew and increased in volume, water entered the cell in exact balance with the synthesis of new molecules during S and $G_2$, but the relationship between water structure and the kind and conformation of the intracellular molecules changed. Figure 1 shows the correlation between the chromatin condensation cycle in these cells and water nmr properties. The effect of chromatin was further explored and the results given in Section III. This study showed that in cells that are grown as monolayers, the ratio of water to dry solids changed during the cell cycle even in a constant osmolarity medium. The pattern of water content and water property changes was related to the genetic programming of the

**Table II   nmr Relaxation Times of Water Protons and Water Content during the Synchronized HeLa Cell Cycle[a,b]**

| Cell cycle stage | $T_1$ (msec) | $T_2$ (msec) | $H_2O$ (%) |
|---|---|---|---|
| Mitosis | | | |
| M (0 minutes) | 1020 ± 84 | 130 ± 13 | 88.25 ± 0.31 |
| M (30 minutes) | 817 ± 76 | 127 ± 18 | 87.47 ± 0.10 |
| Early: $G_1$ (4 hours) | 638 ± 110 | 110 ± 11 | 85.84 ± 0.30 |
| Late: $G_1$ (8 hours) | 570 ± 56 | 117 ± 11 | 85.47 ± 0.74 |
| S Phase: S (12 hours) | 534 ± 43 | 100 ± 9 | 84.42 ± 0.75 |
| Early | | | |
| $G_2$ (18 hours) | 621 ± 25 | 96 ± 8 | 84.54 ± 0.20 |
| $G_2$ (19 hours) | 690 ± 4 | — | 84.28 ± 0.63 |
| Late: $G_2$ (20 hours) | 739 ± 59 | 116 ± 7 | 84.27 ± 1.07 |
| Mitosis: M (22 hours) | 1020 ± 84 | 130 ± 13 | 88.25 ± 0.31 |

[a] From Beall (1979).
[b] Each value is the mean of eight to ten experiments ± standard deviation.

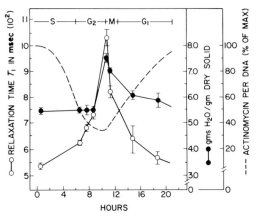

**Fig. 1.** Spin-lattice relaxation time, $T_1$, for water protons and water content as a function of the synchronized HeLa cell cycle. (—O—O—), $T_1$ (msec) $\pm$ the standard deviation of eight to ten experiments; (—●—●—), grams $H_2O$ per gram dry solids of cell pellets $\pm$ standard deviation; (- - - -), actinomycin D binding ability of chromatin during the cell cycle (Pederson and Robins, 1972).

cell since it was reproducible in every cell cycle and that chromatin structure had an influence on water.

## D.   Chinese Hamster Ovary Cells

In order to test the general applicability of the pattern of changes seen in the cell cycle of the HeLa cell to other cell systems, the above experiments were repeated for monolayer Chinese Hamster Ovary (CHO) cells grown on McCoy's medium and synchronized with colcemid and thymidine (Beall and Robinson, 1979). Table III summarizes the results of these experiments. Mitotic CHO cells have the highest $T_1$ values (889 msec) of the cell cycle and the highest water contents (89.8 %). The interphase CHO cells have a lower $T_1$ value of 681 msec and a lower water content of 85%. The most mobile water molecules of the cell cycle occur in mitosis for CHO cells as they did for HeLa cells. Water content and nmr relaxation time values have been measured for a number of cell lines by these authors and their results are shown in Table IV. To date, only the CHO and HeLa cell cycles have been examined in detail. Cercek *et al.* (1973a,b) have utilized the techniques of fluorescence polarization and fluorochromasia to measure the structuredness of the cytoplasmic matrix in CHO cells as an indirect measure of water–macromolecular interactions. Their results suggest the least organization of the cytoplasmic matrix in mitosis and the most organization in S phase, substantiating the nmr results of Beall *et al.* (1976).

Table III   nmr Water Relaxation Times of Mitotic and Interphase Chinese Hamster Ovary Cells

| Sample | $T_1$ (msec) | $T_2$ (msec) | $H_2O$ (%) |
|---|---|---|---|
| Mitotic CHO cells | 888 | 114 | 92.4 |
| | 945 | 127 | 93.3 |
| | 925 | 116 | 89.0 |
| | 819 | 106 | 89.2 |
| | 923 | 115 | 87.8 |
| | 836 | 102 | 87.0 |
| Mean | 889 | 113 | 89.8 |
| Interphase CHO cells | 632 | 96 | |
| | 666 | 92 | |
| | 711 | 102 | |
| | 757 | 106 | |
| | 740 | 111 | 84.2 |
| | 630 | 99 | 85.2 |
| | 623 | 95 | 84.8 |
| | 711 | 103 | 85.2 |
| | 673 | 82 | 85.3 |
| | 655 | 96 | 85.0 |
| | 689 | 100 | 85.0 |
| Mean | 681 | 98 | 85.0 |

## E.   Ehrlich Ascites Tumor Cells

Dupre and Hempling (1978) have examined the osmotic properties of the Ehrlich ascites tumor cell grown in *suspension* culture. Cells grown in suspension tend to be rounded and swollen compared to counterparts grown as monolayers; however, these studies revealed very interesting facts about the properties of water during the cell cycle. Total cell water in suspension culture cells remained constant throughout the cell cycle at 82% by weight, although the cells grew and doubled in total volume before division. The electrolyte concentration remained constant when expressed as milliequivalents per liter cell water. For a given surface area, temperature, and osmotic gradient, the membrane permeability to water decreases from M to S. In agreement with this, freeze-fracture experiments (Scott *et al.*, 1971) have shown that intramembranous parti-cles become diluted out during mitosis and early $G_1$ only to be restored in S and late $G_2$. They show that the heat of activation for water permea-bility peaks in late S or early $G_2$ as well. Their data show that the fraction of cell water which is osmotically active (Ponder's R) decreases from 0.75 at S to about 0.56 following mitosis. In these cells it seems that at the time when the membrane is most permeable to water, a large fraction of cell

**Table IV   nmr Relaxation Times for Cultured Cell Systems**

| Cell type | $T_1$ (msec) | $T_2$ (msec) | $H_2O$ (%) |
|---|---|---|---|
| I.[a] | | | |
| HeLa | 651 | 116 | 84.8 |
| CHO | 630 | 97 | 85.0 |
| Established mouse | | | |
| Mammary cancer | | | |
| Lines | | | |
| ESD/BALB CL3 | 650 | 128 | |
| MTV-L/BALB CL2 | 782 | 115 | |
| DMBA/BALB CL2 | 670 | 95 | |
| Primary mouse mammary | | | |
| Normal | 916 | 158 | 90.8 |
| Preneoplastic HAN D2 | 1029 | 187 | 90.0 |
| Mammary adenocarcinoma | 1155 | 206 | 91.4 |
| WI 38 Human fibroblasts | 1154 | 165 | |
| Human breast cancer | | | |
| Established Lines MDA-MB | | | |
| 231 | 934 | 123 | 86.5 |
| 157 | 907 | 135 | 87.4 |
| 361 | 849 | — | 88.4 |
| 453 | 770 | 113 | 87.4 |
| 330 | 752 | — | 88.5 |
| 435 | 607 | 112 | 87.3 |
| 331 | 549 | 75 | 88.5 |
| 431 | 521 | 126 | — |
| 436 | 499 | 100 | 84.3 |
| Scott and White | | | |
| SW 527 | 729 | 110 | 86.2 |
| SW 613 | 606 | 104 | 85.0 |

*(continued)*

water is not available to dissolve solutes or is associated with cellular macromolecules.

## F.   Friend Leukemia Cells

Haran *et al.* (1979) have studied water permeability changes of friend leukemia cells (FLC) during differentiation by $O^{17}$ nmr. In this case water containing $O^{17}$ isotope was used in the growth medium and taken up by the cells. $O^{17}$ nmr offers some advantages in interpretation. The rate at which $O^{17}$ leaves the external environment and enters the cell can be determined and the permeability calculated. Friend leukemia cells were induced to enter division by 2% $Me_2SO$ and monitored for stage of the cell cycle. The results showed that during the differentiation of FLC there are two periods in which the water permeability is greatly in-

**Table IV**—*Continued*

| Cell type | $T_1$ (msec) | $T_2$ (msec) | $H_2O$ (%) |
|---|---|---|---|
| Swiss Mouse | | | |
| 3T3 | 818 | 113 | 85.4 |
| SV 3T3 | 757 | ′85 | 83.2 |
| II[b] | | | |
| Chinese Hamster lung fibroblasts V79-S171 | 800 | | |
| Human liposarcoma HLS-2 | 914 | | |
| Human amnionic VRC-4 | 749 | | |
| Monkey kidney VERO | 724 | | |
| 9-Day chick embryo CEC | 733 | | |
| Ascites tumor cells | | | |
| Rat | 1911 | | |
| Rat | 1150 | | |
| Mouse hepatoma | 600 | | |
| Mouse melanoma | 620 | | |
| Mouse fibrosarcoma | 620 | | |
| Chicken red blood cells | 470 (with virus $T_1$ = 560) | | |
| Human leukemic cells | | | |
| Before treatment | 1031 | | |
| After treatment | 612 | | |
| Human red blood cells | | | |
| Normal oxy | 382 | | |
| Normal deoxy | 260 | | |
| Sickle cell oxy | 879 | | |
| Sickle cell deoxy | 864 | | |

[a] Data from laboratory of Beall and Hazlewood.
[b] Data from Hazlewood (1979).

creased. The first peak in permeability is after approximately one cell division near mitosis, similar to the Dupre experiments on Ehrlich ascites cells. The second peak of permeability increase occur after 6 days of cell growth when the majority of cells have not only divided but are differentiated to produce hemoglobin. Interestingly Loritz *et al.* (1977) observed two distinct changes in cell volume of FLC during $Me_2SO$ induction which correspond to a 70% decrease in cell volume after one division and a decrease to 50% volume at about 6 days. Water membrane permeability and volume change correspondence was attributed to membrane structure changes by Haran *et al.* (1979); however, no consideration of cytoplasmic properties was given.

## G. Plant Cells

The role of water in the regulation of plant cell growth has been studied by several indirect experiments which can at best only point to the importance of consideration of this variable in interpretation of plant cell growth.

The effect of the osmolarity of the external solution on the growing tips of plants has been studied extensively. One such example is the effect of sucrose solutions of various osmolarities on the number of mitoses per thousand cells in growing onion root tips. Olmstead (1966) showed that the addition of as little as 10 mOsm of sucrose to the growth medium could suppress mitosis by 10% and the addition of 200 mOsm of sucrose could suppress mitosis by 90%. In sucrose solutions of 0.2–1.2 M, Greenfeld (1942) showed suppression of the growth of *Chorella vulgaris* giant algae in proportion to the sucrose present. The disruption of mitotic activity in plants and animals by osmotic abnormalities should signal the close association of water and electrolytes with the control of division.

Etzler and Drost-Hansen (1979) have examined the germination of turnip seeds and the growth of the green thermophilic alga, *Cyanidium caldarium,* as a function of temperature. They have found minimums and maximums in germination and growth near the same temperatures where water between quartz plates displays thermal anomolies in its structure. For example, turnip seeds show a minimum amount of germination at 15°C and then increase their germination rate with temperature until about 22°C, after which the germination rate decreases until 30°C and then starts to increase once again. The algae plant displays growth maximums at 28°–30°C, 38°–40°C, and 48°–50°C but almost no growth in between these narrow ranges. The temperatures that affect growth the most seem to correspond to temperatures where there are maximums in water viscosity, ion selectivity, and specific heat for water in several systems. This type of indirect evidence should be considered carefully, but it does indicate a correlation between growth and water properties.

## III. INTERACTIONS OF WATER WITH CELLULAR MACROMOLECULES

### A.  Chromatin

The results of Beall *et al.* (1976, 1978a) on the nmr properties of water during the HeLa cell cycle were shown at that time to bear a striking correlation to the chromatin condensation cycle of the cell as defined by Pederson and Robins (1972). When the chromatin was condensed into the chromosomes water molecules seemed to be able to move about most freely and when the chromatin was diffuse in the nucleus and open to bind water on its surface the $T_1$ values were lowest (see Fig. 1). Since the HeLa cell has a large nuclear to cytoplasmic ratio it seemed reasonable that this group of cellular macromolecules might have a profound influence on the properties of water near their surface. Of course, all ions, metabolites, and macromolecules in the cell will affect water structure, but in some systems one group may predominate and be easily studied. In order to directly study water–chromatin interactions it would have been most logical to remove chromatin from the cell. However, as biochemists have discovered, this causes condensation of the material which cannot be reversed except in high molarity salts. Therefore, the decision was made to study chromatin in the viable cell or isolated nucleus.

S phase HeLa cells were treated with spermine, a polyamine having chromatin-condensing ability. Visible chromatin condensation was accompanied by an increase in the $T_1$ or water motion without an increase in the water-to-dry-solids ratio in the cell (see Table V). These data indicated a correlation between chromatin structure in the nucleus and the pattern of $T_1$ changes seen for the whole cell; however, they were not conclusive.

Nuclei were isotonically isolated from S and $G_2$ synchronized HeLa cells by an adaptation of the cytochalasin B method of Prescott *et al.* (1972). The double membrane-bound nuclei were treated with spermine. S phase nuclei had the most restricted water motion of the cell cycle at 470 msec for $T_1$, even lower than the whole S phase cell (see Table VI), and $G_2$ nuclei, which already showed some chromatin condensation, had higher $T_1$ values at 690 msec. S phase karyoplasts (isolated nuclei) were treated with spermine and showed chromatin condensation accompanied by an increase in the mobility of water molecules without an increase in water content (see Table VII). These experiments suggested that a good deal of the reproducible pattern of phase-specific

**Table V  Treatment of Isolated S Phase Cells with Spermine**

| Measurement | Control S phase nuclei | 0.02 M Spermine-treated nuclei |
|---|---|---|
| $T_1$ (msec) | | |
| 1 hour | 538 | 638 |
| 2 hour | 546 | 701 |
| $H_2O$ (%) | | |
| 1 hour | 84.4 | 84.4 |
| 2 hour | 84.6 | 85.0 |

Table VI   Water Properties of Isolated S and $G_2$ Nuclei of HeLa Cells

| Variable | Whole cells | Isolated nuclei |
|---|---|---|
| S Phase Cells (12 hours) | | |
| $T_1$ (msec) | $534 \pm 43$ | $457 \pm 29$ |
| $T_2$ (msec) | $100 \pm 9$ | 53 |
| $H_2O$ (%) | 84.4 | 84.0 |
| $G_2$ Phase Cells (19 hours) | | |
| $T_1$ (msec) | $690 \pm 4$ | $692 \pm 30$ |
| $T_2$ (msec) | — | $114 \pm 15$ |
| $H_2O$ (%) | 84.2 | 80.0 |

changes in water $T_1$ values during the HeLa cell cycle was due to the massive conformational changes in the cell chromatin which resulted in water structure changes in its vicinity. The similar pattern for CHO cells may indicate a similar mechanism for these monolayer grown cells as well.

## B.   Microtubules

Another good candidate for a group of abundant cellular macromolecules which could have a profound effect on cellular water would be the filamentous proteins of the cytoskeleton. Considering first the microtubule proteins, Beall et al. (1978b, 1979a,b) have studied several cell systems where microtubules seem to play an important role in cell water structure.

The nmr relaxation times of water have been found to vary among a series of human breast cancer cell lines according to the rate of division of the cells and the macromolecular organization of the cytoskeleton (Beall et al., 1978b, 1979a). Cell population doubling times from 1 to 18 days correlated with $T_1$ values from 934 to 499 msec. Fast dividing cells (1–2 days) demonstrated high $T_1$ values (750 msec) and a diminished cytoplasmic microtubule complex by antibody immunofluorescence. Moderately fast dividing cells (3–7 days) demonstrated medium values for $T_1$ (750–600 msec) and had approximately 50% of a full microtubule

Table VII   Treatment of S Phase Nuclei with Spermine

| Sample | $T_1$ (msec) | $H_2O$ (%) |
|---|---|---|
| Isolated S phase nuclei—controls | $457 \pm 29$ | $84.0 \pm 0.9$ |
| Isolated S phase nuclei—treated with 0.02 M Spermine | $617 \pm 43$ | $84.0 \pm 0.5$ |

Table VIII   nmr Relaxation Times, Doubling Times, and Microtubule Complexes in Established Cell Lines of Human Mammary Carcinoma[a]

| Cell line MDA-MB | Doubling time (days) | $T_1$ (msec) $\pm$ SD | Full microtubule complex (%) |
|---|---|---|---|
| 231 | 1 | 934 ± 78 | 89 |
| 157 | 1–1½ | 907 ± 10 | 7 |
| 361 | 1½ | 849 ± 25 | 0 |
| 134 | 1½–2 | 717 ± 64 | 0 |
| 453 | 1½–2 | 770 ± 15 | 1.5 |
| 330 | 1½–3 | 752 ± 39 | 55 |
| 435 | 6–7 | 607 ± 9 | 34 |
| 331 | 5–7 | 549 ± 136 | 48 |
| 431 | 12–14 | 521 | 83 |
| 436 | 16–18 | 499 ± 49 | 79 |

[a] From Beall *et al.* (1979b).

complex. Slowly dividing cells of the ten lines in the series demonstrated low $T_1$ values (600–500 msec) and had abundant polymerized cytoplasmic microtubules. $T_1$ and $T_2$ values did not correlate with cellular hydration. A three-way correlation between cellular doubling time, water proton relaxation times, and the system of microtubules in these cells suggests a close interaction between water and microtubules (see Table VIII).

Interphase HeLa and CHO cells treated for 30 minutes with colcemid, a drug that depolymerizes microtubules, also show an increase in $T_1$ values upon the destruction of one of the anchoring structures of the cytoplasmic matrix (see Table IX).

Table IX   HeLa and CHO Cells Treated with Colemid to Depolymerize Microtubules

| Cell type | Variable (msec) | Control | Colcemid treated[a] |
|---|---|---|---|
| HeLa cells | $T_1$ | 620 | 797 |
| | | 593 | 781 |
| | $T_2$ | 104 | 122 |
| | | 102 | 129 |
| CHO cells | $T_1$ | 630 | 707 |
| | | 623 | 732 |
| | $T_2$ | 99 | 130 |
| | | 95 | 136 |

[a] Cells were incubated in the usual medium containing 0.10 µg/ml colcemid for 1 hour.

A further test of the role of microtubules in the determination of water properties in cells was accomplished in an *in vitro* system of purified dog brain microtubule protein. In the presence of GTP and $Mg^{2+}$, the microtubule protein was capable of undergoing a temperature-dependent polymerization which could be monitored by nmr (Beall *et al.*, 1978b, 1979b,c) (see Fig. 2). The water molecules in the microtubule solution showed a phase transition-like behavior as a function of temperature. At low temperatures when the protein was in a globular form, water relaxation time behavior was a linear function of temperature as would be predicted for any protein solution. However, in the range of 15°–18°C when the microtubules polymerized into long fibers the behavior of the water molecules underwent an inflection and movement to another linear region typical of another state for water. This behavior was also seen in a pellet of WI-38 cells containing large amounts of cytoplasmic microtubules (see Fig. 3).

Preliminary data from experiments on HeLa and CHO cells, which have measured the self-diffusion coefficient of water in these cells as a

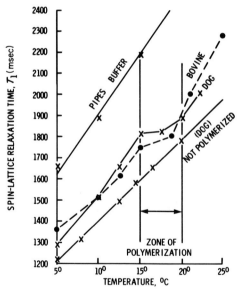

**Fig. 2.** Spin-lattice relaxation time, $T_1$, for water protons in a purified solution of dog (-×-×-×-) and bovine (—●—●—) microtubule protein (2% solution), as a function of temperature. Between 5°–15°C $T_1$ varies linearly with temperature as predicted. At 15°–18°C $T_1$ deviates from linear behavior and returns along a different temperature relationship after 18°C indicating a significant change in water–macromolecular relationships at the point of microtubule polymerization. Pipes buffer follows a linear relationship throughout the temperature range.

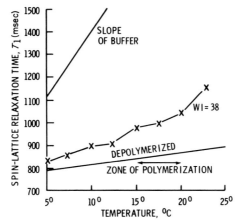

**Fig. 3.** WI-38 human fibroblast cells $T_1$ versus temperature. $T_1$ deviates from a linear relationship with temperature in the 15°–20°C range when microtubules depolymerize.

function of temperature, show the same type of inflection near the polymerization temperature for microtubules (Beall *et al.*, 1979c).

The ability of a 2% microtubule protein solution or the low concentrations of microtubules found in cells to affect the relaxation times of water in a manner that can be detected by present equipment indicate that more than a few layers of water molecules are affected by the protein surface and that microtubules may be the anchoring structures for even greater cytoplasmic organization such as the "microtrabeculae" proposed by Wolosewick and Porter (1976). If the entire filamentous structure of the cytoplasm is integrated with the water structure which surrounds and supports it, then the role of water becomes even more crucial in understanding a cell cycle where this matrix is formed and breaks down in a patterned way.

## C. Actin Filaments

Actin and actinlike filamentous proteins are becoming more and more important to our understanding of cell motility, membrane structure, and cellular function. The relationship between water and actin can mostly be speculation at this point. We have shown that in S and $G_2$ phase HeLa cells treated with the drug cytochalasin B, an increase in water mobility accompanies the depolymerization of the actin filaments (see Table X). Much work remains to be done on the interactions of actin and water, and this area will probably be one of the most productive to increase our knowledge of water's role in physiological functioning.

**Table X   HeLa Cells Treated with Cytochalasin B to Depolymerize Actin Filaments**

| Stage of the cell cycle | Variable | Control cells | Cytochalasin B-treated cells[a] |
|---|---|---|---|
| S phase cells | $T_1$ (msec) | $534 \pm 43$ | $681 \pm 14$ |
| | $H_2O$ (%) | $84.4 \pm 0.7$ | $82.9 \pm 2.0$ |
| $G_2$ phase cells | $T_1$ (msec) | $690 \pm 4$ | $795 \pm 2$ |
| | $H_2O$ (%) | $84.3 \pm 0.6$ | $84.0 \pm 0.5$ |

[a] Cells were incubated in the usual medium with 10 $\mu$g/ml cytochalasin B for 1 hour at 37°C.

## IV.  CONCLUSIONS

A base of knowledge exists concerning the role of water in cell division. The importance of water is recognized in all fields of biology, but a systematic study of the role of water–macromolecular interactions in physiological function has just begun. The ability to apply the tools of biophysics to this complex problem holds hope for its solution. With the growing awareness among cell biologists that it is the integration of all the molecules of the cell, including water, which produce the whole, there will be greater interest in the study of water. The thermodynamics of cellular processes require the consideration of water properties, and the new cellular anatomy has a large role for water.

Studies of the role of water in the cell cycle are just beginning. Only a few systems have been studied and although they offer tantalizing glimpses of what water may be doing in the real cell, no conclusions can be reached at this time. Although we have neglected the close association of water and electrolytes in this discussion, one should note that alterations in the properties of water during the cell cycle may be the driving force for the movements of electrolytes which have a profound effect on the control of mitosis. The ability of minor changes in external osmolarity to halt the division cycle without killing the cell suggest a delicate balance between water and electrolyte concentrations and the control of division.

A good guess at this time would be that the genetic programming of the cell cycle results in the production of and conformational changes in large groups of cellular macromolecules which influence the structure of water in their vicinity. The cyclic changes in water permeability of the membrane and water properties of the cytoplasm may in turn affect changes in intracellular ion concentrations which act as controls on the genome. Although speculative, this scheme offers a proposal to be tested by experimentation.

## ACKNOWLEDGMENTS

The author wishes to thank Carlton Hazlewood for his advice and collaboration in many of the experiments reported in this chapter.

The work was supported in part by the National Institutes of Health grants GM-20154 and CA-21624, the Robert Welch Foundation, and the Office of Naval Research Contracts N00014-76-C-0100 and N00014-78-C-0068.

## REFERENCES

Andronikashvili, A., and Mrevlishvili, G. (1976). *In* "L'eau et Les Systemes Biologiques" (A. Alfsen and A. J. Berteaud, eds.), p. 275. CNRS, Paris.

Beall, P. T. (1979). *In* "Cell-Associated Water" (W. Drost-Hansen and J. Clegg, eds.), pp. 271–292. Academic Press, New York.

Beall, P. T., and Robinson, D. (1979). *Tissue Culture Abstr.* **16**, 228.

Beall, P. T., Hazlewood, C. F., and Rao, P. N. (1976). *Science* **192**, 904–906.

Beall, P. T., Chang, D. C., and Hazlewood, C. F. (1978a). *In* "Biomolecular Structure and Function" (P. Agris, ed.), pp. 233–237. Academic Press, New York.

Beall, P. T., Bohmfalk, J., Fuller, G., and Hazlewood, C. F. (1978b). *J. Cell Biol.* **79**, 281a.

Beall, P. T., Asch, B. B., Chang, D. C., Medina, D., and Hazlewood, C. F. (1979a). *Biophys. J.* **25**, 238a.

Beall, P. T., Brinkley, B. R., Chang, D. C., and Hazlewood, C. F. 1979b). *Cancer Res.*, accepted for publication.

Beall, P. T., Chang, D. C., and Hazlewood, C. F. (1979c). *Am. Soc. Cell. Biol. Abstr.* **1979**, accepted for publication.

Belagyi, J. (1975). *Acta Biochem. Biophys. Acad. Sci. Hung.* **10**, 63.

Bernstein, J. (1902). *Arch. Gesamte Physiol. Menschen Tiere* **92**, 521–562.

Bloch, F. (1946). *Phys. Rev.* **70**, 460–462.

Bratton, C. B., Hopkins, A. L., and Weinberg, J. W. (1965). *Science* **147**, 738–739.

Bull, H. B. (1943). "Biophysical Biochemistry." Wiley, New York.

Cercek, L., Cercek, B., and Ockey, C. H. (1973a). *Biophysics* **10**, 187–194.

Cercek, L., Cercek, B., and Ockey, C. H. (1973b). *Biophysics* **10**, 195–197.

Chambers, R., and Hale, H. P. (1932). *Proc. R. Soc. London Ser.* **110**, 336.

Chang, D. C., Hazlewood, C. F., Nichols, B. L., and Rorschach, H. E. (1972). *Nature (London)* **235**, 170–172.

Clegg, J. (1979). *In* "Cell-Associated Water" (W. Drost Hansen and J. Clegg, eds.), pp. 363–414. Academic Press, New York.

Cooke, R., and Wein, J. (1973). *Ann. N. Y. Acad. Sci.* **204**, 197–210.

Cooke, R. (1976). *In* "L'eau et les Systemes Biologiques" (A. Alfsen and A. J. Berteaud, eds.), p. 283. CNRS, Paris.

Cope, F. W. (1969). *Biophys. J.* **9**, 303–319.

Crick, F. H., and Hughes, A. F. W. (1950). *Exp. Cell Res.* **1**, 37–39.

Damadian, R. (1973a). *Ann. N. Y. Acad. Sci.* **204**, 211–244.

Damadian, R. (1973b). *Crit. Rev. Microbiol.*, 377–422.

Dick, D. A. T. (1959). *Int. Rev. Cytol.* **8**, 392.

Drost-Hansen, W. (1971). *In* "Chemistry of the Cell Interface, Part B" (H. D. Brown, ed.), p. 1. Academic Press, New York.

Drost-Hansen, W., and Clegg, J., eds. (1979). "Cell-Associated Water." Academic Press, New York.

Dupre, A. M., and Hempling, H. G. (1978). *J. Cell Physiol.* **97**, 381–395.

Ernst, E. (1970). *Acta Biochem. Biophys. Acad. Sci. Hung.* **5**, 57–68.

Etzler, F. M., and Drost-Hansen, W. (1979). *In* "Cell-Associated Water" (W. Drost-Hansen and J. Clegg, eds.), pp. 125–144. Academic Press, New York.

Foter, M. J., and Rahn, O. (1936). *J. Bacteriol.* **32**, 485–487.

Fricke, H., and Jacobsen, L. E. (1939). *J. Phys. Chem.* **43**, 781–785.

Gerson, D. F. (1979). *In* "Cell Cycle Regulation" (J. R. Jeter, I. L. Cameron, G. M. Padilla, and A. M. Zimmerman, eds.), pp. 105–122. Academic Press, New York.

Gerson, D. F., and Burton, A. C. (1977). *J. Cell Physiol.* **91**, 297–304.

Gey, G. O., Coffman, W. D., and Kubicek, M. T. (1952). *Cancer Res.* **12**, 264–269.

Giese, A. C. (1957). "Cell Physiology." Saunders, Philadelphia.

Goldman, D. E. (1943). *J. Gen. Physiol.* **27**, 37–39.

Gortner, R. A. (1932). *Annu. Rev. Biochem.* **1**, 21–54.

Greenfield (1942). *J. Gen. Physiol.* **26**, 44–49.

Haak, R. A., Kleinhaus, F. W., and Ochs, S. (1976). *J. Physiol.* **263**, 115–121.

Haran, N., Malik, Z., and Lapidot, A. (1979). *Proc. Natl. Acad. Sci. U.S.A.* **76**, 3363–3366.

Hazlewood, C. F., Nichols, B. L., and Chamberlain, N. F. (1969). *Nature (London)* **222**, 747–750.

Hazlewood, C. F. (1973). *Ann. N. Y. Acad. Sci.* **204**, 593–606.

Hazlewood, C. F. (1979). *In* "Cell-Associated Water" (W. Drost-Hansen and J. Clegg, eds.), pp. 165–260. Academic Press, New York.

Hodgkin, A. L., and Huxley, A. F. (1939). *Nature (London)* **144**, 710–715.

Hodgkin, A. L., and Katz, S. R. (1949). *J. Physiol. (London)* **108**, 37–45.

Hoeve, C. A. J., and Kakivaja, S. R. (1976). *J. Phys. Chem.* **80**, 745–770.

House, C. R. (1974). "Water Transport in Cells and Tissues." Williams & Wilkins, Baltimore, Maryland.

Jungreis, A. M., Hodges, T. K., Kleinzeller, A., and Schultz, S. G. (1977). "Water Relations in Membrane Transport in Plants and Animals." Academic Press, New York.

Ling, G. N. (1962). "Physical Theory of the Living State." Blaisdell, Philadelphia, Pennsylvania.

Ling, G. N., and Ochsenfeld, M. M. (1965). *Biophys. J.* **5**, 77–84.

Ling, G. N. (1979). *In* "Cell-Associated Water" (W. Drost-Hansen and J. Clegg, eds.), pp. 261–270. Academic Press, New York.

Loritz, F., Bernstein, A., and Miller, R. G. (1977). *J. Cell Physiol.* **90**, 423–437.

Lucke, B., and McCutcheon, M. (1952). *Physiol. Rev.* **12**, 68–70.

Miller, C., and Ling, G. N. (1970). *Physiol. Chem. Phys.* **2**, 495–524.

Minkoff, L., and Damadian, R. (1976). *Physiol. Chem. Phys.* **8**, 349–387.

Odeblad, E., Bahr, B. N., and Lindstrom, G. (1956). *Arch. Biochem. Biophys.* **63**, 221–225.

Olmstead, E. G. (1966). "Mammalian Cell Water." Lea & Febiger, Philadelphia, Pennsylvania.

Pederson, T., and Robins, E. (1972). *J. Cell Biol.* **55**, 322–326.

Ponder, E. (1948). "Hemolysis and Related Phenomena." Grune & Stratton, New York.

Prescott, D. M., Myerson, D., and Wallace, J. (1972). *Exp. Cell Res.* **71**, 480–487.

Rao, P. N. (1968). *Science* **160**, 774–776.

Sachs, F., and Latorre, R. (1974). *Biophys. J.* **14**, 316–319.

Schwan, H. P., and Foster, K. R. (1977). *Biophys. J.* **17**, 193–201.

Scott, R. E., Carter, R. L., and Kidwell, W. R. (1971). *Nature (London)* **233**, 219–220.

Slayman, C. L. (1977). *In* "Water Relations in Membrane Transport in Plants and Animals" (A. M. Jungreis and T. K. Hodges, eds.), pp. 69–86. Academic Press, New York.

Sponsler, O. L., and Bath, J. D. (1942). *In* "The Structure of Protoplasm" (W. Seifiz, ed.), pp. 22–24. Iowa State College Press, Ames, Iowa.

Swift, T. J., and Fritz, O. G. (1969). *Biophys. J.* **9,** 54–59.
Troshin, A. B. (1966). "Problems of Cell Permeability." Permagon, Oxford.
Walter, J. A., and Hope, A. B. (1960). *Aust. J. Biol. Sci.* **24,** 497–502.
Wilson, W. L., and Heilbrunn, L. V. (1960). *Q. J. Microsc. Sci.* **101,** 95–98.
Wolosewick, J. J., and Porter, K. R. (1976). *Am. J. Anat.* **147,** 303–323.
Zaner, K. (1973). Doctoral Thesis, unpublished data quoted in Damadian (1973a).
Zimmerman, J. R., and Britton, W. E. (1957). *J. Phys. Chem.* **61,** 1328–1334.

# 9

# Regulation of Cell Reproduction in Normal and Cancer Cells: The Role of Na, Mg, Cl, K, and Ca

I. L. CAMERON, N. K. R. SMITH, T. B. POOL, B. G. GRUBBS, R. L. SPARKS, and J. R. JETER, JR.

**249**

NUCLEAR–CYTOPLASMIC INTERACTIONS
IN THE CELL CYCLE

## I.  INTRODUCTION

A recurring idea stemming from the late nineteenth and early twentieth century is that inorganic substances or small ions govern the processes of growth, embryonic induction, cell differentiation, and cancer. As evidence for this idea, Beebe in 1905 showed that the ratio of potassium (K) concentration over sodium (Na) concentration is less than one for cancerous and embryonic tissues but that this ratio is greater than one for most mature animal tissues (Beebe, 1905). Our recent electron microprobe X-ray data on the intracellular ratio of K/Na concentration in tumor and nontumor tissues show similar results to this early report (see below).

Considerable interest in the regulatory role of inorganic ions continued until World War II when general interest in inorganic ions seemed to decline for a time. This decline of interest in the subject of inorganic ions as regulators of cell processes was in part due to a shift in interest to macromolecules with their greater diversity. Such diversity of macromolecular species, such as the specific enzymes, allowed for a more diverse and precise regulation of differential cell function, while it was not clear how a small number of ion species could exert such diverse and precise regulation of differential cell function.

Interest in the role of small inorganic substances or small ions in controlling cell reproduction and in tumor cells has increased over the last 10–15 years. It now seems reasonable to suggest that changes in the intracellular concentration of inorganic substances and small ions can cause a cascade of biochemical, physiological, and morphological events leading to a coordinated and appropriate response on the part of the cell without the need to control each and every event in the cell's response. In other words, the inorganic substances and small ions may act to trigger and generally coordinate cellular response, while enzymes and the like may act to fine tune the specific steps in the cell's response.

Before continuing, a note on the use of terms such as ion, element, and electrolyte seems appropriate. Because techniques for the analysis of the intracellular state of specific ions or elements are increasingly available, the terminology used should be as precise as possible. These newer methods for analysis of the state of elements include: ion-selective electrodes (Hinke, 1969; Palmer *et al.*, 1978; Palmer and Civan, 1977; Gupta *et al.*, 1978), nmr spectroscopy (Czlisler and Smith, 1973; Ling and Cope, 1969), and X-ray absorption edge spectroscopy (Huang *et al.*, 1979). These new methods for determining the chemical state of ions (such as Na, Mg, Cl, K, and Ca) demonstrate that elements are not always in their free ionic state in cells. Intracellular ions are said to be free when their chemical state is similar to that in a water solution at equivalent ionic strengths; otherwise they are regarded as bound. Whether or not an ion is free or bound will certainly influence its cellular function. For these reasons, it is important to be as precise in terminology as possible. Without knowing the state of ions and their transmembrane concentration gradients, one cannot make accurate calculations of a cell's membrane potential using Goldman's constant field equation (1943). Thus, the activity coefficients of intracellular and extracellular ions, not the total element concentration values, are required for use in this constant field equation in order to calculate a cell's membrane potential.

## II. METHODS TO MEASURE INTRACELLULAR CONCENTRATION OF ELEMENTS

From 1931 to the early 1950s the gravimetric methods, for example, using zinc uranyl acetate precipitation for measurement of sodium, were tedious and difficult. Before and during this period, the titrimetric methods for measurement of chloride were substantially simpler than the method for sodium. After 1950, the introduction of the flame photometer reversed the situation and made Na easier to measure than Cl. ·

Flame photometry is the method now most commonly used for small cation (i.e., Na, Mg, K, Ca) analysis. Preparation for analysis involves washing the cells several times, using centrifugation and resuspension in an appropriate wash solution which is often free of the specific element(s) to be measured. Cell water content is determined by centrifuging the cell suspension at 1500 g for 4 minutes in preweighed plastic tubes. The wet weight of the packed cells is measured and the cells are then dried for a day at 100°C. Water content is the calculated difference between wet weight (corrected for extracellular water in the packed cells) and dry weight. The correction for extracellular water is usually determined by employing a nonpermeable, nonmetabolizable radioactive measurable marker of the extracellular space, such as [14C]inulin. The element content is then determined by flame photometry. Data can be expressed in millimoles per kilogram dry weight or per liter of cell water.

If cell number is known, the data can also be expressed on a per cell basis.

This method is subject to potential criticisms or questions such as: does the washing and handling of the cells perturb the element concentration determinations; what portion of the cell water remains after the drying procedure; how accurate or reliable is the measure of extracellular space? The above method is sometimes attempted on subcellular fractions such as nuclei, mitochondria, etc., to get at the element concentration in subcellular compartments. The major criticism of this approach is the question of translocation of elements during the cell fractionation and isolation procedures (Jones *et al.*, 1979). One must remember that membranes are permeable to Na (certainly less so than to other elements, but nevertheless, *are* permeable). Consider the following mobilities of ions in aqueous systems: Na = 5.2, K = 7.6, and Cl = 7.9 $\mu$m/sec/V/cm, respectively. These are absolute mobilities for ions in pure water with a field of 1 cm and a field strength of 1 V. When adjusted for conditions found in cytosol, mobilities are reduced to approximately 1–2 $\mu$m/sec. It can be seen then that these ions are far from being static and it becomes impossible (even using nonaqueous solvents) to isolate fast enough to assay without major translocations or redistributions. So fractionation procedures are far from ideal for determining element concentrations at a subcellular level. Additionally, these methods are only valid for pure populations of cells. This greatly restricts the types of systems that can be approached.

A summary of several methods for measuring intracellular calcium has recently been reviewed by Caswell (1979). The methods include metallochromic dyes, luminescent proteins, fluorescent chelate probe, radioautography, and electron microprobe procedures. Proton and ion microprobe methods are just being developed for measurement of elements in cells.

The introduction of electron excitation of characteristic X rays by an electron probe microanalysis system now allows spatial resolution in the micron and submicron range for appropriately prepared tissue sections. The advantages of energy dispersive X-ray microanalysis are as follows: it can measure cells as they exist *in situ;* the method is both qualitative and quantitative; one can be sure of the cell type since it is an analytical–morphological procedure; values can be readily obtained for the various compartments of a single cell, therefore, the technique does not necessarily have to rely on large averages of cytoplasm versus large averages for nuclei, etc.; and multiple elemental data are retrieved during one analysis. The method is valid but is worthless unless the proper methods of tissue preparation are employed. Liquefied propane is an excellent freezing fluid for cells and small pieces of tissues because its freezing rate is rapid enough to prevent large ice crystals from forming and to prevent translocation of diffusible materials (Brown *et al.*, 1969).

## III. SUMMARY AND EVALUATION OF THEORIES ON THE ROLE OF NA, MG, CL, K, AND CA IN REGULATION OF CELL REPRODUCTION IN NORMAL AND CANCER CELLS

### A. Membrane Cation Transport and the Regulation of Cell Reproduction

Pardee (1964) postulated that the plasma membrane could regulate cell reproduction. He suggested that cell membranes can exert selective control on the entry of molecules which themselves would be stimulatory or inhibitory to cell growth. Thus, an irreversible change in the plasma membrane permeability properties could lead to initiation of cell reproduction and transformation of the cell to a neoplastic state with independence from organismal controls that regulate cell reproduction of normal cells.

It follows from the above hypothesis that to understand how cell reproduction is controlled one must learn what triggers cell reproduction and that to understand neoplasia one must learn what control mechanism has failed.

One suggestion of how the cell membrane could regulate cell growth and reproduction comes from the finding that transport of certain amino acids or other nutrients essential for growth is coupled to the cotransport of $Na^+$ via the $Na^+$-$K^+$-ATPase membrane pump system (Schultz and Curran, 1970). Such an energy-requiring ion pump in the cell membrane is thought to be responsible for the asymmetric distribution of $Na^+$ and $K^+$ between the inside and outside of the cell. An electrochemical gradient can serve to move glucose, amino acids, or other nutrients into the cell when it is coupled to the active transport of ions like $Na^+$ and $K^+$. One has only to assume that nutrients are limiting to cell growth to see how growth can be controlled by such an ion transport mechanism (Holley, 1972).

Accordingly, the cell membrane potential (also called the transmembrane potential) is seen by some as providing most of the energy to drive the transport of nutrients in the $Na^+$-dependent transport systems. Some recent proponents of this membrane potential gradient hypothesis cite as key evidence two reports which purport to show membrane hyperpolarization when quiescent cells are stimulated to divide (Villereal and Cook, 1977, 1978). In fact, membrane potential was not directly measured in either of these two reports but was calculated from data on the accumulation of the nonmetabolizable amino acid, $\alpha$-aminoisobutyric acid, in fibroblasts in the presence of different concentrations of the ionophore, valinomycin. On the other hand, the direct measurements of membrane potential and simultaneous measurements of mitotic activity and cell density in both 3T3 and Chinese hamster ovary cell cultures reveal a 5- to 6-fold increase in the membrane potential when cells progress from the mitotically active logarithmic growth phase to mitotic arrest at saturation cell densities (Cone and Tongier, 1973). Thus, the data on hyperpolarization of the cell membrane in rapidly dividing cells seem in controversy.

Again, according to the membrane potential gradient hypothesis for nutrient accumulation by cells, the driving force for nutrient transport is the magnitude of the electrochemical gradient across the cell membrane. Thus, the transport of nutrient would be sensitive to the $Na^+$ electrochemical activity as reflected by the intracellular $Na^+$ concentration or membrane potential. If an increased membrane potential or $Na^+$ concentration gradient is causally related to the neoplastic process, one might predict that rapidly dividing tumor cells would have a hyperpolarized membrane and/or a large $Na^+$ concentration gradient compared to their normal cell types of origin. A review of the literature on the membrane potential of paired mammalian normal and tumor cell types measured *in situ* is summarized in Table I. The data presented in Table I make it clear that a hypopolarized and not a hyperpolarized membrane is a characteristic feature of rapidly proliferating neoplastic cells. Although the rapidly dividing tumor cells *in vivo* have a hypopolarized membrane, it does not seem wise to extrapolate this ob-

Table I    Membrane Potential ($-mV$) of Paired Mammalian Normal and Tumor Cell Types *in Situ*

| Cell type | Normal | Tumor | Reference |
|---|---|---|---|
| Liver | $-51$ | $-44$ | Limberger, 1963 |
| Muscle | $-89$ | $-16$ | Balitsky and Shuba, 1964 |
| Gastric | $-22$ | $-15$ | Kanno and Masui, 1968 |
| Thyroid[a] | $-47$ | $-23$ | Jamakosmanovic and |
|  | $-39$ | $-23$ | Loewenstein, 1968 |
| Liver | $-37$ | $-20$ | Binggeli and Cameron (1980) |
| Fibroblasts | $-43$ | $-14$ | Binggeli and Cameron (1980) |

[a] Measured in two different species.

servation to rapidly proliferating normal cells *in vivo* without experimental support.

It, therefore, seems that membrane hyperpolarization is not causally related to increased cell reproduction in the neoplastic cells but, on the contrary, membrane hyperpolarization appears to be a feature of the nontumor cells which demonstrate little if any cell reproduction *in vivo*. What then is the relationship between membrane potential, intracellular Na concentration, $Na^+$ flux, and nutrient transport in the regulation of cell reproduction and neoplasia? Shen *et al.* (1978) have attempted to answer this question by doing electrophysiological studies on the apical membrane ionic permeabilities of primary cell cultures of mouse mammary gland in midpregnant, preneoplastic, and neoplastic states. They report that $Na^+$ permeability increases with tumorigenesis but that $K^+$ and $Cl^-$ permeabilities were unchanged. The results suggest both an increased $Na^+$ permeability and an increased $Na^+,K^+$-ATPase activity in the transformed cells. These authors propose that an increased $Na^+$ permeability might lead to increased $Na^+,K^+$-ATPase activity. This increased $Na^+$ transport in the tumor cell could facilitate an increased nutrient transport and permit or cause growth and cell proliferation without a need for membrane hyperpolarization. The role of cation transport and cell reproduction as relates to lymphocyte stimulation has been reviewed by Kaplan (1978). He concludes that an increased $Na^+$-$K^+$-ATPase activity is also involved in stimulation of lymphocytes to proliferate but doubts that the uptake of several amino acids is caused by the increased $Na^+$-$K^+$-ATPase activity because the uptake of these amino acids was not inhibited by ouabain at concentrations of 5 m$M$ (van den Berg, 1974).

Cone (1969, 1971) and co-workers have put forth a theory that the cell surface acts to exert control over intracellular events leading to cell reproduction. The theory correlates cell reproductive activity with ionic concentrations associated with membrane potential. His "unified theory" states that "changes in the intracellular ionic concentration levels resulting from changes induced in active and passive ion transport through the plasma membrane by various surface conditions, are hypothesized to constitute a basic controlling influence by modulating, either directly or indirectly, one or more key metabolic events required for the initiation of mitogenesis" (Cone and Tongier, 1973).

Cone's main support for his theory is that the lowering of membrane potential will initiate the mitotic process. His theory fits with the data reported in Table I. In a test of his theory, nondividing (postmitotic) neurons in culture were treated with ouabain (a poison to the $Na^+,K^+$-

ATPase membrane pump) and an increase in $Na^+$ in the culture medium to cause a sustained depolarization of the neurons (Stillwell *et al.*, 1973; Cone and Cone, 1976). The membrane of the neurons was found to hypopolarize from –60 m V to –12 m V under these conditions, and the treatment did result in initiation of DNA synthesis and mitosis; however, cytokinesis did not occur, which consequently resulted in the production of binucleate neurons. Cone's ideas and experiments support the contention that the intracellular $Na^+$ concentration controls cell reproduction. Accordingly, high intracellular $Na^+$ concentrations are mitogenic. Any treatment which causes a sustained high intracellular $Na^+$ concentration (by changing the $Na^+,K^+$-ATPase pump activity or in any other way which leads to high intracellular $Na^+$ concentration) will initiate and sustain cell reproduction. The conclusion of Kaplan (1978) that the activation and the continuous function of the $Na^+,K^+$-ATPase pump was an absolute requirement for initiation and maintenance of the activated proliferative state of lymphocytes (a conclusion based mainly on use of ouabain at concentrations of $10^{-4}$ $M$ which strongly inhibits the $Na^+$-$K^+$-ATPase pump) might appear to be in conflict with the experimental use of ouabain at concentrations of $10^{-5}$ and $10^{-6}$ $M$ to initiate cell proliferation in nonproliferating neurons (Cone and Cone, 1976). Perhaps it is not the increased activity of the $Na^+,K^+$-ATPase pump itself which initiates and maintains cell reproduction but a change in the intracellular environment, which can be influenced by "the pump." The data of Stillwell *et al.* (1973) and Cone and Cone (1976) on the ouabain stimulation of DNA synthesis and mitosis in neurons suggest that this may be the case for the following reasons: (1) the effective concentration of ouabain could be reduced from $10^{-5}$ $M$ to $10^{-6}$ $M$ when the medium was supplemented with Na Cl: (2) both veratridine and the ionophore gramicidin were also effective in stimulating the neurons; and (3) the only basic commonality of action of these different treatments is to effect an increase in intracellular Na. Thus, an increase in intracellular Na seems to be directly or indirectly related to the cell proliferation state.

Cone's "unified theory" on control of cell proliferation is an extension of classical membrane theory derived from dilute-solution theory. For this reason Cone's theory and other theories based on membrane transport have been challenged by those who do not believe that the assumptions used in the derivation of the classical membrane theory obtain for cells (Hazlewood, 1972, 1980). Briefly, the criticisms of the classical membrane theory are: (1) that the energy requirements to operate membrane pumps are greater than the energy available to the cell; (2) that the cytoplasm is not a dilute solution; (3) that a large proportion of cellular Na and K are adsorbed or "bound"; (4) that membrane potential is not a function of extracellular K concentration as it should be according to theory; and (5) that the cell membrane is not rate-limiting to the diffusion of water or Na.

Although Hazlewood criticizes the classical membrane theory basis of Cone's unified theory, he does agree with Cone that an increase in the concentration of Na in the cytoplasmic water will increase the chances of cell reproduction. Hazlewood (1980) has put forth his own theory which states that the concentration of Na in cell water is regulatory to cell reproduction. Hazlewood's theory differs from Cone's theory in that, instead of the cell membrane, the tertiary structure of cellular macromolecules, like proteins and nucleic acids, are postulated as the sites of regulation of intracellular Na content. According to Hazlewood, change in the tertiary structure of a macromolecule (brought about by association of ions, drugs, hormones, ATP, carcinogens, or metabolites) "induces long-range changes in the macromolecules which, in turn, alter the physical state of the cytoplasmic water favoring the accumulation of sodium in the cytoplasmic water." In order to test Hazlewood's theory, more experiments which measure the physical state of ions and water will be required. Such measurements are becoming more available with development of modern equipment and technology.

## B.   Other Theories and Observations on the Role of Na, Mg, K, and Ca in Regulation of Cell Reproduction

Withdrawal followed by the appropriately timed replacement of any specific element essential to a cell will influence and control the cell's metabolism. Thus, any element can be considered the most important if it is the sole and limiting element to a cell's growth. This fact is sometimes forgotten in our quest to find *the* most important element or ion controlling initiation of cell reproduction.

With this note of caution in mind, two recent reports are discussed which support the idea that the divalent cations, $Ca^{2+}$ and $Mg^{2+}$, have a major regulatory role in the initiation of cell replication (Balk *et al.*, 1979; Rubin *et al.*, 1979). Balk *et al.* suggest that either the failure of cellular divalent cation homeostasis or perhaps a bypass of a divalent cation-dependent initiation mechanism is involved in the neoplastic state. Their ideas are based on the following experimental findings: Normal chicken fibroblasts and Rous sarcoma virus-infected chicken fibroblasts proliferate rapidly in normal medium containing Ca (1.2 m$M$) and Mg (0.7 m$M$). Reduction of Ca to 0.125 m$M$ decreases cell reproduction only of the normal fibroblasts but not the virus-transformed fibroblasts. Likewise, reduction of Mg to 0.05 m$M$ had a similar effect. When both Mg and Ca were reduced to 0.05 m$M$ and 0.20 m$M$, respectively, the normal fibroblasts stopped cell reproduction but the neoplastic fibroblasts continued to proliferate. These findings suggested to them that cell reproduction is initiated by $Mg^{2+}$ and $Ca^{2+}$. They do not indicate if the neoplastic fibroblasts' divalent cation homeostatis mechanisms have failed or if there has been a bypass of the divalent cation-dependent initiation mechanism. Balk *et al.* (1979) speculate that the virus in the neoplastic fibroblasts may code for a transformation-specific protein such as the src protein $pp60^{src}$ (Brugge *et al.*, 1979), which may be a protein independent of divalent cation control and may substitute for a normal host cell protein such as the sarc protein $pp60^{sarc}$ whose activity might be dependent upon divalent cation. Rubin *et al.* (1979) suggest from their experiments that $Mg^{2+}$ is more directly involved than $Ca^{2+}$ in the regulation of cell reproduction, because protein synthesis is very sensitive to small changes in intracellular $Mg^{2+}$ in the physiological range.

Other cations such as $Na^+$ and $K^+$ have not been as extensively studied as $Ca^{2+}$ and $Mg^{2+}$. There is, however, some evidence of a regulatory role in cell reproduction for $K^+$ (Gunther and Averdunk, 1970; Lubin, 1976; Cameron *et al.*, 1979) and for $Na^+$ (Shank and Smith, 1976).

## IV. INTRACELLULAR ELEMENT CONCENTRATION CHANGES ASSOCIATED WITH REGULATION OF CELL REPRODUCTION: ENERGY DISPERSIVE X-RAY MICROANALYSIS STUDIES OF CELL POPULATIONS *IN VIVO*

Our group is collecting data on the relations between cell reproduction and the intracellular concentration of several elements. The electron microprobe has provided the opportunity to measure the concentration of several elements simultaneously at the subcellular level (nucleus and cytoplasm). The procedures currently used to prepare tissues permit measurement of diffusible and nondiffusible tissue elements without major translocation.

The types of studies we have done are summarized below and include: (1) stimulation of quiescent cells to enter the cell cycle (for example, estrogen stimulation of vaginal epithelium); (2) transition from a mitotically active to a mitotically inactive state (for example, postnatal changes in cardiac myocytes and fibroblasts as they reach confluence in culture); (3)

element concentration differences between paired tumor and nontumor cell types and between rapidly and slowly dividing cell types; and (4) element changes and redistributions in cells during the cell cycle (for example, redistribution of elements at mitosis of an intestinal crypt cell and cell cycle changes in the cytoplasm and chromatin of the acellular slime mold *Physarum*).

All of our microprobe studies have used the same procedures. Briefly, we kill our animals by decapitation to avoid potential element redistribution problems associated with killing with a general anesthetic. In less than 1 minute, 1–3 mm³ pieces of the desired tissue are removed from the animal, mounted on a stub using minced liver as an adhesive, and frozen at a rapid rate in liquid propane cooled in a liquid nitrogen bath. The frozen tissue is then cryosectioned at $-30°$ to $-40°C$. Two- to 4-$\mu$m thick sections are then freeze-dried overnight in a cryosorption apparatus. This procedure was originally designed for radioautographic localization of diffusible elements without translocation and has been validated (Brown *et al.*, 1969). The dried sections are placed across a 2-mm hole in a carbon planchet using graphite adhesive to tack the edges. The sections are then examined at 15 kV in a scanning electron microscope equipped with A Si(Li) X-ray detector. X-Ray pulse-height distribution is measured for selected areas of the cytoplasm or nucleus of specific cells. The X-ray energy spectra are subjected to a multiple least squares fitting program to deconvolute the spectra and to calculate element peak/continuum (P/C) ratios for each element of each spectrum. Quantification of data is done by referencing the P/C ratios to that of a series of standards consisting of salts in bovine serum albumin. It should be noted that the X-ray data are reported in millimoles per kilogram dry weight and not per liter of cell water.

## A. Stimulation of Quiescent Cells to Enter the Cell Cycle—Estrogen-Stimulated Vaginal Epithelium

The basal layer of epithelial cells in the vagina of ovariectomized rats was chosen for analysis because the reproduction of these cells is suppressed by decreasing the level of estrogen in the rat and can be specifically increased by the injection of estrogen (Cameron *et al.*, 1980). Microprobe measurements were made on the cytoplasm of quiescent cells (before estrogen injection) and at 2, 17, and 24 hours after estrogen injection. Mitotic figures were first noticed in the basal cell layer at 24 hours after estrogen stimulation. Sodium, P, S, and Cl concentrations all initially decreased after estrogen stimulation but returned to near the nonstimulated concentration at 24 hours (Table II). Potassium and Mg

**Table II**   Cytoplasmic Element Concentration Changes Following Estradiol Stimulation of the Basal Layer of Cells in the Vaginal Epithelium[a,b]

| Time after estradiol administration | Number cells measured | Na | Mg | P | S | Cl | K |
|---|---|---|---|---|---|---|---|
| 0 Hours | 20 | 574 | 45 | 554 | 233 | 297 | 220 |
| | | ±15 | ±3 | ±32 | ±5 | ±8 | ±5 |
| 2 Hours | 10 | 358 | 50 | 445 | 242 | 212 | 272 |
| | | ±14 | ±3 | ±13 | ±4' | ±6 | ±6 |
| 17 Hours | 21 | 285 | 54 | 458 | 156 | 227 | 344 |
| | | ±24 | ±4 | ±15 | ±4 | ±7 | ±10 |
| 24 Hours | 19 | 418 | 92 | 689 | 214 | 257 | 500 |
| | | ±20 | ±7 | ±14 | ±3 | ±3 | ±10 |
| Statistical analysis of variance | | | | | | | |
| F value | | 40.2 | 21.5 | 25.5 | 4.34 | 31.2 | 204.4 |
| Probability | | <0.001 | <0.001 | <0.001 | <0.001 | <0.001 | <0.001 |
| Critical difference between means[c] | | 61 | 14 | 65 | 18 | 19 | 26 |

[a] Adapted from Cameron *et al.* (1980b).

[b] Data expressed as millimoles per kilogram dry weight ±SE of the mean.

[c] Critical differences between means in each element column. Means in each column which differ by these values are significantly different at the $p < 0.05$ level of probability.

showed an early, substantial, and continued increase in concentration after estrogen stimulation. Thus, the marked increase in cytoplasmic concentration of K and Mg and the concomitantly decreased ratio of Na to K concentration are either directly involved in growth stimulation of these cells or are reflective of the stimulated growth state. Intracellular Na concentration, although somewhat decreased after stimulation, was still relatively high compared to the slowly or rapidly dividing normal cell populations listed in Table IV.

## B. Transition from a Mitotically Active to a Mitotically Inactive State—Postnatal Changes in Cardiac Myocytes and Changes as Fibroblasts Reach Confluence

Radioautographic studies (Klinge and Stocker, 1968; I. L. Cameron, unpublished) show that the proliferative activity of heart muscle cells is high shortly after birth in the rat and in the mouse but that this proliferative activity decreases rapidly thereafter. The proliferative activity ceases by 2 months after birth. Based on these observations (Dykes *et al.*, 1979), microprobe measurements were made on the cytoplasm and nucleus of cardiac myocytes from groups of mice killed at 2, 4, 8, 16, and 32 days of age and in adults 3 to 6 months old. A significant decrease with age was seen in the concentrations of Na, K, P, and Cl while the concentration of S increased significantly with age (Table III). The cytoplasm and the nucleus both showed the same general pattern of element concentration change with age. In this study, higher intracellular concentrations of Na, K, P, and Cl can be correlated with high proliferative activity shortly after birth and the decrease in concentration of Na, K, P, and Cl are correlated to a decrease in cell proliferation and the progressive cytodifferentiation of the cardiac myocytes.

Proll *et al.* (1979) have investigated the elemental concentration changes that occur as rapidly growing BALB C/3T3 fibroblasts become mitotically inactive as they reach confluence in culture. With the use of energy dispersive X-ray microanalysis, it was found that the intracellular concentrations of Na, P, K, S, Mg, and Cl were all significantly greater ($p < 0.001$) in subconfluent 3T3 cell (15,170 cells/cm$^2$) than in confluent 3T3 cells (76,000 cells/cm$^2$). Determinations of mitotic indices (4% of subconfluent and 0% of confluent cells) and labeling indices following a 45-minute pulse of $^3$H-TdR (56% of subconfluent and 24 of confluent cells) confirmed that mitosis and DNA synthesis were also greater in subconfluent versus confluent 3T3 cells. Elemental concentrations within any one cell were not significantly different between the nucleus and cytoplasm in either the subconfluent or confluent cells, thus indicating no preferential sequestering of elements in either cell compartment.

**Table III   Least Squares Linear Regression Analysis of Changes in Elemental Concentrations of the Mouse Cardiac Myocyte Cytoplasm from 2 Days to 64 Days of Age**[a,b]

| Parameter | Na | K | Cl | Mg | P | S |
|---|---|---|---|---|---|---|
| Slope[c] | −0.64 | −0.26 | −0.87 | −0.25 | −0.19 | 3.0 |
| Intercept at birth[c] | 200 | 470 | 190 | 67 | 470 | 310 |
| Correlation coefficient[c] | −0.88 | −0.82 | −0.96 | −0.77 | −0.93 | 0.91 |
| Student's $t$[d] | 3.78 | 2.85 | 6.80 | 2.39 | 5.25 | 4.30 |
| Probability significance of slope versus a slope of zero | <0.05 | <0.05 | <0.01 | NS | <0.01 | <0.02 |

[a] Data adapted from Dykes *et al.* (1979).

[b] Groups of four to five mice were killed and their cardiac myocytes microprobes at 2, 4, 8, 16, 32, and 64 days after birth. Data expressed in millimoles per kilogram dry weight per day.

[c] Rounded to two significant figures.

[d] Calculated against a slope of zero.

The decrease in elemental concentrations after 3T3 cells reach conflu-ence is 2.5-fold for Na, 4.3-fold for P, 3.1-fold for S, 1.5-fold for Cl, 6.6-fold for K, and 16.3-fold for Mg. These data, plus other studies done with transformed cells which characteristically have a rapid rate of pro-liferation, suggest that Na and perhaps other elements are involved in mitogenesis. We are currently studying SV40 virus-transformed 3T3 fibroblasts to determine the magnitude of elemental changes under the same conditions used in this investigation.

## C. Element Concentration Differences between Paired Tumor and Nontumor Cell Types and between Rapidly and Slowly Dividing Cell Types

Normal and tumor tissue cell types were classified into four categories: tumor cell types, nontumor counterparts of the tumor cell types, rapidly dividing cell types, and slowly dividing cell types (Smith *et al.,* 1978; Cameron *et al.,* 1979, 1980a). Examples of specific cell types which are included in each of the four categories are listed in Table IV. Energy dispersive X-ray microanalysis was done to determine element concen-tration (millimoles per kilogram dry weight) for Na, Mg, P, S, Cl, and K. Calcium was at the margin of detectability in some of these tissue cells and therefore was not included in the overall analysis. Table IV lists the intracellular elemental concentrations for the four categories of cell types studied. A statistical analysis of variance of the data is given at the bottom of the table.

Tumor cells and their nontumor counterparts showed significant dif-ferences. The Na and Cl concentrations of the tumor cells are markedly higher compared to the nontumor cells. Other elements show no ele-ment concentration difference between these two cell type categories. Rapidly dividing cells show significant increases in the concentration of five elements (Na, Mg, P, Cl, and K) compared to the slowly dividing cells. The magnitude of the Na and Cl concentration differences between the slowly and rapidly dividing cells was less than between the tumor cells and their nontumor counterparts. The nuclear and the cytoplasmic concentrations of both Na and Cl showed parallel changes in the indi-vidual cell types. The statistical analysis of the tumor cells and the rapidly dividing cells show several significant differences in concentra-tion of elements. Sodium and Cl are both shown to be in higher concen-tration in the tumor cells while Mg, P, and K are in higher concentration in the rapidly dividing cells.

As mentioned in the Introduction, in 1905 Beebe reported that the weight ratio of K/Na was less than one for tumor tissue and greater than one in most mature animal tissues. The intracellular weight ratio of

Table IV    Element Concentration Differences between Tumor and Nontumor Cells and between Rapidly and Slowly Dividing Cells[a,b]

| Group number and cell type category | Na | Mg | P | S | Cl | K |
|---|---|---|---|---|---|---|
| Group I, nontumor cells[c] | 138 | 44 | 495 | 173 | 142 | 339 |
| | (±11) | (±3) | (±28) | (±17) | (±9) | (±13) |
| Group II, tumor cells[d] | 451 | 44 | 527 | 215 | 329 | 348 |
| | (±6) | (±4) | (±30) | (±17) | (±17) | (±13) |
| $p$ value of difference between groups I and II | <0.001 | NS | NS | NS | <0.001 | NS |
| Group III, rapidly dividing cells[e] | 196 | 63 | 626 | 242 | 206 | 549 |
| | (±12) | (±3) | (±42) | (±30) | (±8) | (±15) |
| Group IV, slowly dividing cells[f] | 140 | 50 | 515 | 213 | 159 | 389 |
| | (±8) | (±3) | (±30) | (±22) | (±8) | (±21) |
| $p$ value of difference between groups III and IV | <0.001 | <0.005 | <0.10 | NS | <0.001 | <0.001 |
| $p$ value of difference between groups II and III | <0.005 | <0.005 | <0.10 | NS | <0.001 | <0.001 |

[a] Adapted from Cameron *et al.* (1980a).

[b] Rapidly dividing cell populations are defined as cell populations with a turnover time of 7 days or less (Cameron, 1971), or with a mitotic index of 1.5% or greater (Altman and Katz, 1976). Concentration in millmoles per kilogram dry weight ±SE.

[c] Nucleus and cytoplasm of rat and mouse hepatocytes, rat and mouse lactating mammary epithelium.

[d] Nucleus and cytoplasm of rat Morris hepatoma 7777, mouse H6 hepatoma, mouse mammary adenocarcinoma $C_3H$, rat mammary adenocarcinoma 13762NF.

[e] Nucleus and cytoplasm of mouse duodenal interphase crypt cells, cardiac myocytes from 2- and 4-day-old mice, rat colon crypt cells, nucleus of rat cortical thymocytes.

[f] Nucleus and cytoplasm of cardiac myocytes from 32- and 64-day-old mice, mouse and rat hepatocytes, mouse and rat lactating epithelium, rat medulary thymocytes and the cytoplasm of rat smooth muscle cells, rat and mouse pancreatic acinar cell (ergastoplasm), mouse transitional epithelium (bladder).

K/Na in tumor and nontumor cells can be calculated from data in Table IV. In agreement with Beebe's findings, the intracellular ratio of K/Na in the tumor cells is 0.77 versus 2.47 in the nontumor cells, 2.80 in the rapidly dividing normal cells and 2.78 in the slowly dividing normal cells.

A prime reason for undertaking this study on different categories of cell types was to test Cone's and Hazlewood's predictions that the intracellular Na concentration should be greater in transformed and rapidly dividing cell types than in nontransformed or slowly dividing cell types. Cone's unified theory should be applicable to rapidly dividing normal and rapidly dividing transformed cells. The data of Table IV show significant increases in the intracellular concentration of Na in both rapidly dividing normal and rapidly dividing transformed cells.

Thus, Cone's and Hazlewood's predictions about Na are substantiated. However, the large and significant difference in intracellular Na concentration between rapidly dividing normal and transformed cell types shows that although an increased intracellular concentration of Na is related to mitogenesis it is more dramatically related to neoplastic cell transformation than to rapid cell proliferation per se. Intracellular Cl concentrations show the same significant pattern as Na in Table IV.

Potassium in the cell is significantly higher in the rapidly dividing normal cells than in the rapidly dividing tumor cells or the other two groups of cells. Apparently an increased K concentration in the normal cells is in some way related to normal mitotic activity in the rapidly dividing cells but not directly related to mitotic activity in the rapidly dividing tumor cells. All of the tumor cells we have microprobed are known to be rapidly dividing cell populations. Cone's prediction that a decrease in the intracellular $K^+$ concentration would be associated with mitotic activity per se is not supported by our comparative data on rapidly dividing normal cells *in vivo*. Thus, our microprobe data on intracellular K concentration show another significant difference between rapidly dividing normal cells and rapidly dividing tumor cells, specifically that the tumor cells have lower $K^+$ concentration than the normal cells.

Magnesium is also in significantly higher concentration in the rapidly dividing normal cells than in the rapidly dividing tumor cells (Table IV). This finding is consistent with the data of Balk *et al.* (1979) that neoplastic cells have bypassed a divalent cation-dependent initiation mechanism or have a failure in divalent cation homeostasis mechanism.

Such a pattern of difference between the four categories of cell types has suggested to us some experiments that can be designed to control cell proliferation in the rapidly dividing tumor cells as compared to rapidly dividing normal cell *in vivo*. Such experimental protocols involve use of drugs and procedures which may differentially effect Na concentration in the different cell types. Awareness of clear-cut differences between normal and tumor cells may provide a unique and effective basis for drug therapy of cancer in the future (see below).

### D. Cell Cycle Changes—Redistribution of Elements at Mitosis and Chromatin and Cytoplasmic Changes in *Physarum*

A major advantage of the microprobe procedure for measurement of elements is the opportunity that this method offers for analyses of different subcellular compartments or structures within a cell. With this in mind, the cellular redistribution of elements was studied at the time of mitosis in the rapidly proliferating crypt cells of the mouse small intestine (Cameron *et al.,* 1979). Several significant elemental concentration differences were found between the interphase nucleus and the cytoplasm and the chromatin of late anaphase–early telophase stage of mitosis in the cryptal cells. Calcium, S, and Cl were all significantly concentrated in the mitotic chromatin when compared to the cytoplasm of the same cell or to the cytoplasm or nucleus of interphase cryptal cells. Calcium was concentrated 2.5- to 8-fold in the mitotic chromatin. Conversely, the late anaphase–early telophase mitotic chromatin had a much lower concentration of Na, Mg, and P than the cytoplasm. These data on P concentration were interpreted as a loss of P by dephosphorylation of chromosomal proteins at the late anaphase–early telophase stage of mitosis. It was speculated that the sequestering of Ca in the metabolically inactive mitotic chromatin might be directly involved in chromosome condensation and gene inactivation. At the same time, the sequestering of Ca in the chromatin would lower $Ca^{2+}$ in the environment of the chromosomes, which might permit microtubule assembly in the spindle apparatus. That changes in chromosome condensation which are associated with changes in the concentration of Na or K (or specific combinations of elements or ions) might regulate gene expression of chromatin has been presented by Kroeger (1963) and Lezzi and Roberts (1972). Further microprobe studies should help identify the role of Na, K, Mg, and Ca in regulation of chromosome condensation and expression of chromosomal genes.

The plasmodial myxomycete *Physarum polycephalum* is an organism well suited to cell cycle studies. All of the nuclei in a large plasmodium divide in synchrony every 8–10 hours. The element concentration in the cytoplasm and in the chromatin of this multinucleated giant cell have been analyzed by electron microprobe by sampling the plasmodium throughout the cell cycle (Jeter *et al.,* 1979). Significant differences were found in both the chromatin and the cytoplasm during the cell cycle. Briefly, the concentration of Na, Mg, P, S, K, and Ca increased about twofold at the time of mitosis in both the chromatin and the cytoplasm while the concentration of Cl showed little if any increase at the time of mitosis. Mitotic chromatin showed marked increases of both Mg and Ca as compared to the chromatin and the cytoplasm of the interphase plasmodium. The increase in divalent cations might be involved in chromosome condensation. The increase in concentration of several elements in the cytoplasm at the time of mitosis might possibly be explained by a loss of intracellular water at the time of mitosis. However, analysis of the wet and dry weight of the plasmodium during the cell cycle did not show any significant change in the water content of the plasmodium during the cell cycle. This leads one to conclude that *Physarum* shows a major pulsatile loss and gain of elements over the cell cycle. Such a major change in element concentration may be correlated with the striking change in the intracellular pH which occurs during the cell cycle of this cell (Gerson and Burton, 1977). These pH changes are cyclical in nature and have a periodicity which duplicates the mitotic cycle. The maximum pH of about 6.4 occurs at mitosis and a minimum pH of about 5.8 occurs at midinterphase, suggesting an efflux of $H^+$ at the time of mitosis. From Gerson and Burton's data it can be predicted that at the time when the intracellular pH is increasing there might also be an influx of positively charged ions into the cell. In agreement with this, our results show that at mitosis, when the pH is maximal and $H^+$ is minimal, the levels of several potential cations (Ca, Na, K, Mg) within the cell are significantly increased in the cytoplasm. This suggests an exchange of $H^+$ for other positively charged ions during the *Physarum* cell cycle. This change in element concentration during the unperturbed cell cycle of *Physarum* can also be correlated to a striking ultrastructural change which is seen as "a cytosolic phase separation" as revealed by transmission electron microscopy of the cytoplasm at the time of mitotic division. The element change can also be correlated to the striking cessation of plasmodial streaming which occurs at the time of mitotic division. Clearly, intracellular element concentration changes occur during the cell cycle of *Physarum*.

## V.  A NEW APPROACH TO REGULATION OF CELL REPRODUCTION IN NORMAL AND CANCER CELLS

A common observation from the microprobe studies done on *in vivo* cell populations as reported above is that rapidly dividing nontumor cells tend to have comparatively high intracellular Na and Cl concentrations of about 200 mmole/kg dry wt. in comparison to slowly dividing or nontumor cells with intracellular Na and Cl concentrations of about 145 mmole/kg dry wt. (Table IV). It also appears that tumor cells have especially elevated concentrations of Na and Cl, averaging 451 and 329 mmole/kg dry wt., respectively.

Assuming that Na and Cl concentrations are especially elevated in tumor cells, one might hypothesize that lowering intracellular Na levels in tumor cells might inhibit their rapid cell division. R. L. Sparks has initiated pilot studies to test this hypothesis in a rapidly dividing tumor cell population *in vivo* by administration of a drug, amiloride, which is purported to block passive diffusion of $Na^+$ into cells without effecting $Na^+,K^+$-ATPase and cause a subsequent decrease in intracellular $Na^+$ (Bentley, 1968; Baer *et al.*, 1967; Finn, 1976; Nagel and Dorge, 1970). Shen *et al.* (1978) have already shown that amiloride hyperpolarizes the membrane of mammary adenocarcinoma cells *in vitro* but not the membrane of mammary epithelium from preneoplastic nodules and mammary epithelium from midpregnant mice.

To determine the effects of amiloride *in vivo,* several preliminary experiments on male A/J mice have been done to determine nonlethal doses. In a series of three experiments, doses ranging from 0.1 to 10 $\mu$g/gm body wt. given in a series of five injections were not fatal. Amiloride was subsequently tested on male A/J mice bearing transplantable H6 hepatomas. Five groups, including saline controls, received injections of amiloride every 8 hours for 104 hours at 0.1, 1, 10, and 100 $\mu$g/gm body wt., respectively. Tumor measurements were made daily. Within 48 hours all the mice in the two highest dose groups had died. All animals at the 0.1 and 1.0 $\mu$g/gm doses received colchicine 3 hours prior to sacrifice in order to collect mitotic cells in metaphase. At sacrifice, liver, duodenum, and tumor were removed and processed for routine histology. Not enough tumor measurements were made on the two highest dose groups to include them in tumor growth rate analysis. It was found by one-way analysis of variance (ANOVA) of tumor growth rates that amiloride at 1 $\mu$g/gm (slope of tumor growth in square centimeters cross-sectional area per day was 0.109) suppressed growth compared to control (slope, 0.317; significant $p < 0.001$) and compared to 0.1 $\mu$g/gm (slope, 0.276; $p < 0.01$). Histological examination of the hepatoma to determine the colchicine mitotic index (MI) revealed a suppression at 1 $\mu$g/gm (MI = 0.05) compared to control (MI = 0.074) at the $p < 0.01$ level of significance, as determined by ANOVA. Analysis of the colchicine mitotic index in the duodenal crypts revealed no significant differences in mitotic index (duodenum was used as a control because it is a rapidly proliferating tissue). This indicates a differential drug effect on the tumor cells vs. rapidly dividing normal cells. Future microprobe analysis should enable determination of elemental changes due to amiloride in such experiments and allow measurements that will further substantiate the role of sodium and chlorine in cell proliferation.

Such studies should allow a broader understanding of the role of ions in the regulation of cell proliferation. It seems likely that methods such as amilioride treatment constitute a new and promising approach to the regulation of cell reproduction in cancer cells.

## VI.  CONCLUSIONS FROM ENERGY DISPERSIVE X-RAY MICROANALYSIS MEASUREMENTS OF THE CONCENTRATION OF ELEMENTS IN RELATION TO CELL REPRODUCTION IN NORMAL AND CANCER CELLS *IN VIVO*

All cell populations with a high rate of mitotic activity have in common an elevated concentration of Na and Cl compared to cell populations with a low rate of mitotic activity. Tumor cells with a high rate of mitotic activity have an especially elevated concentration of Na and of Cl.

An elevated concentration of K and of Mg is associated with the initiation and the maintenance of a high rate of mitotic activity in nontumor cell populations *in vivo;* however, an elevated intracellular concentration of K and of Mg is not necessary for the maintenance of a high rate of mitotic activity in tumor cell populations.

Therefore, an especially elevated intracellular concentration of Na and Cl combined with a relatively low concentration of K and Mg (in comparison to rapidly dividing nontumor cell populations) give tumor cell types a unique element concentration characteristic. More data on other normal and tumor cell populations are now needed to strengthen these conclusions.

Working under the assumption that our conclusions about element concentration differences between normal and tumor cells are valid and that these element concentration differences are involved in the regulation of cell reproduction, we have begun to think of ways to modify the intracellular concentration of the various elements in tumor-bearing animals. Thus, our aim is to change the intracellular concentration of rapidly dividing tumor cell types towards that of the nontumor cell types so as to lower their high rate of mitotic activity. One might, for example, consider ways of lowering the elevated concentration of Na or of Cl in tumor cells as a means of reducing their high rate of mitotic activity. Ideally, one would like to find a way to preferentially slow or stop the reproduction of tumor cells without slowing or stopping the reproduction of the rapidly dividing normal cell populations in the animal.

Specifically our current and initial approach is to lower the intracellular concentration of Na and Cl in the tumor cells by the systemic administration of a drug (amiloride) which is known to block the passive diffusion of $Na^+$ into cells *in vivo*. Pilot studies on animals with rapidly growing tumors show that amiloride significantly decreases the rate of mitotic activity of the tumor cells but does not change the rate of mitotic activity in rapidly dividing intestinal crypt cells *in vivo*. Simultaneous measurements of the concentration of the various elements in the normal and tumor cell types are underway to see if the predicted changes in element concentration have occurred due to amiloride treatment.

## REFERENCES

Altman, P. L., and Katz, D. D. (1976). *In* "Cell Biology". Fed. Am. Soc. Exp. Biol., Bethesda, Maryland.
Baer, J. E., Jones, C. B., Spitzer, S. A., and Russo, H. F. (1967). *J. Pharmacol. Exp. Ther.* **157**, 472–485.
Balitsky, K. P., and Shuba, E. P. (1964). *Acta Unio Int. Contra Cancrum* **20**, 1391.
Balk, S. D., Polimeni, P. I., Hook, B. S., Le Stourgeon, D. N., and Mitchell, R. S. (1979). *Proc. Natl. Acad. Sci. U.S.A.* **76**, 3913.
Beebe, (1905). *Am. J. Physiol.* **12**, 167.
Bentley, P. J. (1968). *J. Physiol.* **195**, 317–330.
Binggeli, R., and Cameron, I. L. (1980). *Cancer Res.* **40**, 1830.
Brown, D. A., Strumpf, W. E., and Roth, L. J. (1969). *J. Cell Sci.* **4**, 265.
Brugge, J. S., Collett, M. S., Siddiqui, A., Marezynska, B., Deinhardt, F., and Erikson, R. L. (1979). *J. Virol.* **29**, 1193.
Cameron, I. L. (1971). *In* "Cellular and Molecular Renewal in the Mammalian Body" (I. L. Cameron and J. D. Thrasher, eds.), pp. 45–85. Academic Press, New York.
Cameron, I. L., Smith, N. K. R., Pool, T. B., and Sparks, R. L. (1979). *J. Cell Biol.* **83**, 7a. (Abstr.)
Cameron, I. L., Smith, N. K. R., Pool, T. B., and Sparks, R. L. (1980a). *Cancer Res.* **40**, 1493.
Cameron, I. L., Pool, T. B., and Smith, N. K. R. (1980b). *J. Cell Physiol.* (in press).
Caswell, A. H. (1979). *Int. Rev. Cytol.* **56**, 145.
Cone, C. D. (1969). *Trans. N. Y. Acad. Sci.* **31**, 404.
Cone, C. D. (1971). *J. Theoret. Biol.* **30**, 151.
Cone, C. D., and Cone, C. M. (1976). *Science* **192**, 155.
Cone, C. D., and Tongier, M., Jr. (1973). *J. Cell Physiol.* **82**, 373.
Czeisler, J. L., and Smith, T. J. (1973). *Ann. N. Y. Acad. Sci.* **204**, 261.
Dykes, S. G., Cameron, I. L., and Smith, N. K. R. (1979). *J. Cell. Physiol.* **100**, 305.
Finn, A. L. (1976). *Physiol. Rev.* **56**, 453–464.
Gerson, D. F., and Burton, A. C. (1977). *J. Cell. Physiol.* **91**, 297.
Goldman, D. E. (1943). *J.Gen. Physiol.* **27**, 37.
Gunther, T., and Averdunk, R. (1970). *Z. Klin. Chem. Klin. Biochem.* **8**, 621.
Gupta, B. L., Berridge, M. J., Hall, T. A., and Moreton, R. B. (1978). *J. Exp. Biol.* **72**, 261.
Hazlewood, C. F. (1972). *Science* **177**, 815.
Hazlewood, C. F. (1980). *In* "Nuclear Resonance Effects in Cancer" (R. Damadian, ed.). Pacific Press, (in press).
Hinke, J. A. M. (1969). *In* "Glass Microelectrodes" (M. Lavallee, O. F. Schanne, and N. C. Hebert, eds.), pp. 349–375. Wiley, New York.
Holley, R. W. (1972). *Proc. Natl. Acad. Sci. U.S.A.* **69**, 2840.
Huang, H., Hunter, S. H., Warburton, W. K., and Moss, S. C. (1979). *Science* **204**, 191.
Jamakosmanovic, A., and Loewenstein, W. R. (1968). *J. Cell Biol.* **38**, 556.
Jeter, J. R., Jr., Smith, N. K. R., Wille, J. J., and Cameron, I. L. (1979). *J. Cell Biol.* **83**, 8a. (Abstr.).
Jones, R. T., Johnson, R. T., Gupta, B. L., and Hall, T. A. (1979). *J. Cell Sci.* **35**, 67.

Kanno, Y., and Matsui, Y. (1968). *Nature (London)* **218**, 775.

Kaplan, J. G. (1978). *Annu. Rev. Physiol.* **40**, 19.

Klinge, O., and Stocker, E. (1968). *Experientia* **24**, 269.

Kroeger, H. (1963). *Nature (London)* **200**, 1234.

Lezzi, M., and Roberts, M. (1972). *In* "Developmental Studies on Giant Chromosomes" (W. Beerman, ed.), pp. 35–57. Springer-Verlag, Berlin and New York.

Limberger, J. (1963). *Z. Krebsforsch.* **65**, 590.

Ling, G. N., and Cope, F. W. (1969). *Science* **163**, 1335.

Lubin, M. (1976). *Nature (London)* **213**, 451.

Nagel, W., and Dorge, A. (1970). *Pflügers Arch.* **317**, 84–92.

Palmer, L. G., and Civan, M. M. (1977). *J. Membrane Biol.* **33**, 41.

Palmer, L. G., Century, T. J., and Civan, M. (1978). *J. Membrane Biol.* **40**, 25.

Pardee, A. B. (1964). *Natl. Cancer Inst. Monogr.* **14**, 7.

Proll, M. A., Pool, T. B., and Smith, N. K. R. (1979). *J. Cell Biol.* **83**, 13a.

Rubin, A. H., Terasaki, M., and Sanui, H. (1979). *Proc. Natl. Acad. Sci. U.S.A.* **76**, 3917.

Schultz, J., and Curran, P.F. (1970). *Physiol. Rev.* **50**, 637.

Shank, B. B., and Smith, N. E. (1976). *J. Cell Physiol.* **87**, 377.

Shen, S. S., Hamamoto, S. T., Bern, H. A., and Steinhardt, R.A. (1978). *Cancer Res.* **38**, 1356.

Smith, N. R., Sparks, R. L., Pool, T. B., and Cameron, I. L. (1978). *Cancer Res.* **38**, 1952.

Stillwell, E. F., Cone, C. M., and Cone, C. D., Jr. (1973). *Nature (London) New Biol.* **246**, 110.

van den Berg, K. J. (1974). *In* "The Role of Amino Acids in the Mitogenic Activation of Lymphocytes", p. 103. Radiobiological Inst. TNO, Rysiwijk (ZH), Netherlands.

Villereal, M. L., and Cook, J. S. (1977). *J. Supramol. Struct.* **6**, 179.

Villereal, M. L., and Cook, J. S. (1978). *J. Biol. Chem.* **253**, 8257.

# 10

# The Significance and Regulation of Calcium during Mitotic Events

## JESSE E. SISKEN

## I. INTRODUCTION

Nucleocytoplasmic interactions are generally visualized as involving the transport of materials across the nuclear envelope. This interaction, typified by interphase cells, is quite different during the course of cell division. For example, except in organisms which undergo intranuclear division, the nuclear envelope is not present during cell division and no longer serves to partition the components of the nucleus from those of the cytoplasm. Second, genetic activity is greatly reduced during division—only some 4 and 5 S RNA is made during this period (Zylber and Penman, 1971)—so that there is a reduction in molecular interactions between chromatin and the remainder of the cell at this time. In

NUCLEAR–CYTOPLASMIC INTERACTIONS
IN THE CELL CYCLE

addition, the most obvious interaction between the genetic material and the rest of the cell during division is a physical one in the sense that (1) the mitotic or meiotic spindle, whose components are mainly, if not entirely, cytoplasmic in most organisms, functions to partition the chromosomes into two daughter nuclei, and (2) the cytokinetic mechanisms operate to separate the nuclei into two separate cells. This chapter is concerned with these physical interactions and deals mainly with karyokinesis and cytokinesis in animal cells.

Many studies have suggested that at least some of the complex events of mitosis involve the regulation of calcium ions (see, e.g., Mazia, 1961; Arnold, 1976; Rebhun *et al.*, 1976a; Hepler, 1977; Rebhun, 1977; Harris, 1978) which in itself is a very complex matter that has been the subject of a large amount of work and a vast literature. A discussion of this field is beyond the scope of this chapter but an inkling of the complexity of this topic can be gathered from the following points:

1. Mammalian cells live in a milieu containing about $10^{-3}$ $M$ calcium but maintain cytosolic concentrations of this ion in the range of $10^{-7}$ to $10^{-5}$ $M$ depending upon the state of the cell.

2. In various cells, calcium may be bound to the plasma membrane and is sequestered at high concentrations within membrane-bound organelles such as mitochondria, the perinuclear space, endoplasmic reticulum, and perhaps other populations of vesicular elements.

3. Movement of calcium between various pools and its effects in particular intracellular systems may be influenced by a number of factors including the electrical state of the plasma membrane, the activity of calcium pumps, cyclic nucleotide levels, ATP, and calcium-binding proteins.

These topics have been discussed in detail in a number of recent reviews and symposia (e.g., Rasmussen and Goodman, 1977; Scarpa and Carafoli, 1978; Bolton, 1979).

The aim of this chapter is to briefly discuss some work from our own and other laboratories which may provide further insight into the mechanisms of cell division, the role of calcium in this process, and the manner in which this process is regulated. These topics have previously been considered by others whose interests have centered around the mechanisms of movement per se, i.e., the application of force to move chromosomes during prometaphase and anaphase, the effects of calcium on microtubules, and the mechanisms of cytokinesis. The focus in this chapter will be somewhat different and will be divided into two parts. The first section will deal mainly with metaphase, a stage about which little is known, and to a small extent with anaphase. Some questions about the nature of metaphase will be raised and evidence reviewed which indicates (a) that metaphase is sensitive to agents which modify intracellular calcium levels, and (b) that for, as yet, unknown reasons, the duration of this stage is prolonged in many transformed cells.

The second part of this chapter will discuss the role of calcium ions in cytokinesis and review some quantitative data that support the idea of a role for calcium in the furrowing process and are consistent with the existence of a calcium pool associated with the plasma membrane.

## II. METAPHASE

### A. Visible Events of Metaphase

The known events of metaphase have been thoroughly reviewed by Mazia (1961) and more recently by Nicklas (1971) and McIntosh et al. (1975) so there is no need to do so here. However, a few points relevant to later discussion need to be made.

The metaphase cell by definition is one which contains a mitotic spindle with the kinetochores of the chromosomes aligned upon its equator. The chromosomes are not totally static on the metaphase plate but undergo oscillations along the long axis of the spindle. However, one sees no progression of movements during this period which might represent a continuous transition from the earlier events of prophase and prometaphase to the movements of anaphase. As stated by Mazia (1961), metaphase "... strikes us as an interruption of the flow of events during which the mitotic apparatus is waiting for something to happen! That visible event it is waiting for is the parting of the sister chromosomes. Much else may be happening." McIntosh et al. (1975) described it as a wait of variable duration. It is not an obligatory step in the mitotic process for all organisms (Mazia, 1977) since there are a number of cells in which this phase does not occur under normal (Belar, 1926 cited by Mazia, 1961; Heath, 1974) or experimental (Galinsky, 1949) conditions. In grasshopper neuroblasts, anaphase movement begins immediately after the last chromosome arrives on the spindle equator (Carlson, 1956), whereas in other organisms, there is a finite period between the alignment of chromosomes on the plate and the initiation of anaphase movement (see Zirkle, 1970). Thus, in cells in which metaphase does occur, one would suppose, as did Zirkle (1970), that it represents a period during which something must be happening by way of preparation for the onset of anaphase movement.

The major events that lead to the metaphase configuration include: (a) the replication of the chromosomes and their condensation into shorter, denser forms which are observably doubled; (b) the assembly of the mitotic apparatus including the separation of mitotic centers and the growth of microtubules; and (c) the movement of the chromosomes to the equatorial region of the mitotic spindle, i.e., the metakinetic movements. The events involved in its termination are discussed in the next section.

## B. Duration of Metaphase and Its Termination

By definition, metaphase begins when the chromosomes are aligned on the equator of the spindle and terminates when sister chromatids (during mitosis and meiosis II) or homologous chromosomes (during meiosis I) begin movement toward their respective poles. Some questions we would like to be able to answer about this stage are:

1. What determines how long a cell will remain in the metaphase configuration?
2. What is happening during this period?
3. What events occur to start the (usually) synchronous, poleward movement of the chromosomes?

As a point of departure for considering these questions, it should first be recognized that the mean duration of metaphase is relatively constant for a particular cell type under a given set of conditions. For example, Fig. 1 shows the duration of metaphase as a function of population doubling level for a line of human skin fibroblast cells which undergoes *in vitro* senescence of the type reported by Hayflick (1965) for WI-38 and since observed by many others. Although the generation times get longer and growth fractions get smaller in the later passages of such populations (Macieira-Coelho *et al.,* 1966; Cristofalo and Scharf, 1973; Bowman *et al.,* 1975; Mitsui and Schneider, 1976), the mean duration of metaphase in the cells which do divide remains essentially constant in the range between 10.5 and 11.5 minutes (Sisken and Bonner, 1979). HeLa cells also appear to have a fairly characteristic metaphase duration which is, however, about twice as long as that of the skin fibroblasts. Over the years we have measured the duration of metaphase in HeLa in many time lapse cinemicrographic films made as controls for a variety of experiments. As shown in Fig. 2a, the mean values for the individual films are distributed over a discrete range with most of them falling between 18 and 25 minutes. It should be noted that there is no overlap at all between the mean values for the HeLa cells and the fibroblasts.

The distribution of metaphase durations within populations also va-

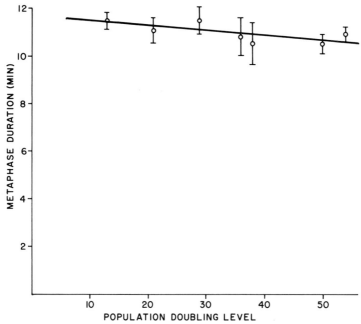

**Fig. 1.** The duration of metaphase in human skin fibroblasts as a function of population doubling level. The calculated regression coefficient ± SE is $-0.030 \pm 0.016$ with a $p >$ 0.10 (modified from Sisken and Bonner, 1978).

ries and in the specific population shown in Fig. 2b, the distribution is unimodal with some skewness towards the longer end of the distribution. Interestingly, the degree of skewness resembles that of generation times first reported for mammalian cells by Killander and Zetterberg (1965) and Sisken and Morasca (1965), but further work will be required to determine whether this is a consistent feature of dividing cells.

We have also measured the durations of metaphase in a number of other normal and transformed cells and have found that the mean duration of metaphase in the transformed cells we have studied is significantly longer than the metaphases in nontransformed cells in nearly all populations examined to date. To cite another example, the duration of metaphase in WI-38 cells was determined to be $7.6 \pm 0.48$ (SE) minutes while in SV40-transformed WI-38 cells grown under the same conditions it is $35.5 \pm 1.67$ minutes, i.e., approximately five times longer (Sisken and Bonner, 1978). The reasons for these differences are unclear but two possibilities are (1) that the transformed cells we have examined are mostly hyperploid and the duration of metaphase is, for some reason, longer when the number of chromosomes is increased, and

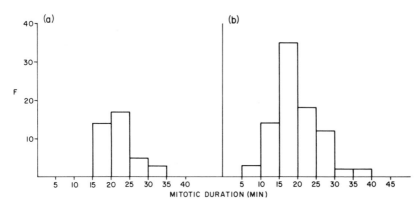

**Fig. 2.** a, Frequency distribution of average metaphase durations in HeLa cells obtained from individual control experiments performed over a period of 5 years. b, Frequency distribution of metaphase durations of 88 individual HeLa cells within a single experiment which yielded a mean (± SE) metaphase duration of 20.0 ± 0.86.

(2) that some metabolic or structural alterations which occur in cells as a result of transformation cause metaphase duration to be prolonged. It may be particularly pertinent in view of later discussion to note that calcium regulation and metabolism are often altered when cells become transformed (e.g., see Mikkelsen, 1978) and that the cytoskeletal network also appears to be reduced in some (Brinkley *et al.*, 1975) but not all tumor cells [see Asch *et al.* (1979) for discussion].

Further studies of differences in metaphase duration between normal and tumor cells and attempts to understand why the differences occur are now in progress in our laboratory. In any event, even though the duration of metaphase may vary somewhat within and between populations of a given cell type, the above data indicate that it is a functional characteristic of the cell and not simply a random variable. This being the case then, what is it that determines when metaphase terminates, i.e., when anaphase movement will begin?

The onset of anaphase movement involves two main events: The physical separation of the attached chromatids (or chromosomes) from each other and their poleward movement. A key question for our understanding of mitotic or meiotic events is whether these are coordinately regulated but separate occurrences or whether chromatid or chromosome separation per se are causally related to the onset of chromosome movements. Put another way, we may ask, as Mazia (1961) has done, whether the spindle is fully prepared to function from the beginning of metaphase and is only waiting for chromatid separation to occur or whether something else must occur during metaphase before the spindle is ready to function? (The following paragraphs pertain to both mitosis and meiosis but for ease of expression, I will refer only to mitosis.)

A detailed discussion of the various models of mitosis is not appropriate for this chapter but it may be said that according to many views of mitosis, the metaphase alignment is considered to be an equilibrium situation in which poleward tensions are exerted on the still attached chromatids (e.g., see Mazia, 1961; McIntosh *et al.*, 1969, 1975; Margolis *et al.*, 1978). That is, during prometaphase the chromosomes behave as though they are under the influence of attractive forces applied to the kinetochores from both poles and they become situated on the metaphase plate when these forces are equalized. As long as the chromatids remain bound to each other, the equilibrium persists and the spindle remains under tension. However, once the chromatids separate from each other, a phenomenon which is not dependent upon a functional spindle (see Mazia, 1961), the equilibrium is broken and the chromatids are free to begin movement toward their respective poles. According to this concept, it is a chromosomal event, separation of chromatids, which allows the spindle to carry out anaphase movements.

One piece of evidence in favor of this idea is the observation by Upcott (1939) that the asynchronous separation of chromatids in polyploid *Tulipa* species seems to allow some premature movement to the poles.

Izutzu's (1959) work, discussed below, is also cited as support for this hypothesis. The implication, then, is that the termination of metaphase is controlled by the mechanisms involved in separating the chromatids. As pointed out by Mazia (1961), one of the niceties of this idea is that all of the movements of mitosis, i.e., metakinesis, premetaphase stretch, metaphase chromosome oscillations, and anaphase can be explained on the basis of a single concept.

Whether this is all that is involved is open to some question. More recently, this concept was included in a model of mitosis developed by Margolis *et al.* (1978) which was derived from their observations that the native microtubule exists in a steady-state, opposite end, assembly/disassembly equilibrium (Margolis and Wilson, 1978). They also proposed, however, that a cessation of assembly of kinetochore-to-pole microtubules occurred simultaneously with chromosome separation, suggesting, therefore, that at least two events occur to initiate anaphase movement. Observations of Inoué and Ritter (1975) do, in fact, indicate that kinetochore microtubules begin disassembly at this same time.

Other arguments exist contrary to the idea that the initiation of anaphase movement requires only the physical separation of chromatids. For example, there is evidence that suggests that poleward tensions on chromosomes are actually lower during metaphase than during prometaphase and anaphase in both meiosis and mitosis (see review by Luykx, 1970). This conclusion is based mainly on observations that chromosomes are deformed to a greater extent by tensions at the kinetochore during prometaphase and anaphase than during metaphase.

Other evidence that the presumed equilibrium of metaphase is characterized by low tensions on the kinetochores comes from observations of chromosome movements after irradiation of kinetochores and spindle fibers of the chromosomes of grasshopper spermatocytes with microbeams of ultraviolet light. Izutzu (1959) observed that, following irradiation of the kinetochore of one chromosome of a bivalent during metaphase, there was a period of about 25 minutes during which the bivalent did not move. At the end of the time the bivalent began gradual movement in the direction of the spindle pole toward which the unirradiated kinetochore was oriented. Thus, the magnitude of any poleward tensions on these metaphase chromosomes must be small relative to the other stages. This does not imply that the spindle is static during metaphase since zones of reduced birefringence produced by such irradiation of spindle fibers (Forer, 1965) and small particles or aggregates of material associated with the spindle (Allen *et al.,* 1969) are observed to move poleward during the period. The observations of Forer (1966) that crane fly spermatocyte spindles responded to ultraviolet microbeam irradiation differently during metaphase than during anaphase also suggests that some fundamental changes must occur in the spindle during metaphase or at the metaphase–anaphase boundary. There is at least one other piece of evidence in the literature which suggests that chromosome or chromatid separation per se is not sufficient to permit or trigger spindle movements. Schrader (1936) showed that in *Amphiuma,* the first part of poleward movement of chromosomes is not led by the kinetochores. It also is not led by the ends of the chromosomes, which might indicate a type of neocentric activity that has been observed in some cases (Rhoades, 1952). According to Schrader (1936), the distal ends moved as rapidly as the kinetochores as though the early movement of anaphase was due to a mutual repulsion of the sister chromatids. Later in anaphase, the kinetochores again take the lead. Such observations would seem to argue that chromosome separation and spindle activation might be independent though normally simultaneous events. In the case of *Amphiuma,* chromosome separation seems to occur prior to spindle activation.

With respect to the question of what is happening during metaphase, although there is little morphological activity occurring at this time, it is not a metabolically inert period. In addition to the reduction in RNA synthesis mentioned above, the rate of protein synthesis is decreased and the level of free polysomes is reduced but significant protein synthesis still occurs [see reviews by Prescott (1976) and Baserga (1976) for further details]. Energy is required by the cell during this period since treatment with inhibitors of respiration can block cells in metaphase (Epel, 1963; Tobey *et al.*, 1969). In addition, the duration of metaphase in cultured mammalian cells is as sensitive to low and high temperatures as any stage of the mitotic cycle and more sensitive than anaphase (Sisken *et al.*, 1965). It could be that abnormal temperatures affect some rate-limiting metabolic step or perhaps the assembly–disassembly equilibrium of microtubules (Fuseler,1975).

There are a number of other metabolic systems which differ between mitotic cells and those in interphase, including changes in the cell surface, calcium ion content, and cyclic nucleotide levels. For example, Clothier and Timourian (1972) described several peaks in the rate of uptake of calcium into dividing sea urchin eggs, one of which was during metaphase, and Huot *et al.* (1976) reported maximal levels of calcium in HeLa cells during mitosis. Further, a peak of a $Ca^{2+}$-dependent ATPase activity occurs during mitosis in sea urchin eggs (Petzelt, 1972) and in mouse mastocytoma cells (Petzelt and Auel, 1977) with the peak occurring during metaphase in the sea urchin (Petzelt, 1972).

Observations also indicate that cAMP levels are generally at their lowest in the cell cycle during division (Burger *et al.*, 1972; Millis *et al.*, 1972; Sheppard and Prescott, 1972); of particular interest are reports that in Novikoff hepatoma cells (Zeilig and Goldberg, 1977) and cleaving *Mactra solidissima* eggs (Geilenkirchen *et al.*, 1977), there is a decrease in cAMP and an increase in cGMP at about the time of onset of anaphase movement. Thus, some interesting metabolic events take place during metaphase and since calcium and cyclic nucleotide metabolism are closely linked (see e.g., Rasmussen and Goodman, 1977) and there are dynamic changes in intracellular levels of these substances during the course of mitosis, it is reasonable to think that they might be involved in the regulation of some aspects of this process (but see Rebhun, 1977).

Taken together, then, these findings indicate there is more to the initiation of chromosome movement than a simple separation of the chromatids and that metaphase is more complex than a tug of war between two tractors pulling on a pair of joined plates whose glue is about to come apart. They also suggest that calcium ions and/or cyclic nucleotides may somehow be involved in the onset of anaphase movement.

## C.  Effects of Calcium on the Duration of Metaphase and Anaphase

### 1.  General Considerations

As indicated earlier, there has been considerable interest in the role of calcium ions in mitosis for a long time but the most recent surge derives in part from two sets of findings. The first is the observation by Weisenberg (1972) that brain tubulin can be made to polymerize *in vitro* only when $Ca^{2+}$ levels are maintained at a low level. This was followed by numerous other experiments which have demonstrated that calcium can inhibit the assembly of microtubules (see Bryan, 1975; Olmsted, 1976; Kirschner, 1978 for reviews) and that the experimental manipulation of intracellular calcium levels appears to affect the integrity of the microtubules of the mitotic spindle (Rebhun *et al.*, 1975; Kiehart and Inoué, 1976)—the cytoplasmic microtubule complex (Fuller and Brinkley, 1976) and the axonemes of the heliozoan *Actinosphaerium* (Schliwa, 1976).

The second set of observations, as discussed in a later section, suggests that the contractile protein-containing, cortical microfilaments also depend on the presence of calcium ions in order to form and/or to function. Such findings have led a number of investigators to consider the idea that calcium might be involved in the regulation of mitotic events and that this regulation is dependent upon the calcium-sequestering capacities of membrane-bound organelles (Hepler and Palevitz, 1974; Harris, 1976, 1978; Hepler, 1977; Rebhun, 1977). As was expressed by Rebhun (1977), for example, the coordination of mitotic events "... involves changes in distribution and activity of a calcium-sequestering system which must first reduce free calcium levels in the cell center where MA (mitotic apparatus) is formed and then deliver calcium to the ... cortex where it is required for ... " induction of cleavage furrow microfilaments. The discovery that a calcium-dependent regulator protein is localized in the spindle zone (Welsh *et al.*, 1978), where it may act to modulate the sensitivity of microtubules to low levels of calcium ions (Marcum *et al.*, 1978) is an additional factor that must be considered. In any case, the general concept is that calcium plays a key role in the mitotic cell where its availability in time and location must be regulated in a highly specific manner.

### 2.  Effects on the Mitotic Spindle

Many experiments have been done, mainly with cleaving invertebrate and amphibian embryos, which indicate that alteration of calcium levels in live cells can affect the mitotic spindle. Baker and Warner (1972) showed that injection of the calcium-chelating agent EGTA into cleaving *Xenopus* embryos caused relaxation of the cell cortex and a slowdown in onset or failure of cytokinesis without affecting nuclear division. Immersion of embryos in calcium-free medium containing EGTA had no effect on cleavage. On the other hand, Timourian *et al.* (1972) found that EDTA, which chelates both magnesium and calcium, when applied directly to the surface of cleaving sea urchin eggs, had somewhat different effects. Treatment prior to midmetaphase caused reversible "fading" of the mitotic apparatus and the arrest of cleavage. On the other hand, treatment after furrowing had begun had no effect on the completion of the process. Timourian *et al.* (1972) suggested that calcium might be required for the determination of the cleavage furrow. Their data, taken together with those of Baker and Warner (1972), also suggest that low levels of calcium do not affect spindle function but that magnesium ions might be required to maintain mitotic spindle integrity *in vivo*.

That increased levels of calcium might disorganize the mitotic spindle is supported by the findings of Rebhun and co-workers (e.g., 1975, 1976a,b) who observed that caffeine can cause the spindle of marine eggs to degenerate. Their work indicates that this could be due to a caffeine-induced release of intracellular calcium as occurs in muscle (Naylor, 1963; Weber and Herz, 1968) or to an alteration in the disulfide-sulfhydryl status of tubulin (see Rebhun, 1976a).

More recently, Kiehart and Inoué (1976) found that the injection of calcium into sea urchin eggs next to the mitotic spindle could cause a local, temporary loss of birefringence. They interpreted this to mean that the cell could rapidly sequester calcium ions and that calcium might be involved in the local control of microtubule assembly and disassembly during mitosis.

In our own work we have used time lapse cinemicrography to study the effects of agents that alter the ionic content of the cell or its milieu on mitotic stages of mammalian cells (Sisken and VedBrat, 1976; VedBrat *et al.*, 1979). These include nicotine, caffeine, and the ionophore A23187, all of which are known to increase the availability of free cytosolic calcium ions (see e.g., Ahmad and Lewis, 1962; Naylor, 1963; Sandow and Brust, 1966; Weber and Herz, 1968; Weiss, 1966, 1968; Reed and Lardy, 1972; Bianchi, 1975; Pressman, 1976) in other systems. However, at the biochemical level they have all been noted to have other effects as well. For example, nicotine is known to cause cells to lose potassium and take up sodium (Ahmad and Lewis, 1962), and to either increase [at $10^{-3}$ $M$ (Johnson and Epel, 1976)] or decrease [at $3 \times 10^{\times 5}$ $M$ (Brown and Halliwell, 1972)] intracellular pH. Caffeine can inhibit cyclic AMP phospodiesterase and, therefore, increase cyclic AMP levels (Butcher and Sutherland, 1962) and A23187 could increase the diffusion of magnesium in cells (Pfeiffer *et al.*, 1978). Thus, although it seems probable that the effect of these agents with respect to cell motility are related to their ability to increase the cytosolic content of free calcium ions (see above references), such effects have to be cautiously interpreted. For example, increased free calcium concentrations could alter cyclic nucleotide levels (see Rasmussen and Goodman, 1977) and, in fact, it has been reported that a 2-minute exposure to 10 $\mu M$ A23187 causes a significant decrease in intracellular cAMP thought to be due to an inhibition of adenyl cyclase (Zonefrati *et al.*, 1978). In addition, it appears that increased free calcium levels in invertebrate eggs can lead to an increase in intracellular pH (see Epel, 1978).

With regard to the effects relevant to the mitotic spindle, caffeine has previously been noted to inhibit mitosis (Cheney, 1948) and cause degeneration of the spindle in invertebrate eggs (Rebhun *et al.*, 1975), and A23187 has been shown to disorganize the cytoplasmic microtubular complex but not the mitotic spindle of mammalian cells as seen by fluorescent antibody techniques (Fuller and Brinkley, 1976). It has also been shown to cause breakdown of microtubules in axonemes of an *Actinosphaerium* (Schliwa, 1976) but not affect mitosis in a plant cell (Wick and Hepler, 1976).

We have found that, at low concentrations, all three agents increase

**Table I    Effects of Agents Which Increase Cytosolic Levels of Calcium Ions on the Duration of Metaphase in HeLa Cells**[a]

| Agent | Concentration ($M$) | Metaphase duration (min) | $\pm SE$[b] | n | p | References[c] |
|---|---|---|---|---|---|---|
| Nicotine | 0 | 24.79 | 1.26 | 630 | | 1 |
| | $6.17 \times 10^{-5}$ | 25.78 | 2.09 | 322 | — | |
| | $6.17 \times 10^{-4}$ | 28.92 | 2.00 | 309 | 0.11 | |
| Caffeine | 0 | 19.7 | 2.12 | 160 | | 2, 3 |
| | $1 \times 10^{-3}$ | 27.5 | 2.26 | 104 | 0.032 | |
| A23187 | 0 | 21.67 | 0.85 | 485 | | 2, 3 |
| | $1 \times 10^{-6}$ | 26.93 | 1.38 | 210 | 0.01 | |

[a] Obtained from Microbiological Associates.

[b] Each value is a mean derived from a number of separate experiments. The calculation of standard error included variance components for both intraexperiment and interexperiment variation (VedBrat *et al.*, 1979).

[c] Key to references: (1) VedBrat *et al.* (1979); (2) Sisken and VedBrat (1976); (3) Sisken *et al.* (1980).

Table II    Effects of A23187 on the Duration of Anaphase in HeLa Cells[a]

| Concentration ($M$) | Anaphase duration | | | | References |
| | Minutes | ±SE | n | p | |
| --- | --- | --- | --- | --- | --- |
| 0 | 5.02 | 0.13 | 507 | | 2, 3 |
| $1 \times 10^{-6}$ | 5.39 | 0.22 | 200 | 0.18 | |

[a] Calculations and references as in Table I.

the duration of metaphase (Sisken and VedBrat, 1976; VedBrat et al., 1979) (Table I). This is statistically significant only for caffeine and A23187, and it should be noted that the significance is greatest for A23187 which is probably the most calcium-selective agent in the group. The fact that all of the agents increase calcium levels, though by different mechanisms, and have similar effects on metaphase, suggests that it is the change in calcium levels which is causing the prolongation. That this is not a nonspecific toxic effect comes from measurements which show that all of these treatments speed up cytokinesis (Sisken and VedBrat, 1976; VedBrat et al., 1979) (see below). Since calcium ions decrease the stability of microtubules and since microtubule instability might affect the organization of the spindle and the onset of anaphase, it is reasonable to expect that such treatments might prolong metaphase. However, as noted above, since changes in calcium levels in cells can have other effects, these results have to be interpreted with some caution. On the other hand, these findings do lend credence to hypotheses (Hepler and Palevitz, 1974; Harris, 1976; Rebhun, 1977) which suggest that the concentration of free calcium in the spindle zone is normally maintained at a low level during this period.

The effects of nicotine, caffeine, and A23187 on the duration of the period from the onset of anaphase movement to the initiation of furrowing (Sisken and VedBrat, 1976; VedBrat et al., 1979) were also measured. As explained elsewhere (Sisken, 1973), this is an approximation of the duration of anaphase in the cultured cells we have studied. As indicated by data on the effects of A23187 shown in Table II, none of these treatments produced any significant effect on this stage of mitosis.

## D.    Relevance of the Calcium Effect to Events of Metaphase

That increased levels of cytosolic calcium can prolong metaphase supports the idea that levels of available calcium must be reduced in the spindle zone for a cell to have a normal metaphase. Since anaphase durations do not appear to be affected by these treatments, the data suggest either that the state of the spindle undergoes some change to become less sensitive to calcium between metaphase and anaphase or that the capacity of the cell to regulate its level of cytosolic calcium is greater during anaphase.

The first possibility fits with earlier observations that anaphase cells are more resistant than metaphase cells to a variety of treatments (Mazia, 1961), including hyper- and hypothermia (Sisken *et al.*, 1965), and the same may hold true for a calcium-related effect as well.

At the present time, I am unaware of any evidence in favor of the idea that anaphase cells have a greater capacity to regulate their levels of available calcium than do metaphase cells; however, there is some evidence against it since the activity of a spindle-associated calcium-dependent ATPase reaches a peak in metaphase rather than in anaphase in sea urchin embryos (Petzelt, 1972). The suggestion, then, is that the cell may have greater calcium-sequestering potential during metaphase, or at the metaphase–anaphase border, which appears to lessen the sensitivity of the mitotic spindle to manipulation of intracellular calcium levels. Therefore, as suggested by Mazia (1961), metaphase may well be more than just a waiting period until chromatids separate.

Although none of the questions posed at the beginning of this section is answerable at this time, the following statements summarize the discussion up to this point:

1. The duration of metaphase is not a random variable but a characteristic of a particular cell type.

2. The metaphase period seems to be concerned with something more than the separation of chromosomes or chromatids.

3. The regulation of calcium levels appears to be important for normal spindle function and that an excess of calcium can lead to a prolongation of metaphase.

4. The anaphase spindle appears to be much less sensitive to increased calcium levels than does the metaphase spindle.

## III. CYTOKINESIS

### A. Brief Description

The process of cytokinesis in animal cells involves a constriction in a plane previously occupied by the metaphase plate at the equator of the mitotic spindle. In cultured cells we have studied, this occurs when the chromosomes have completed about 80% of their movement apart, irrespective of the velocity of anaphase chromosome movement (Sisken, 1973). A number of elegant studies have shown that the position of the plane of the furrow is dependent upon the astral zones of the spindle and that up to a point in anaphase this plane can be shifted by moving the spindle. However, once furrow progression passes beyond a critical point, it is no longer dependent upon the spindle which can then be removed from the cell without effect on the continuation of cleavage (for a detailed treatment and references, see Rappaport, 1971; Beams and Kessel, 1976).

From the mechanical point of view, cytokinesis is brought about by a contractile ring or band containing microfilaments 4–7 nm in diameter (e.g., see Schroeder, 1970). Various methods have been used to demonstrate that this band contains actin (Schroeder, 1973; Perry *et al.*, 1971; Sanger, 1975; Herman and Pollard, 1978), and more recently, fluorescent antibody techniques have indicated that it also contains myosin (Fujiwara and Pollard, 1976), and $\alpha$-actinin (Fujiwara, *et al.*, 1978). It thus resembles a muscle fiber with respect to its content and has been compared to muscle from a functional point of view for some time (see, e.g., Wolpert, 1960).

## B.  Involvement of Calcium

### 1.  Effects of Increased Levels of Calcium

As previously indicated, a number of studies, done mostly with invertebrate and amphibian oocytes, suggest that cytokinesis is a calcium-dependent process and, given the composition of the contractile ring, one might expect a priori that this would be so. Some of the main experimental findings which support this idea are:

1.  Agents which chelate calcium when applied either internally (Baker and Warner, 1972) or externally (Timourian *et al.*, 1972) may block cytokinesis.

2.  Injection of calcium into amphibian eggs can induce cortical contractile responses (Gingell, 1970; Hollinger and Schmetz, 1976).

3.  Treatment of frog oocytes with the ionophore A23187 induces cortical contractions (Schroeder and Strickland, 1974) and can speed up the rate of travel of the furrow in the cleavage divisions of squid embryos (Arnold, 1975). Since the stimulatory effects of the ionophore occurred in the presence or absence of external calcium, it was proposed that

**Table III  Effects of Agents Which Increase Cytosolic Levels of Calcium Ions on the Duration of Cytokinesis in HeLa Cells[a]**

| Agent | Concentration (M) | Cytokinesis duration (minutes) | ±SE | n | p | References |
|-------|-------------------|--------------------------------|------|-----|--------|------------|
| Nicotine | 0 | 3.06 | 0.074 | 290 | | 1 |
| | $6.17 \times 10^{-5}$ | 2.33 | 0.127 | 96 | 0.0008 | |
| | $6.17 \times 10^{-4}$ | 2.32 | 0.146 | 45 | 0.0015 | |
| Caffeine | 0 | 2.80 | 0.09 | 180 | | 2, 3 |
| | $1 \times 10^{-3}$ | 2.18 | 0.09 | 140 | 0.0009 | |
| A23187 | 0 | 3.00 | 0.08 | 514 | | 2, 3 |
| | $1 \times 10^{-6}$ | 2.01 | 0.15 | 200 | 0.0001 | |

[a] Calculations and references as in Table I.

internal stores of calcium may be released by the inophore (Schroeder and Strickland, 1974; Arnold, 1975) as occurs in other systems (Babcock et al., 1976).

In our own work we have measured the effects of nicotine, caffeine, and A23187 on cytokinesis in HeLa cells (Sisken and VedBrat, 1976; VedBrat et al., 1979). Examples of data obtained (Table III) show that all of these agents speed up the process of cytokinesis, i.e., they shorten its duration by about 22–33%. This is not a large difference but it should be noted that these treatments cause a process, which normally would have occurred anyway, to proceed at a faster than normal rate.

## 2. Effects of Reduced External Levels of Calcium

We also studied the effects of a series of treatments which limit the availability of external calcium (Sisken et al., 1980). These included (a) growth in synthetic, calcium-free medium supplemented with dialyzed serum, (b) growth in standard medium containing the calcium-chelating agent EGTA, (c) growth in standard medium pretreated with Chelex 100 (Bio-Rad) (an ion exchange resin which removes both calcium and magnesium ions) to which magnesium was added back, and (d) growth in medium containing lanthanum ion which is thought to bind to calcium-binding sites on the surface of the plasma membrane and to inhibit the transport of calcium across it (Takata et al., 1966).

The unexpected result obtained was that these treatments also caused a speed up of cytokinesis (Table IV) to an extent quantitatively similar to the treatments shown in Table III. These findings were also contrary to observations of Timourian et al. (1972) who found that EDTA applied externally to sea urchin eggs inhibited cytokinesis. The situation for

*Xenopus* eggs is not clear since Gingell's experiments appear to demonstrate a requirement for extracellular calcium in order for cytokinesis to occur (Gingell, 1970), while Baker and Warner (1972) found that a calcium-free, EGTA-containing buffer had no effect on cleavage.

The fact that cytokinesis is stimulated by deprivation of extracellular calcium to about the same extent as it is when treated with nicotine, caffeine, and A23187 seems paradoxical but a similar phenomenon has been reported in other cell types. For example, it has been observed that amphibian cardiac (Bianchi, 1969) and mammalian smooth (Hurwitz *et al.*, 1967) muscle will contract spontaneously when placed in a calcium-free environment. The explanation offered for these observations was that a pool of calcium exists bound to the inner surface of the plasma membrane and that this pool is stabilized in position by calcium bound to the outer surface. It was suggested that removal of calcium from the external surface causes a conformational change in the membrane which in turn releases into the cell the calcium bound to the inner surface. The existence of such a pool was also suggested by Gingell and Palmer (1968) as a possible explanation for a cortical response obtained when nondividing frog eggs were exposed to polycations. Evidence concerning the existence of a similar pool in other systems has been reviewed recently by Bolton (1979).

Such a pool could account for the effects of chelating agents and calcium-free media on mitotic cells as well. One could speculate, for example, that the polar zones of the spindle determine the position of the cleavage furrow by establishing a localized, circumferential pool of calcium ions bound to the inner surface of the plasma membrane and that some alteration of the membrane, possibly resulting from spindle activity, would allow release from this pool into the cortical region during anaphase–telophase where it could establish the position of the cleavage furrow and participate in the process. Experimental removal of external calcium might have speeded up contraction by increasing the rate or amount of release of calcium from this pool. Whether such a

**Table IV    Effects of Ethylene Glycol bis ($\beta$-Aminoethyl Ether)-$N,N'$-Tetraacetic Acid (EGTA) on the Duration of Cytokinesis in HeLa Cells[a]**

| Concentration (mM) | Cytokinesis duration (minutes) | ±SE | n | p | Reference |
|---|---|---|---|---|---|
| 0 | 2.91 | 0.06 | 432 | | 2, 3 |
| 2.0 | 2.49 | 0.08 | 114 | 0.001 | |

[a] Calculations and references as in Table I.

model could be contemplated for cells other than cultured mammalian cells is not clear in view of the results cited above.

## C. Summary of Cytokinesis

Based upon the observations of others and our own quantitative data, it appears that cytokinesis is a calcium-dependent process and, as we have previously mentioned (VedBrat *et al.*, 1979), supports the idea that calcium availability might be rate-limiting for cytokinesis in the normal cell. Among the questions that remain open are: (a) What is the molecular role of calcium in cytokinesis? (b) What is the normal source of calcium? (c) How is the availability of calcium regulated throughout the course of mitosis? Some discussion relevant to these questions is presented in the next section.

## IV.  REGULATION OF CALCIUM LEVELS DURING THE COURSE OF CELL DIVISION

It is clear from the work of others as well as from our own studies that increased levels of calcium can prolong metaphase (i.e., delay the onset of chromosome movement) but stimulate cytokinesis in the same cell minutes later. Such findings support the suggestions made by others that the cell might control some of the major events of cell division by regulating the temporal and spatial availability of calcium ions. This would include reducing calcium levels in the spindle zone during metaphase and perhaps early anaphase and making calcium available in the region of the cortical microfilaments during late anaphase and telophase.

As previously noted, calcium regulation in cells is a complex phenomenon and may involve several membrane-bound organellar systems. Particles visible in the light microscope and membrane-bound elements as seen in the electron microscope have often been observed to be present in the polar zones of the spindle during metaphase and early anaphase (see Rebhun, 1960; Mazia, 1961; Rappaport, 1971; Harris, 1975, 1978; Hepler, 1977) and to move into the interzonal region during late anaphase and telophase (Rappaport, 1971). These elements, which could well contain calcium-sequestering systems, would thus fit the hypothesis of Rebhun (1977), i.e., they might sequester calcium in the polar zones during metaphase and release it in the interzonal region during late anaphase–telophase. The discovery of Petzelt (1972) that a peak in activity of a spindle-associated, calcium-dependent ATPase occurs during metaphase, and the observation of Kiehart and Inoué (1976) that the local loss of spindle birefringence induced by the microinjection of calcium in the region of the spindle poles is reversed in a matter of minutes, supports this hypothesis. The idea that mobile, calcium-sequestering organelles move into the interzonal region in late anaphase where they might release their calcium and thus stimulate cleavage is also consistent with the findings that the spindle poles determine the plane of cleavage.

As reasonable as the above suggestions might be, they are not necessarily correct. The experiments in which we have limited the availability of extracellular calcium to dividing HeLa cells are consistent with the possibility that a pool of calcium exists on the inner surface of the plasma membrane in mitotic cells. This pool, if it really exists, is strategically placed with respect to the cortical microfilaments for which it could serve as a local source of calcium for cytokinesis. Thus, while calcium sequestration at the polar zones of the spindle might occur and be important for assembly of the spindle, a different pool of calcium might be involved in cytokinesis.

A final point to be discussed concerns the prolongation of metaphase in certain tumor cells. The mechanisms behind this are unknown and could be related to one of two categories of causes. The first involves the fact that most transformed cells are hyperploid which suggests that the increased duration of metaphase in these cells might be due to some karyotypic change, either because the cell simply has to deal with more chromosomes or because the specific gene content of the genome is altered as a result of the heteroploidy. On the other hand, since heteroploidy generally develops late in tumor evolution (Nowell, 1976), it is possible that the increased metaphase duration is related to a cellular change which occurred prior to the development of heteroploidy, possibly as a result of the initial transforming event. Since calcium regulation is known to be altered in many tumor cells (Mikkelsen, 1978), further studies of differences between normal and tumor cells with respect to these parameters might provide some useful leads with respect to the mechanisms that determine the duration of metaphase and the regulation of the division process as a whole.

## ACKNOWLEDGMENT

The original research included in this chapter was supported in part by the University of Kentucky Tobacco and Health Research Institute Project KTRB 24111 and a grant from the Sanders-Brown Kentucky Research Center on Aging. The author wishes to express his sincere appreciation to Drs. Stanley Blecher, Louis Boyarsky, Joseph Engelberg, Arthur Forer, and Peter Hepler for their helpful comments and criticisms of the manuscript.

## REFERENCES

Ahmad, K., and Lewis, J. J. (1962). *J. Pharmacol Exp. Ther.* **136,** 298.
Allen, R. D., Bajer, A., and LaFountain, J. (1969). *J. Cell Biol.* **43,** 4a.
Arnold, J. M. (1975). *Cytobiologie* **11,** 1.
Arnold, J. M. (1976). *In* "The Cell Surface in Animal Embryogenesis and Development" (G. Poste and G. L. Nicolson), eds. Elsevier/North-Holland, Amsterdam.
Asch, B. B., Medina, D., and Brinkley, B. R. (1979). *Cancer Res.* **39,** 893.
Babcock, D. F., First, N. L., and Lardy, H. A. (1976). *J. Biol. Chem.* **251,** 3881.
Baker, P. F., and Warner, A. E. (1972). *J. Cell Biol.* **53,** 579.
Baserga, R. (1976). "Multiplication and Division of Mammalian Cells." Dekker, New York.
Beams, H. W., and Kessel, R. G. (1976). *Am. Sci.* **64,** 279.
Belar, K. (1926). *Ergeb. Forsch. Zool.,* **6,** 255 (cited by Mazia, 1961).
Bianchi, C. P. (1969). *Fed. Proc. Fed. Am. Soc. Exp. Biol.* **28,** 1624.
Bianchi, C. P. (1975). *In* "Cellular Pharmacology of Exitable Tissues" (T. Naŕahashi, ed.) Thomas, Springfield, Illinois.
Bolton, T. B. (1979). *Physiol. Rev.* **59,** 606.
Bowman, P. D., Meek, R. L., and Daniel, C. W. (1975). *Exp. Cell Res.* **93,** 184.
Brinkley, B. R., Fuller, G. M., and Highland, D. P. (1975). *Proc. Natl. Acad. Sci. U.S.A.* **72,** 4981.
Brown, D. A., and Halliwell, J. V. (1972). *Br. J. Pharmacol.* **45,** 349.
Bryan, J. (1975). *Am. Zool.* **15,** 649.
Burger, M. M., Bombik, B. M., Breckenridge, B., Sheppard, J. R. (1972). *Nature (London) New Biol.* **239,** 161.
Butcher, R. W., and Sutherland, E. W. (1962). *J. Biol. Chem.* **237,** 1244.
Carlson, J. G. (1956). *Science* **124,** 203.
Cheney, R. H. (1948). *Biol. Bull. (Woods Hole, Mass.)* **94,** 16.
Clothier, G., and Timourian, H. (1972). *Exp. Cell Res.* **75,** 105.
Cristofalo, V. J., and Sharf, B. B. (1973). *Exp. Cell Res.* **76,** 419.
Epel, D. (1963). *J. Cell Biol.* **17,** 315.
Epel, D. (1978). *In* "Cell Reproduction" (E. R. Dirksen, D. M. Prescott, and C. F. Fox, eds.), p. 367. Academic Press, New York.
Forer, A. (1965). *J. Cell Biol.* **25,** 95.
Forer, A. (1966). *Chromosoma* **19,** 44.
Fugiwara, K., and Pollard, T. D. (1976). *J. Cell Biol.* **71,** 848.
Fugiwara, K., Porter, M. E., and Pollard, T. D. (1978). *J. Cell Biol.* **79,** 268.
Fuller, G. M., and Brinkley, B. R. (1976). *J. Supramol. Struct.* **5,** 497.
Fuseler, J. W. (1975). *J. Cell Biol.* **67,** 789.
Galinsky, I. (1949). *J. Hered.* **40,** 289.
Geilenkirchen, W. L. M., Jansen, J., Coosen, R., and Van Kijk, R. (1977). *Cell Biol. Int. Rep.* **1,** 419.
Gingell, D. (1970). *J. Embryol. Exp. Morphol.* **23,** 583.
Gingell, D., and Palmer, J. F. (1968). *Nature (London)* **217,** 98.
Harris, P. (1975). *Exp. Cell Res.* **94,** 409.
Harris, P. (1976). *Exp. Cell Res.* **97,** 63.
Harris, P. (1978). *In* "Cell Cycle Regulation" (J. R. Jeter, Jr., I. L. Cameron, G. M. Padilla, and A. M. Zimmerman, eds.), p. 75. Academic Press, New York.
Hayflick, L. (1965). *Exp. Cell Res.* **36,** 614.
Heath, I. B. (1974). *In* "The Cell Nucleus" (H. Busch, ed.), Vol. II, p. 487. Academic Press, New York.

Hepler, P. K. (1977). *In* "Mechanisms and Control of Cell Division" (T. L. Rost, and E. M. Gifford, Jr., eds.), p. 212. Dowden, Hutchinson and Ross, Stroudsburg, Pennsylvania.

Hepler, P. K., and Palevitz, B. A. (1974). *Annu. Rev. Plant Physiol.* **25,** 309.

Herman, I. M., and Pollard, T. D. (1978). *Exp. Cell Res.* **114,** 15.

Hollinger, T. G., and Schuetz, A. W. (1976). *J. Cell Biol.* **71,** 395.

Huot, J., Landry, J. C., and Bergeron, C. (1976). *J. Cell Biol.* **70,** 398a.

Hurwitz, L., P. D. Joiner, and Von Hagen, S. (1967). *Am. J. Physiol.* **213,** 1299.

Inoué, S., and Ritter, H. Jr. (1975). *In* "Molecules and Cell Movement" (S. Inoué, R. E. Stepens, eds.), p. 3. Raven, New York.

Izuztu, K. (1959). *Mie Med. J.* **9,** 15.

Johnson, J. D., and Epel, D. (1976). *Nature (London)* **262,** 661.

Kiehart, D. P., and Inoué, S. (1976). *J. Cell Biol.* **70,** 230a.

Killander, D., and Zetterberg, A. (1965). *Exp. Cell Res.* **40,** 12.

Kirschner, M. W. (1978). *Int. Rev. Cytol.* **54,** 1.

Luykx, P. (1970). *Int. Rev. Cytol. Suppl.* **2,** 173 pp.

Macieira-Coelho, A., Ponten, J., and Phillipson, L. (1966). *Exp. Cell Res.* **42,** 673.

McIntosh, J. R., Hepler, P. K., and Van Wie, D. G. (1969). *Nature (London)* **224,** 659.

McIntosh, J. R., Cande, W. Z., and Snyder, J. A. (1975). *In* "Molecules and Cell Movement" (S. Inoué and R. E. Stephens, eds.), p. 31. Raven, New York.

Marcum, J. M., Dedman, J. R., Brinkley, B. E., and Means, A. R. (1978). *Proc. Natl. Acad. Sci. U.S.A.* **75,** 3771.

Margolis, R. L., and Wilson, L. (1978). *Cell* **13,** 1.

Margolis, R. L., Wilson, L., and Kiefer, B. I. (1978). *Nature (London)* **272,** 450.

Mazia, D. (1961). *In* "The Cell" (J. Brachet and A. E. Mirsky, eds.), Vol. III, p. 77. Academic Press, New York.

Mazia, D. (1977). *In* "Mitosis: Facts and Questions" (M. Little, N. Paweletz, C. Petzelt, H. Ponstingl, D. Schroeter and H. P. Zimmerman, eds.), p. 196. Springer-Verlag, Berlin and New York.

Mikkelsen, R. B. (1978). *Prog. Exp. Tumor Res.* **22,** 123.

Millis, A. J. T., Forrest, G., and Pious, D. A. (1972). *Biochem. Biophys. Res. Commun.* **49,** 1645.

Mitsui, Y., and Schneider, E. L. (1976). *Mech. Ageing Dev.* **5,** 45.

Naylor, W. G. (1963). *Am. J. Physiol.* **204,** 969.

Nicklas, R. B. (1971). *Adv. Cell Biol.* **2,** 225.

Nowell, P. C. (1976). *Science* **194,** 23.

Olmsted, J. B. (1976). *In* "Cell Motility" (R. Goldman, T. Pollard and J. Rosenbaum, eds.), p. 1081. Cold Spring Harbor Lab., Cold Spring Harbor, New York.

Perry, M. M., John, H. A., and Thomas, N. S. T. (1971). *Exp. Cell Res.* **65,** 249.

Petzelt, C. (1972). *Exp. Cell Res.* **70,** 333.

Petzelt, C., and Auel, D. (1977). *Proc. Natl. Acad. Sci. U.S.A.* **74,** 1610.

Pfeiffer, D. R., Taylor, R. W., and Lardy, H. A. (1978). *Ann. N.Y. Acad. Sci.* **307,** 402.

Prescott, D. M. (1976). "Reproduction of Eukaryotic Cells." Academic Press, New York.

Pressman, B. C. (1976). *Annu. Rev. Biochem.* **45,** 501.

Rappaport, R. (1971). *Int. Rev. Cytol.* **31,** 169.

Rasmussen, H., and Goodman, D. B. P. (1977). *Physiol. Rev.* **57,** 509.

Rebhun, L. (1960). *Ann. N.Y. Acad. Sci.* **90,** 357.

Rebhun, L. I. (1977). *Int. Rev. Cytol.* **49,** 1.

Rebhun, L. I., Jemiolo, D., Ivy, N., Mellon, M., and Nath, J. (1975). *Ann. N. Y. Acad. Sci.* **253,** 362.

Rebhun, L. I., Nath, J., and Remillard (1976a). *In* "Cell Motility" (R. Goldman, T. Pollard, and J. Rosenbaum, eds.) Cold Spring Harbor Lab., Cold Spring Harbor, New York.

Rebhun, L. I., Miller, M., Schnaitman, T. C., Nath, J., and Mellon, M. (1976b). *J. Supramol. Struct.* **5,** 199.

Reed, P. W., and Lardy, H. A. (1972). *J. Biol. Chem.* **247,** 6970.

Rhoades, M. M. (1952). *In* "Heterosis" (J. W. Gowen, ed.), p. 66. Iowa State College Press, Ames, Iowa.

Sandow, A., and Brust, M. (1966). *Biochem. Z.* **345,** 232.

Sanger, J. W. (1975). *Cell Tissue Res.* **161,** 431.

Scarpa, A., and Carafoli, E. (1978). "Calcium Transport and Cell Function" *Ann. N. Y. Acad. Sci.* **307.**

Schliwa, M. (1976). *J. Cell Biol.* **70,** 527.

Schrader, F. (1936). *Biol. Bull. (Woods Hole, Mass.)* **70,** 484.

Schroeder, T. E. (1970). *Z. Zellforsch.* **109,** 431.

Schroeder, T. E. (1973). *Proc. Natl. Acad. Sci. U.S.A.* **70,** 1688.

Schroeder, T. E., and Strickland, D. L. (1974). *Exp. Cell Res.* **83,** 139.

Sheppard, J. R., and Prescott, D. M. (1972). *Exp. Cell Res.* **75,** 293.

Sisken, J. E. (1973). *Chromosoma* **44,** 91.

Sisken, J. E., and Bonner, S. V. (1978). *J. Supramol. Struct. Suppl.* **2,** 313.

Sisken, J. E., and Bonner, S. V. (1979). *Mech. Ageing Dev.* **11,** 191.

Sisken, J. E., and Morasca, L. (1965). *J. Cell Biol.* **25,** 179.

Sisken, J. E., and VedBrat, S. S. (1976). *J. Cell Biol.* **70,** 130a.

Sisken, J. E., Morasca, L., and Kibby, S. (1965). *Exp. Cell Res.* **39,** 103.

Sisken, J. E., VedBrat, S. S., and Nasser, M. (1980). Effects of caffeine, ionophores and calcium deficient media on mitosis and cytokinesis of HeLa cells. (in preparation).

Takata, M., Pickard, W. F., Lettvin, J. Y., and Moore, J. W. (1966). *J. Gen. Physiol.* **50,** 461.

Timourian, J., Clothier, G., and Watchmaker, G. (1972). *Exp. Cell Res.* **75,** 296.

Tobey, R. A., Peterson, D. F., and Anderson, E. C. (1969). *In* "Biochemistry of Cell Division" (R. Baserga, ed.), p. 39. Thomas, Springfield, Illinois.

Upcott, M. (1939). *Chromosoma* **1,** 178.

VedBrat, S. S., Sisken, J. E., and Anderson, R. L. (1979). *Eur. J. Cell Biol.* **19,** 250.

Weber, A., and Herz, R. (1968). *J. Gen. Physiol.* **52,** 750.

Weisenberg, R. C. (1972). *Science* **177,** 1104.

Weiss, G. B. (1966). *J. Pharmacol. Exp. Ther.* **154,** 595.

Weiss, G. B. (1968). *J. Pharmacol. Exp. Ther.* **163,** 43.

Welsh, J. J., Dedman, J. R., Brinkley, B. R., and Means, A. R. (1978). *Proc. Natl. Acad. Sci. U.S.A.* **75,** 1867.

Wick, S. M., and Hepler, P. K. (1976). *J. Cell Biol.* **70,** 209a.

Wolpert, L. (1960). *Int. Rev. Cytol.* **10,** 163.

Zeilig, C. E., and Goldberg, N. D. (1977). *Proc. Natl. Acad. Sci.* **74,** 1052.

Zirkle, R. E. (1970). *Radiat. Res.* **41,** 516.

Zonefrati, R., Tanini, A., Rotella, C., and Toccafondi, R. (1978). *In* "Molecular Biology and Pharmacology of Cyclic Nucleotides" (G. Folco and R. Paoletti, eds.), p. 185. Elsevier/North Holland, Amsterdam.

Zylber, E. A., and Penman, S. (1971). *Science* **172,** 947.

11

# Cyclic Nucleotides and the Control
# of Erythroid Cell Proliferation

### W. J. GEORGE, L. A. WHITE, and J. R. JETER, Jr.

## I. INTRODUCTION

Erythropoietin (Ep) is the glycoprotein hormone that regulates the production of red blood cells. This regulation is very important since there are numerous physiological and/or pathophysiological states during which the body must produce new red blood cells. For example, additional erythrocytes must be produced in certain situations where: (1) tissues exhibit enhanced oxygen utilization; (2) there is an increased rate of red blood cell destruction; or (3) there is a decrease in the supply of oxygen available to the red blood cells, as in exposure to hypoxia (Fisher, 1972). However, the mechanism by which Ep produces its effects on the control of red cell proliferation is not well understood. There apparently is a renal cell mechanism which responds to a low oxygen tension and, as a result, causes an enhanced production of Ep (Fisher, 1969). There is evidence that renal cell oxygen tensions are involved in Ep production, such as the finding that hypoxemic perfusion of isolated kidneys will stimulate Ep production (Fisher and Birdwell, 1961; Fisher et al., 1962; Reissmann and Nomura, 1962; Zangheri et al., 1963). Agents such as thyroid hormone (Evans et al., 1957, 1961) and ACTH (Fisher and

**293**

NUCLEAR–CYTOPLASMIC INTERACTIONS
IN THE CELL CYCLE

Crook, 1962) increase oxygen demand and also stimulate Ep production. Also, agents that cause an enhanced hemolysis of red blood cells (Fisher and Mueller, 1971), thus producing marked reductions in hematocrit, cause enhanced titers of Ep as a compensatory measure to restore normal red cell mass. After the primary stimuli of reduced oxygen tension, increased oxygen utilization, or increased red cell breakdown, an enhanced plasma titer of Ep results. Ep, as the primary hormone, then acts on erythroid cell precursors in the bone marrow to accelerate the development of mature erythrocytes.

The precise mechanism by which Ep acts on these target cells to cause differentiation and proliferation has not been completely defined. It has been suggested by Stohlman (1964) that the primary site of Ep action is on erythroid stem cells, which differentiate into mature cells of the erythroid line. It has been also demonstrated by others (Alpen and Cranmore, 1959; Alpen *et al.*, 1962; Fried *et al.*, 1969) that Ep causes differentiation of immediate precursor cells into recognizable or differentiated erythroid cells.

In many nonerythroid tissues, differentiated cells no longer have the capacity to divide. In a few tissues (e.g., liver), cells may be triggered to begin proliferation by injury; however, in erythroid and certain other tissues, cells must be continually replaced as they are injured or as they complete their finite life span. The erythrocyte is one such cell that must be continually replaced as its life span is only about 120 days and also because there is frequently a loss of these cells due to hemorrhage (Stohlman, 1971). The erythrocyte is essential to life since it transports oxygen to, and removes carbon dioxide from, other cells in the body. In order to perform this function, the red blood cell is packed with hemoglobin (95% dry wt. and 30% wet wt.) at the expense of other cellular components. The mature erythrocyte is nonnucleated and cannot divide; therefore, it must be constantly replenished from hemopoietic tissue, primarily the bone marrow. To accomplish this, cells flow from an undifferentiated (pluripotent) cell compartment to the erythroid-committed stem cell (unipotent) compartment (Stohlman, 1971). The erythroid-committed cells undergo a series of cell divisions which produce increasing numbers of differentiated daughter cells, until the mature red blood cell is formed (Morse *et al.*, 1970).

Erythrocyte maturation consists of two processes: (1) cell proliferation in which the cells undergo mitosis with the number of cells doubled during each round of cell division; and (2) cell differentiation, which may be defined as the synthesis of specific cellular products that characterize particular types of cells. In this latter case, Ep induces hemoglobin synthesis which distinguished erythroid cells from other cells. As the extent of differentiation increases, the capacity for cell division in these cells decreases. The undifferentiated progenitor cells have the capacity for self-perpetuation, thus allowing their population to be maintained. These cells also possess the ability to produce daughter cells that differentiate (Stohlman, 1971; Hayes *et al.*, 1975). The undifferentiated progenitor is pluripotent and may produce cells that differentiate along three lines: the granulocytic, erythrocytic, or megakaryocytic pathway (Stohlman, 1970a) (Fig. 1). Once a cell is committed to a particular cell type, it undergoes a limited number of cell divisions as it matures (Stohlman, 1970b). The pluripotent stem cells have not been shown to be under humoral control and the stimulus which apparently maintains the necessary population of committed stem cells, by regulating the flux of erythroid cells from the pluripotent compartment, is unclear (Fisher, 1972). Ep does not appear to act on the pluripotent stem cell to stimulate erythropoiesis, but rather it acts on early undifferentiated erythroid-committed stem cells (Bruce and McCulloch, 1964). The morphological identity of the specific target cell for Ep has not yet been established.

There is evidence to indicate that once Ep exerts its effect on the erythroid-committed stem cell, it may not be necessary for Ep to be

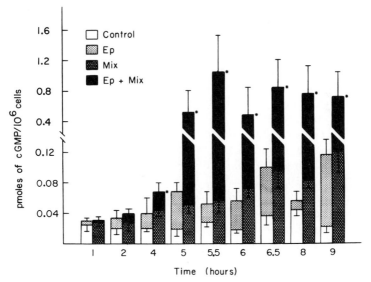

**Fig. 1.** Effect of erythropoietin (Ep) and/or 1-methyl-3-isobutylxanthine (MIX) on cyclic GMP levels in fetal liver cultures. Ep (100 mU) or MIX ($5 \times 10^{-6}\ M$) was added at the beginning of the incubation. Each value represents the mean ± SEM of three individual experiments each performed in triplicate. (*) Indicates significant difference from control ($p < 0.05$).

present for the remainder of the erythropoietic process (i.e., until the red blood cell matures). Experiments conducted by Filmanowicz and Gurney (1961) demonstrated that a single injection of a high dose of Ep stimulates erythropoiesis in hemopoietic tissue. There is also evidence that Ep acts on more highly differentiated cells, such as the proerythroblast or the early normoblast, and may even act on the release of reticulocytes from marrow. Studies employing perfused hindlimbs of rabbits or rats have shown that sustained administration of Ep causes an increase in the release of reticulocytes from bone marrow (Kuna et al., 1959; Gordon et al., 1962; Fisher et al., 1965).

The committed stem cells have been shown to be proliferating cells; that is, they normally proceed through $G_1$, S, $G_2$, and M phases of the cell cycle (DeRobertis et al., 1975). Erythroid-committed stem cells proceed through the cell cycle at either a basal or enhanced rate, depending upon regulator influences, whereas the pluripotent stem cell is predominantly in a "resting state," as shown in experiments conducted by Morse and co-workers (Morse et al., 1969). These investigators employed hydroxyurea and [$^3$H]thymidine to kill all cells that passed through the S phase and, therefore, destroy all cycling cells. When these agents were used, cells capable of responding to Ep were reduced in number, whereas undifferentiated colony-forming stem cells were unaffected with respect to their ability to restore all lines of hemopoiesis (Morse et al., 1969).

When cells enter the erythroid-committed compartment, they begin to differentiate, developing morphological characteristics that distinguish them from other hemopoietic cells (Fig. 1). The first identifiable cells are the proerythroblasts, followed by the large basophilic normoblasts, small basophilic normoblasts, and polychromatic normoblasts. These cells are capable of cell division and are thought to respond to Ep. These cells mature into late normoblasts, reticulocytes, and then erythrocytes, all of which are no longer capable of dividing and therefore do not respond to Ep (Stohlman, 1970a).

The primary action of Ep is to stimulate erythrocyte proliferation and thus produce an increase in the number of circulating cells. This stimulation of erythrocyte proliferation could be accomplished by: (1) accelerating the cell cycle of the developing erythroid cells, which results in a decreased time of development for individual cells; (2) increasing the number of stem cells entering the erythroid-committed stem cell compartment; (3) increased percentage of cells successfully completing maturation; or (4) a combination of points (1), (2), and (3). However, it has been demonstrated that Ep does not affect the uncommitted stem cell (Bruce and McCulloch, 1964) and, therefore, it is unlikely that Ep acts to increase the flow of stem cells into the erythroid cell line. It is more plausable that Ep acts to accelerate division of the erythroid-committed cells (Curry *et al.*, 1967; O'Grady *et al.*, 1968; O'Grady and Lewis, 1970). By increasing the number of cells in mitosis during a given period, the maturation time necessary to produce erythrocytes will be shortened. Furthermore, a greater number of cells will be produced since each mitosis results in a doubling of the number of cells available for maturation. In addition, it has been proposed that a small fraction of cells in the proerythroblast stage may proceed directly to the nondividing pool through what has been described as the upper pathway of skipped division. This pathway is believed to occur more frequently when erythropoiesis is stressed (Stohlman 1970b). However, the pathway results in a lower yield of erythrocytes than would occur through cell division.

## II. EFFECTS OF EP STIMULATION ON DNA, RNA, AND PROTEIN SYNTHESIS

The interaction of Ep with Ep-sensitive cells results in the initiation of a series of biochemical events. Information obtained from short-term cultures of both fetal liver and adult erythropoietic tissue has contributed to the elucidation of the sequence of intracellular events induced by Ep. When erythroid cells are placed in culture, the rates of synthesis of RNA, DNA, and hemoglobin begin to decay unless Ep is added (Krantz and Goldwasser, 1965, 1970; Cole and Paul, 1966). This indicates that Ep is necessary for the maintenance of functional erythroid cells in culture.

The use of biochemical inhibitors has further contributed to the elucidation of the sequence of events induced by Ep and the dependence of these events upon prior reactions. Stimulation of RNA synthesis is the first event observed in bone marrow and fetal liver cultures after Ep is added to cultures (Krantz and Goldwasser, 1965; Gross and Goldwasser, 1969; Djaldetti *et al.*, 1972; Handler *et al.*, 1974). Increase in RNA synthesis occurs 20 to 60 minutes after the addition of Ep to cultures. The RNA produced has been shown to be a 150 S RNA species and is not present in control cultures which do not contain Ep. The RNA that is produced is insensitive to protease, DNAase, urea, and detergent disaggregation, which indicates that it is a single molecule (Djaldetti *et al.*, 1972). Metabolic blockers that inhibit DNA synthesis [5'fluorodeoxyuridine (FdUrd), cytosine arabinoside, or hydroxy urea] (Djaldetti *et al.*, 1972; Datta and Dukes, 1975), or which inhibit protein synthesis (cycloheximide) (Maniatis *et al.*, 1974), do not affect this early

RNA synthesis. Protein synthesis occurs immediately following the Ep-induced increase in RNA synthesis. This protein synthesis is prevented if the early RNA synthesis is inhibited by actinomycin D (Cole and Paul, 1966; Paul *et al.*, 1973). Although specific proteins have not been identified, it is known that the protein synthesized is not hemoglobin.

Stimulation of DNA synthesis occurs 4 to 12 hours after the addition of Ep to both fetal liver and bone marrow cultures (Paul and Hunter, 1968; Gross and Goldwasser, 1970; Paul *et al.*, 1973). Inhibitors of RNA synthesis (actinomycin D) block both the effect of Ep on DNA synthesis and subsequent hemoglobin synthesis. DNA synthesis is completely abolished when actinomycin D is added 1 hour prior to Ep addition (Paul and Hunter, 1969). Also, puromycin (Paul and Hunter, 1969) and cycloheximide (Gross and Goldwasser, 1970) inhibit the Ep-induced increase in DNA synthesis, suggesting that Ep-stimulated DNA synthesis is dependent on prior synthesis of RNA and protein.

A second wave of RNA synthesis occurs 10 to 20 hours after Ep addition. This stimulation represents the peak effect of Ep on RNA synthesis and accounts for the majority of RNA synthesized (Gross and Goldwasser, 1969; Conkie *et al.*, 1975). A number of species of RNA which sediment at 55–64 S, 45 S, 9 S, 6 S, and 4 S are produced during this second period of RNA synthesis (Nicol *et al.*, 1972). This RNA includes ribosomal precursor RNA (45 S), ribosomal RNA (28 S and 18 S), and transfer RNA (Nicol *et al.*, 1972; Maniatis *et al.*, 1974). Stimulation of the 9 S fraction is thought to represent messenger RNA, which may be the messenger-specifying globin formation (Gross and Goldwasser, 1970), or another unrelated protein (e.g., histone) (Terada *et al.*, 1975). This stimulation of RNA synthesis is blocked by the DNA synthesis inhibitor, FdUrd, which suggests that the majority of RNA synthesis is dependent upon prior DNA synthesis (Paul and Hunter, 1969).

Ep stimulates iron uptake and hemoglobin synthesis in bone marrow cultures (Krantz *et al.*, 1963; Gross and Goldwasser, 1970) and fetal liver cultures (Paul and Hunter, 1968, 1969; Terada *et al.*, 1975). The amount of iron uptake peaks at 24 to 48 hr after Ep addition. Cell division may be required for this stimulation of iron uptake which is thought to be the result of an increased number of cells producing a basal amount of hemoglobin rather than an absolute increase in the amount of hemoglobin synthesized in each individual cell (Chui *et al.*, 1971). Nonetheless, there is an overall increase in the amount of hemoglobin synthesized which is dependent upon both prior DNA synthesis and RNA synthesis. Treatment with FdUrd, as well as other DNA synthesis inhibitors, prevents the Ep-induced increase in hemoglobin synthesis (Paul and Hunter, 1968, 1969; Gross and Goldwasser, 1970;

Ortega and Dukes, 1970). From these studies, it may be concluded that Ep-induced DNA synthesis is a necessary requirement for Ep-induced hemoglobin synthesis. Addition of actinomycin D completely abolishes Ep-stimulated heme synthesis, indicating that RNA synthesis is necessary for the effect of Ep. These results suggest that the sequence of events induced by Ep, which is eventually expressed as hemoglobin synthesis, include: (1) early transcription of the heterogenous 150 S RNA; (2) translation of necessary protein; (3) DNA synthesis; (4) synthesis of a 9 S RNA (possibly the mRNA for globin); and (5) hemoglobin synthesis.

A bone marrow cytoplasmic factor that stimulates RNA synthesis in isolated nuclei has also been described (Chang and Goldwasser, 1973). In these studies, a cytoplasmic extract was obtained from cultured erythroid cells incubated with Ep. This cytoplasmic factor produces a 25–29% stimulation of RNA synthesis during a 15-minute incubation with nuclei from unstimulated bone marrow cells, as well as nuclei from liver or kidney. Trypsin inactivates this cytoplasmic factor, but ribonuclease has no comparable effect.

## III. CYCLIC NUCLEOTIDES AND ERYTHROID CELL DIFFERENTIATION

Although there are a number of isolated events which occur following the interaction of Ep with a target receptor, the precise mechanism of action of Ep is unknown. Indeed, the mechanism of action of other hormonal or mitogenic agents which act to induce cellular proliferation or differentiation is yet to be described. However, the cyclic nucleotides, cyclic AMP, and cyclic GMP are being implicated as possible mediators of these events, in addition to their well-known actions on other cellular processes.

Reports of the actions of the cyclic nucleotides upon erythropoiesis have been contradictory and have led to much confusion as to what role, if any, the cyclic nucleotides may play in the development of erythrocytes. Experiments conducted to test the effect of cyclic AMP on heme synthesis in bone marrow cultures have yielded diverse results. Stimulation of heme synthesis by cyclic AMP has been observed in human marrow cells (Gorsheim and Gardner, 1970; Brown and Adamson, 1974), rat bone marrow cells (Dukes, 1971), and rabbit, sheep, and dog bone marrow cells (Brown and Adamson, 1974). Other investigators have reported that cyclic AMP produces no stimulatory effect on heme synthesis and some even reported an inhibitory action on this process in mouse bone marrow (Bottomley *et al.,* 1971; Brown and Adamson,

1977a), rat fetal liver (Graber *et al.*, 1972), guinea pig bone marrow (Brown and Adamson, 1974), rat bone marrow (Brown and Adamson, 1974; Olander, 1975), and rabbit bone marrow (Rodgers *et al.*, 1976). Cyclic GMP has no effect on heme synthesis in fetal rat liver cultures (Graber *et al.*, 1972), or in rabbit marrow cultures (Rodgers *et al.*, 1976).

Studies with canine bone marrow cells (Brown and Adamson, 1974) demonstrate that cyclic AMP stimulates heme synthesis in this species. These investigators also suggest that certain canine marrow cells responded to Ep, whereas others responded to cyclic AMP. This distinction between cells was based on the observations that Ep-sensitive cells possessed different sedimentation properties than cyclic AMP-sensitive cells. While there was some overlap in sedimentation, cells responding to Ep sedimented less rapidly than cells sensitive to cyclic AMP (Brown and Adamson, 1977a). This indicates that cyclic AMP-responsive cells are larger or less mature than Ep-responsive cells, or may be in a different stage of the cell cycle (Brown and Adamson, 1977b).

Differences in culture techniques and batches of serum also çóntribute to differences in the effect of cyclic AMP on heme synthesis (Olander, 1975). Olander (1975) compared rat bone marrow cultures containing serum from two separate lots obtained from the same company. Ep stimulated heme synthesis in both sets of cultures; however, cyclic AMP stimulated heme synthesis in cultures containing one batch of serum and not in cultures containing the other batch. It was suggested (Olander, 1975) that an unknown factor in serum could modify both the direction and degree of the cyclic AMP effect on heme synthesis.

Studies designed to test the effects of Ep on cyclic nucleotide content have consistently shown an elevation in cyclic GMP levels. Rodgers *et al.* (1976) found that cyclic GMP concentrations are increased after 1 hour of incubation with Ep in rabbit bone marrow cultures. In fetal rat liver cultures, which have a greater percentage (65–70%) of erythroid cells than does bone marrow (35%), cyclic GMP is also elevated by Ep (Fig. 1). However, the cyclic AMP content of these cells does not change (Fig. 2). The increase in cyclic GMP content is potentiated by the phosphodiesterase inhibitor, 1-methyl-3-isobutylxanthine (MIX) (Graber *et al.*, 1977). In addition, Graber *et al.* (1977) found that the periods of fetal development during which Ep was able to stimulate heme synthesis in fetal liver cells coincides with the period in which Ep increases cyclic GMP content. *In vivo* experiments indicate that the Ep-induced increase in cyclic GMP is a physiological event. Cyclic nucleotide content was measured in spleens of mice exposed to hypoxia, an event that elevates titers of Ep (Rodgers *et al.*, 1976). In this study, cyclic GMP content is

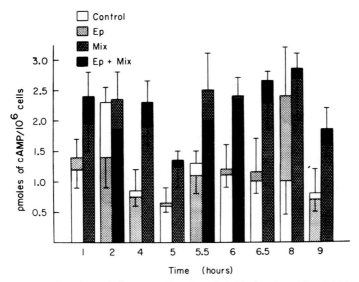

**Fig. 2.** Effect of erythropoietin (Ep) and/or 1-methyl-3-isobutylxanthine (MIX) on cyclic AMP levels in fetal liver cultures. Ep (100 mU) or MIX ($5 \times 10^{-6}$ M) was added at the beginning of the incubation. Each value represents the mean ± SEM of three individual experiments, each performed in triplicate.

increased, and cyclic AMP levels are decreased with as little as 15 minutes of exposure to hypoxia.

The elevation of cyclic GMP occurs at 4 hours (Fig. 1) after the administration of Ep. This time lag suggests that Ep produces its effects on cyclic GMP indirectly by first affecting other cellular events which must occur prior to the increase in cyclic GMP. There is evidence that Ep treatment results in an enhanced rate of synthesis of cyclic GMP since inhibition of phosphodiesterase activity by MIX increases cyclic GMP content. This conclusion is based on our observations that the concentration of MIX used in these studies only increases cyclic GMP content 2- to 3-fold, whereas Ep plus MIX treatment produces 15- to 20-fold increases in cyclic GMP concentrations (Fig. 1).

Since Ep is capable of stimulating iron incorporation into heme and since cyclic GMP content is elevated by Ep administration, it seemed reasonable to postulate that cyclic GMP should also stimulate iron incorporation into heme. This would occur if cyclic GMP were an intracellular mediator of this process. However, this does not appear to be the case. The direct addition of cyclic AMP, cyclic GMP and/or MIX to cultures does not affect $^{59}$Fe incorporation into heme (Table I).

Table I    Effect of Ep, Methylisobutylxanthine (MIX), or the Cyclic Nucleotides on Heme Synthesis in Fetal Liver Cultures

| | $^{59}$Fe incorporation[a] (%) | |
|---|---|---|
| Agent | Before addition of Ep | After addition of Ep |
| Control | $100 \pm 7$ | $246 \pm 29$ |
| db-cAMP $1 \times 10^{-6}$ $M$ | $89 \pm 3^b$ | $218 \pm 50$ |
| 8-Br-cGMP $1 \times 10^{-6}$ $M$ | $125 \pm 17^b$ | $258 \pm 32$ |
| MIX $1 \times 10^{-5}$ $M$ | $129 \pm 12^b$ | $294 \pm 63$ |
| db-cAMP + MIX | $86 \pm 7^b$ | $238 \pm 47$ |
| 8-Br-cGMP + MIX | $134 \pm 23^b$ | $304 \pm 60$ |

[a] Transferrin-bound $^{59}$Fe was added at 20 hours of incubation and was continued for a total of 24 hours ($n = 3$).

[b] Significant difference from value with corresponding designation ($p < 0.05$).

The lag in cyclic GMP accumulation following administration of Ep is unlike that seen with other hormonal systems (e.g., epinephrine stimulation of cyclic AMP or acetylcholine stimulation of cGMP) where there is an immediate increase in cyclic nucleotide content upon administration of the hormone. This indicates that the effect of Ep on the cellular content of cyclic GMP does not occur as a result of classical hormonal responses as seen in other systems.

## IV.    CYCLIC NUCLEOTIDES AND ERYTHROID CELL PROLIFERATION

Both cyclic AMP and cyclic GMP have been suggested as possible components in the regulation of cell proliferation and differentiation in a variety of cells. However, insufficient data are available to completely describe such cellular control mechanisms.

Earlier studies have shown that DNA synthesis, which is a prerequisite for heme synthesis, occurs 4 to 12 hours after the addition of Ep to both fetal liver and bone marrow cultures (Hodgson, 1962; Paul and Hunter, 1968; Gross and Goldwasser, 1970; Paul et al., 1973). In the fetal liver cultures the Ep-induced DNA synthesis occurs coincident with the Ep-induced elevation in the intracellular levels of cyclic GMP (Fig. 1). With this in mind we have undertaken several experiments designed to elucidate the relationship between Ep induction, DNA synthesis, and elevated levels of cyclic GMP. Fetal liver cells were synchronized by double thymidine blockade, and the cyclic nucleotide levels monitored throughout the cell cycle in the presence and absence of Ep. The cell cycle of the control cultures was approximately 16 hours long (Figs. 3 and 4). Interestingly, incubation of these same synchronized cells with Ep shortened the cell cycle to approximately $10\frac{1}{2}$ hours (Figs. 3 and 4). In the control cultures the levels of both cyclic GMP and cyclic AMP oscillated in a reciprocal manner. In these cells the cyclic GMP concentrations were elevated during DNA synthesis and mitosis and cyclic AMP levels increased in $G_2$ and $G_1$. During mitosis cyclic AMP levels were lowest while cyclic GMP concentrations were greatly increased (Figs. 3 and 4).

When Ep was added to the synchronized cultures the cell cycle was shortened by approximately $5\frac{1}{2}$ hours. In addition, the cyclic nucleotide levels retained their same temporal association with the phases of the cell

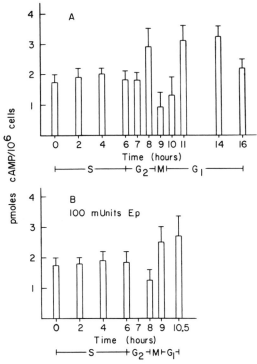

**Fig. 3.** Cyclic AMP concentrations in fetal liver cultures at various stages of the cell cycle. Cyclic AMP levels are expressed as the average ± SEM of six individual determinations. The various phases of the cell cycle are S (DNA synthesis), $G_1$ and $G_2$ (gap phases), and M (mitosis). A represents cyclic events in fetal liver cells in the absence of erythropoietin (Ep). B represents cyclic events in fetal liver cells following addition of Ep (100 mU) which decreased the duration of the cell cycle.

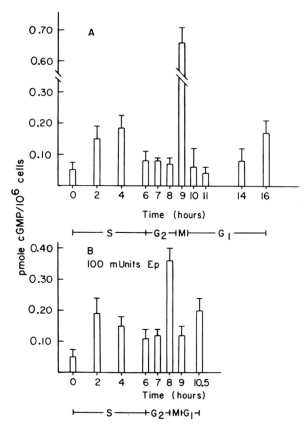

**Fig. 4.** Cyclic GMP concentrations in fetal liver cultures at various stages of the cell cycle. Cyclic AMP levels are expressed as the average ± SEM of six individual determinations. The various phases of the cell cycle are S (DNA synthesis), $G_1$ and $G_2$ (gap phases), and M (mitosis). A represents cyclic events in fetal liver cells in the absence of erythropoietin (Ep). B represents cyclic events in fetal liver cells following addition of Ep (100 mU) which decreased the duration of the cell cycle.

cycle and with each other. This could be interpreted as supporting their involvement in the control of the cell cycle or cell cycle related events. If the cell cycle related fluctuations in both cyclic nucleotides had changed as a result of the shortening of the cell cycle, then that evidence could argue against their being involved in the control of cell proliferation or other cell cycle related events.

Further work was done to assess the possible relationship between the coincident effects of Ep on increasing cyclic GMP concentrations and stimulating DNA synthesis. In one set of experiments fetal liver cells in

**Table II   Effect of FdUrd$^a$ on Ep Stimulation of Cyclic GMP Concentrations**

| Agent | Cyclic GMP/$10^6$ cells (pmole) |
|---|---|
| Control | $0.037 \pm 0.009$ |
| Ep (100 mU) | $0.096 \pm 0.010^b$ |
| Ep + MIX | $0.214 \pm 0.016^b$ |
| MIX ($1 \times 10^{-5}$ $M$) | $0.074 \pm 0.012^b$ |
| FdUrd ($1 \times 10^{-4}$ $M$) | $0.035 \pm 0.008$ |
| Ep + FdUrd | $0.087 \pm 0.009^b$ |
| FdUrd + MIX | $0.062 \pm 0.008^b$ |
| Ep + FdUrd + MIX | $0.234 \pm 0.020^b$ |

$^a$ 5-Fluorodeoxyuridine (FdUrd) = $1 \times 10^{-4}$ $M$.
$^b$ Indicates significant difference from control with $p < 0.05$ ($n = 5$).

exponential growth were incubated for 6 hours with Ep, dibutyryl (db)-cyclic AMP, 8-bromo (8-br)-cyclic GMP, or MIX and DNA synthesis monitored (Table II). Ep stimulated DNA synthesis to 185% of control; however, neither db-cyclic AMP, 8-br-cyclic GMP, nor MIX had an effect on the rate of DNA synthesis at this time period.

In still other experiments DNA synthesis was blocked by 5'-fluordeoxyuridine (FdUrd) and cyclic GMP levels monitered. This was done in both control and Ep-stimulated cultures. FdUrd blocked DNA synthesis in both groups of cultures but had no effect on the elevation of cyclic GMP concentrations in the Ep-stimulated cells (Table III). Cyclic GMP levels were also monitered in fetal liver cultures at 6 hours of incubation with an antibody to Ep (anti-Ep). The antibody alone had no

**Table III   Effect of Ep and Cyclic Nucleotides on DNA Synthesis**

| Agent | Control (%) |
|---|---|
| Control | $100 \pm 4$ |
| Ep (100 mU) | $185 \pm 11^a$ |
| MIX ($1 \times 10^{-5}$ $M$) | $108 \pm 5$ |
| cAMP ($1 \times 10^{-5}$ $M$) | $97 \pm 4$ |
| cGMP ($1 \times 10^{-5}$ $M$) | $106 \pm 7$ |
| Ep + MIX | $176 \pm 10^a$ |
| FdUrd ($1 \times 10^{-4}$ $M$) | $0.05 \pm 0.01$ |
| FdUrd + Ep | $0.08 \pm 0.01$ |

$^a$ Indicates significant difference from control with $p < 0.05$ ($n = 9$).

Table IV   Effect of Ep and Anti-Ep on Cyclic GMP
Concentrations in Fetal Liver Cells at 6 Hours of Incubation

| Agent | Cyclic GMP/$10^6$ cells (pmole) |
|---|---|
| Control | $0.027 \pm 0.007$ |
| Ep (100 mU) | $0.064 \pm 0.014^a$ |
| Ep + MIX | $0.235 \pm 0.045^a$ |
| MIX ($1 \times 10^{-5}$ M) | $0.058 \pm 0.009^a$ |
| Anti-Ep | $0.026 \pm 0.004$ |
| Ep + Anti-Ep | $0.035 \pm 0.009$ |
| Ep + Anti-Ep + MIX | $0.068 \pm 0.009^a$ |
| MIX + Anti-Ep | $0.064 \pm 0.008^a$ |

[a] Indicates significant difference from control with $p < 0.05$
($n = 9$).

effect on cyclic GMP levels. However, it did block the increase in cyclic GMP content induced by Ep (Table IV).

These studies indicate that the Ep-induced increase in cyclic GMP concentrations appears to be neither a cause nor a result of the stimulation of DNA synthesis. The addition of a lipophilic derivative of cyclic GMP, as well as cyclic AMP or MIX, had no effect on DNA synthesis. This indicates that cyclic GMP is not an intermediate in the stimulation of DNA synthesis. If cyclic GMP was acting as a traditional "second messenger," the addition of such a second messenger would be expected to produce the same effect as the primary hormone, Ep. This does not appear to be the case. Furthermore, the inhibition of DNA synthesis by FdUrd has no effect upon the elevation of concentrations of cyclic GMP by Ep. Therefore, it does not seem likely that cyclic GMP is involved in the biochemical events which result in heme synthesis.

Ep does stimulate cyclic GMP accumulation in the fetal liver cells. The fact that the antibody to Ep blocks this increase in cellular cyclic GMP content indicates that Ep may affect other cellular events in addition to those directly responsible for heme synthesis.

## V.  CONCLUSIONS

1. Erythropoietin (Ep) stimulation of fetal rat liver cultures causes an increase in the intracellular levels of cyclic GMP, but not AMP, at the same time that it stimulates DNA synthesis.

2. The Ep-induced increase in cyclic GMP concentrations appears to be neither a cause nor a result of the stimulation of DNA synthesis.

3. Cyclic GMP does not appear to be an intracellular mediator of the increase in heme synthesis seen following Ep stimulation. It does not appear to act as in the manner of a classical "second messenger" following stimulation by Ep.

## REFERENCES

Alpen, E. L., and Cranmore, D. (1959). *Ann. N.Y. Acad. Sci.* **77,** 753-765.

Alpen, E. L., Cranmore, D., and Johnston, M. E. (1962). *In,* "Erythropoiesis" (L. O. Jacobson and M. Doyle, eds.), p. 184. Grune & Stratton: New York.

Bottomley, S. S., Whitcomb, W. H., Smithee, G. A., and Moore, M. Z. (1971). *J. Lab. Clin. Med.* **77,** 793-801.

Brown, J. E., and Adamson, J. W. (1974). *Blood* **44,** 913.

Brown, J. E., and Adamson, J. W. (1977a). *Br. J. Haematol.* **35,** 193-208.

Brown, J. E., and Adamson, J. W. (1977b). *Cell Tissue Kinet.* **10,** 289-298.

Bruce, W. R., and McCulloch, E. A. (1964). *Blood* **23,** 216.

Chang, C. S., and Goldwasser, E. (1973). *Dev. Biol.* **34,** 246-254.

Chui, D. H. K., Djaldetti, M., Marks, P. A., and Rifkind, R. A. (1971). *J. Cell Biol.* **51,** 585-595.

Cole, R. J., and Paul, J. (1966). *J. Embryol. Exp. Morphol.* **15**(2), 245-260.

Conkie, D., Kleiman, L., Harrison, P. F., and Paul, J. (1975). *Exp. Cell Res.* **93,** 315-324.

Curry, J. L., Trentin, J. J., and Wolf, N. (1967). *J. Exp. Med.* **125,** 703-720.

Datta, M. C., and Dukes, P. P. (1975). *Biochem. Biophys. Res. Commun.* **66,** 293-302.

DeRobertis, E. D. P., Salz, F. A., and DeRobertis, E. M. F. (1975). "Cell Biology," 6th Ed., Chapter 2, pp. 13-36. Saunders, Philadelphia.

Djaldetti, M., Preisler, H., Marks, P. A., and Rifkind, R. A. (1972). *J. Biol. Chem.* **247,** 731-735.

Dukes, P. P. (1971). *Blood* **38,** 822.

Evans, E. S., Contopoulos, A. N., and Simpson, M. E. (1957). *Endocrinology* **60,** 403.

Evans, E. S., Rosenberg, L. L., and Simpson, M. E. (1961). *Endocrinology* **68,** 517.

Filmanowicz, E., and Gurney, C. W. (1961). *J. Lab. Clin. Med.* **57,** 65.

Fisher, D. B., and Mueller, G. C. (1971). *Biochim. Biophys. Acta* **248,** 434-448.

Fisher, J. W. (1969). *In* "Biological Basis of Medicine" (E. E. Bittary and N. Bittar, eds.), Vol. 3, pp. 41-71. Academic Press, New York.

Fisher, J. W. (1972). *Pharmacol. Rev.* **24**(3), 459-508.

Fisher, J. W., and Birdwell, B. J. (1961). *Acta Haematol.* **26,** 244.

Fisher, J. W., and Crook, J. J. (1962). *Blood* **19,** 557.

Fisher, J. W., Sanzari, N. P., Birdwell, B. J., and Crook, J. J. (1962). *In* "Erythropoiesis" (L. O. Jacobson and M. Doyle, eds.), p. 78. Grune & Stratton, New York.

Fisher, J. W., Lajtha, L. G., Buttoo, A. S., and Porteous, D. D. (1965). *Br. J. Haematol.* **11,** 342.

Fried, W., Kilbridge, T., Krantz, S., McDonald, T. P., and Lange, R. D. (1969). *J. Lab. Clin. Med.* **73,** 244.

Gordon, A. S., LoBue, J., Dornfest, B. S., and Cooper, G. W. (1962). *In* "Erythropoiesis" (L. O. Jacobson and M. Doyle, eds.), p. 321. Grune & Stratton, New York.

Gorsheim, D., and Gardner, F. H. (1970). *Blood* **36,** 847.

Graber, S. E., Carrillo, M., and Krantz, S. B. (1972). *Proc. Soc. Exp. Biol. Med.* **141,** 206-210.

Gross, M., and Goldwasser, E. (1969). *Biochemistry* **8**(5), 1795–1805.

Gross, M., and Goldwasser, E. (1970). *J. Biochem. Chem.* **245**(7), 1632–1636.

Handler, E. E., Mendelson, N., and Handler, E. S. (1974). *Blood* **44**(4), 535–542.

Hayes, E. F., Firkin, F. C., Koga, Y., and Hays, D. M. (1975). *J. Cell Physiol.* **86**, 213–220.

Krantz, S. B., and Goldwasser, E. (1965). *Biochim. Biophys. Acta* **103**, 325–332.

Krantz, S. B., and Goldwasser, E. (1970).

Kuna, S., Gordon, A. S., Morse, B. S., Lane, F. B., III, and Charipper, H. A. (1959). *Am. J. Physiol.* **196**, 769.

Maniatis, G. M., Rifkind, R. A., Bank, A., and Marks, P. (1974). *Adv. Exp. Med. Biol.* **44**, 221–243.

Morse, B. S., Rencricca, N. J., and Stohlman, F. (1969). *Proc. Soc. Exp. Biol. Med.* **130**, 986–989.

Morse, B. S., Rencricca, N. J., and Stohlman, F. (1970). *Blood* **35**, 761–774.

Nicol, A. G., Conkie, D., Lanyon, W. G., Drewienkiewicz, C. E., Williamson, R., and Paul, J. (1972). *Biochim. Biophys. Acta* **277**, 342–363.

O'Grady, L. F., and Lewis, J. P. (1970). *J. Lab. Clin. Med.* **76**, 445–450.

O'Grady, L. F., Lewis, J. P., and Throbaugh, F. E. (1968). *J. Lab. Clin. Med.* **71**, 693–703.

Olander, C. P. (1975). *Experimentia* **31**(8), 981–983.

Ortega, J. A., and Dukes, P. P. (1970). *Biochim. Biophys. Acta* **204**, 334–339.

Paul, J., and Hunter, J. A. (1968). *Nature (London)* **219**, 1362–1363.

Paul, J., and Hunter, J. A. (1969). *J. Mol. Biol.* **42**, 31–41.

Paul, J., Conkie, D., and Burgos, H. J. (1973). *Embryol. Exp. Morphol.* **29**, 453–472.

Reissmann, K. R., and Nomura, T. (1962). *In* "Erythropoiesis" (L. W. Jacobson and M. Doyle, eds.), p. 71. Grune & Stratton, New York.

Rodgers, G. M., Fisher, J. W., and George, W. J. (1976). *Biochem. Biophys. Res. Commun.* **70**, 287–294.

Stohlman, F., Jr. (1964). *Ann. N.Y. Acad. Sci.* **119**, 578–585.

Stohlman, F., Jr. (1970a). *In* "Formation and Destruction of Blood Cells" (T. J. Greenwalt and G. A. Jamieson, eds.), pp. 65–82. Lippincott, Philadelphia.

Stohlman, F., Jr. (1970b). "Hemopoietic Cellular Proliferation." Grune & Stratton, New York.

Stohlman, F., Jr. (1971). *In* "Kidney Hormones" (J. W. Fisher, ed.), Academic Press, New York.

Terada, M., Raminez, F., Cantor, L., Maniatis, G. M., Bank, A., Rifkind, R. A., and Marks, P. A. (1975). *Proc. Int. Conf. Erythropoiesis, 4th,* pp. 23–32.

Zangheri, E. D., Compana, H., Ponce, F., Silva, J. C., Fernandez, F. D., and Suarez, J. R. E. (1963). *Nature (London)* **199**, 572.

# 12

# Calcium and Cyclic Nucleotide Interactions during the Cell Cycle

PAUL A. CHARP and GARY L. WHITSON

## I. INTRODUCTION*

Since the discovery of the first cyclic nucleotide, cAMP, by Sutherland and Rall (1958), this compound has been shown to play an important

*Key to abbreviations used in this chapter: cAMP, adenosine-3′,5′-monophosphate; cGMP, guanosine-3′,5′-monophosphate; AC, adenyl cyclase; GC, guanyl cyclase; PDE, phosphodiesterase; db-cAMP, dibutyryl cAMP; CDR, calcium-dependent regulatory protein; PK, protein kinase; EHS, end of heat shock.

NUCLEAR–CYTOPLASMIC INTERACTIONS
IN THE CELL CYCLE

role in many physiological processes in most cell types, both prokaryote and eukaryote (Drummond *et al.,* 1975). Although their exact biological role has not yet been elucidated, cyclic nucleotides can effect nucleic acids by either inhibiting (Froehlich and Rachmeler, 1972, 1974) or stimulating both DNA and RNA synthesis (Abell and Monihan, 1973). Cyclic nucleotides also affect growth, causing both developmental and morphological changes (O'Neill *et al.,* 1976). Changes in cyclic nucleotide levels also signal the onset of cell division (Hadden *et al.,* 1972; Charp and Whitson, 1978) as there is a large increase in the intracellular levels of cGMP and a concommitant decrease in cAMP just before or during the division process (Charp and Whitson, 1978; see George *et al.,* this volume, Chapter 11).

More recently, the role of calcium in relation to alteration in cyclic nucleotide levels and its role in cell regulation (Rassmussen, 1970; Rassmussen *et al.,* 1972; Berridge, 1975; Rebhun, 1977) has stimulated an extensive search for calcium-dependent and calcium regulatory proteins (Kretsinger, 1976). What roles these intracellular regulators play in the control of cellular physiological processes remains to be determined.

In this chapter we will discuss changes in cyclic nucleotide levels during the cell cycle as related to cell cycle progression as well as the interactions of calcium and cyclic nucleotides which occur during cell division.


## II.  CYCLIC NUCLEOTIDE CHANGES DURING THE CELL CYCLE

There have been numerous studies that indicate a positive role for cyclic nucleotides as controllers or regulators for cell cycle progression. The majority of these studies involve the measurement of cAMP in conjunction with the ability of certain cell types to traverse a certain phase of the cell cycle. The methods for measuring cyclic nucleotides as well as the enzymatic mechanisms involved in the metabolism of cyclic nucleotides are covered in detail in Brooker *et al.* (1979).


### A.  $G_0$ and $G_1$ Periods

Due to the low percentage of mammalian cells in $G_1$ in an asynchronous population, many investigators have synchronized cells using the mitotic shakeoff procedure to yield large populations of cells in the $G_1$ phase. Sheppard and Prescott (1972) synchronized CHO cells and compared the levels of cAMP occurring during mitosis, early $G_1$, late $G_1$, and S. Their results showed that the intracellular cAMP levels increased beginning with mitosis and reached a peak in early $G_1$, decreased as cells approached late $G_1$, and increased slightly as the cells entered the S phase. Similar results were seen if cells were synchronized by isoleucine deprivation, indicating that cells blocked in $G_1$ show higher levels of cAMP than cells progressing normally through $G_1$.

Tissue cultures on reaching the confluent state of growth, no further cell divisions occurring, are said to be in $G_0$. This stage may also be obtained by depletion of selected nutrients necessary for cell proliferation. Seifert and Rudland (1974a,b) starved mouse 3T3-4a and SV3T3 fibroblasts by either removing phosphate, histidine, glutamine, leucine, methionine, or serum. In 3T3 cells, over 70% of all cells were stalled in $G_1$ with the maximum (94%) stalled after depletion of serum, glutamine, and histidine. Upon replenishment by addition of missing nutrients, cAMP levels decreased as the cells returned to the cycling state except in the case of addition of methionine. Intracellular measurement of cGMP showed that there was a transient increase in this cyclic nucleotide. Just before cells entered S, cAMP increased and cGMP decreased. These results indicate that elevated cGMP is only associated with movement of cells through early $G_1$ and cAMP shows regular variation during the cycle. This may also suggest that cells have the capacity to measure the ratio or absolute levels of cyclic nucleotides and use this capability to regulate cell cycle progression. Supporting data obtained by Moens *et al.* (1975) also showed that, after the addition of serum to mouse cells, cAMP levels decreased as the cGMP levels increased. More recently Yasuda *et al.* (1978) showed that, after addition of serum to cultures of BALB/c3T3 cells, there was a transient increase in cGMP followed by a decrease.

Russell and Stambrook (1975) measured the cAMP levels in V79 cells synchronized by the mitotic selection procedure and showed that 2 hours after cell division in early $G_1$, cAMP levels began to increase and reached a point over twofold higher at the beginning of S phase. At the same time as the increase in cAMP was observed, the activity of ornithine decarboxylase was shown to increase as well as the concentrations of polyamines. Thus, cAMP appears to play a regulatory role in the enzymatic activity of polyamine biosynthetic pathways associated with the $G_1$ phase of the cell cycle.

Burger *et al.* (1972) treated confluent 3T3 cells with proteases to remove growth restrictions and allow the cells to progress through one cell cycle. After treatment, the intracellular levels of cAMP decreased as cells began to reenter the cycle from $G_0$ to $G_1$. They found that cells treated with proteolytic enzymes and given db-cAMP were prevented from further progression. Thus it appears that high intracellular levels of cAMP in some manner restricts progression through the cell cycle. Fluctuations in cAMP content in cells is necessary, particularly the lowering of cAMP in order for progression through the cell cycle and subsequent cell division.

Similar patterns of cyclic nucleotide fluctuations are seen in HeLa cells (Zeilig *et al.*, 1974), sea urchin embryos (Yasumasu *et al.*, 1973), CHO cells (Hsie *et al.*, 1976; Costa *et al.*, 1976), and BALB/c3T3 cells (Yasuda *et al.*, 1978). All of these studies indicate that $G_1$ arrested cells are associated with high intracellular levels of cAMP. The arrest can be induced, as stated earlier, by serum or nutrient depletion or by contact inhibition as shown in studies by Heidrick and Ryan (1971) and Otten *et al.* (1972). Contact inhibition is seen as a consequence of changes in membrane protein conformation which results in decreased PDE activity and accompanied increases in AC activity, all of which results in increased cAMP levels (Anderson *et al.*, 1973). Even in transformed cell lines, those lines that no longer possess normal contact inhibition, cell cycle progression can be stopped by addition of exogenous cAMP in the form of either db-cAMP (Oler, 1973; Grimes and Schroeder, 1973; Robinson *et al.*, 1974; Eker, 1974) or by agents capable of stimulation of AC activity *in vivo* (Brønstad *et al.*, 1971).

## B. S Phase

Whereas decreasing levels of cAMP appear to be necessary for cells to progress through $G_1$, there seems to be an increase in cAMP just before cells enter the S phase. Using mouse 3T3 fibroblasts, Siefert and Rudland (1974a) showed that high concentrations of cGMP ($10^{-6}$ to $10^{-4}$ $M$) stimulated DNA synthesis. Addition of serum to starved cultures also resulted in a rapid increase in intracellular cGMP. They also found that as cells progress through S, cGMP decreased and remained low, whereas cAMP increased before S and remained at this elevated level until the end of S. Using BALB/c3T3 cells, Seifert and Rudland (1974b) showed that as cells approached S, the intracellular level of cGMP decreased and remained at a depressed level as cells entered S and progressed through DNA synthesis.

In human lymphocytes treated with mitogenic agents, marked increases in cGMP were observed with little increase in cAMP, leading to an approximate 100-fold increase in DNA synthesis (Hadden *et al.*, 1972). These results seem to indicate that an increase in cGMP is necessary in order to stimulate DNA synthesis.

Wang *et al.* (1978), using concanavalin A stimulated mouse lymphocytes, reported that after 10 hours posttreatment, an initial increase in cAMP was followed by a subsequent decrease in cAMP and that this was necessary for cells to enter S. If the decrease in cAMP was inhibited by PDE blocking agents or addition of cAMP analogs, the cells were prevented from entering the S phase. Conversely, Russell and Stambrook (1975), using synchronized V79 cells, showed that cAMP levels increased steadily from $G_1$ to S, peaking at mid-S (2.5-fold increase above mid $G_1$), and then began to decrease as cells left S and progressed into $G_2$. Obviously, the interpretations of data are confusing or conflicting. For if a decrease in cAMP is necessary for progression into S, why do some systems require high intracellular levels of cAMP for DNA synthesis to occur? Van Wijk *et al.* (1973), using Reuber H35 cultured hepatoma cells synchronized by a single thymidine block, added db-cAMP and reported that DNA synthesis was reduced as measured by an increase in the length of S (10.5 hours in control to 29 hours in treated cells). Also, the rate of [$^3$H]deoxyadenosine transport was suppressed by the addition of the analog. This was the only phase of the cell cycle that was altered by the addition of exogenous cAMP. Their conclusion was that db-cAMP somehow disrupts the pyrimidine deoxyribonucleotide pathways, thus resulting in a decrease mainly in thymine precursors to DNA synthesis since the decrease in synthesis could be overcome by addition of thymidine to db-cAMP treated cultures. Addition of db-cAMP to 3T3 cells in early $G_1$ also leads to S blockage (Willingham *et al.,* 1972). However, if db-cAMP was added at plating time and then removed after 11 hours, DNA synthesis occurred at the scheduled time but at a decreased rate, leading to an increase in the length of S. They found, however, if db-cAMP was added to cultures 3 to 6 hours after plating, DNA synthesis was accelerated yielding an earlier peak of S. Willingham and co-workers then concluded that cAMP, therefore, had three regulatory roles in cell cycle progression: (1) High cAMP levels block cells in $G_1$; (2) high levels of cAMP stimulate DNA synthesis when added exogenously to cells in early $G_1$; (3) inhibition of mitosis occurs when cAMP is added exogenously to cells in the S phase.

The acellular slime mold *Physarum,* in addition to having no detectable $G_1$ phase, is a naturally synchronous system; all the nuclei share a common cytoplasm and upon receiving some unknown signal, all divide in synchrony and progress directly into the next S phase. This organism, therefore, is an excellent model to study cyclic nucleotide regulation of S. Kuehn (1972), using *Physarum,* reported a cell cycle dependence of kinase activity resulting in first an activation and then an inhibition of PK by cAMP. As the nuclei entered S, the kinase activity was enhanced but as the nuclei approached the $G_2$ stage, the cAMP inhibited kinase activity. Thus cAMP regulates the progression of the cell cycle in *Physarum* by modulating the activity of PK. It is of interest to note that in *Physarum* systems, it has been shown that kinase activity does not phosphorylate histones (Kuehn, 1971). It is possible that *Physarum* kinases are predominately cGMP-dependent as seen in *Tetrahymena* (Murofushi, 1974). These PK's have different affinities for substrates. Two types prefer casein and histone and are activated by low concentrations of cGMP while the third type is, to a lesser degree, stimulated by cGMP and prefers histone and protamines as the substrate. Kuehn (1972) did not investigate the possibility of cGMP as a stimulator of *Physarum* histones.

More recent investigation into the role of protein kinases and cell cycle progression has indicated the presence of two inhibitors of kinase activity (Costa, 1977) in CHO cells. One type is a general inhibitor of kinase activity and the other inhibitor is specific for cAMP-dependent PK. Costa and co-workers (1976) measured the activity of types I and II of protein kinases and showed that during the traverse from $G_1$ to S, the activity of type II rapidly increased and type I activity remained relatively constant. During S, type II activity decreased, while the activity of type I PK increased. Thus, their results indicate that, whereas type I PK remains relatively stable throughout $G_1$ and S, type II activity may be necessary for the progression into S. The role of the PK inhibitors (Costa, 1977) has yet been defined as pertaining to cell cycle progression. A needed experiment would be to assay type II kinase using a thymidine block to halt progression of S or deprive cells of essential nutrients and then determine the relative activity of the protein kinases and inhibitors under these conditions.

The role of cyclic nucleotides in the progression of S is, as indicated, by no means clear. The general trend is that cAMP increases, then decreases as cGMP decreases, and remains at a decreased level during S. The protein kinase story is also far from complete because all we really know is that cAMP has been shown to stimulate and inhibit different PK activity during different phases of the cell cycle. For instance, it is not known what proteins are phosphorylated by these kinases. Also, to our knowledge there have been no studies on cGMP-dependent protein kinases during the cell cycle.

## C.  $G_2$ and Division

Since there are no excellent methods to synchronize cells to obtain cultures with a high degree of $G_2$ specificity, very few studies have been done to study cyclic nucleotide levels in that stage of the cell cycle. Dickinson *et al.* (1977) used *Tetrahymena* synchronized with a single hypoxic shock accumulated cells in the $G_2$ phase. Their results showed that during the hypoxic treatment, the cells accumulated cAMP, resulting in an increase of almost 100-fold as AC activity decreased. However, the increase in cAMP could be due to a decrease of PDE or release of cAMP from PK. Gray *et al.* (1977) used the same system and measured intracellular cGMP levels and showed that cGMP increased rapidly during the initial start of hypoxia and remained at the elevated level during the hypoxia. Once hypoxia was halted, the cGMP levels rapidly decreased, although not as rapidly as cAMP (Dickinson *et al.*, 1977). However, just before cell division, as measured by an increase in cell number, there is a rapid spike of cGMP.

Artificially elevating cAMP can also arrest cells in the $G_2$ phase as seen in human lymphoid cells (Millis *et al.*, 1972). However, the blockage in $G_2$ is not entirely specific since some cells may be blocked in S as seen in rat liver cells (Nose and Katsuta, 1975), melanoma cells (DiPasquale and McGuire, 1977), and 3T3 cells (Willingham *et al.*, 1972). Therefore, there does not seem to be a specific transition point located in early $G_2$ that is under control by cyclic nucleotides.

Cyclic nucleotide levels in relation to mitosis and cell division are well documented. Most data show that cAMP levels decrease while cGMP levels increase simultaneously. The first cleavages of sea urchin eggs after fertilization are synchronous and thus this system provides for precise measurements of cyclic nucleotide changes in conjunction with cell division. Yasumasu *et al.* (1973), using this system, measured cAMP content during the three cleavages after fertilization. The intracellular cAMP content rose during the 50 minutes after fertilization, decreased as division began, and repeated this pattern as the second and third division approached and occurred. It is noted that early embryos do not

**Table I  Concentration of cAMP in Synchronized Cultures of *Tetrahymena pyriformis*[a]**

| | pM/$10^6$ cells ±SE | |
|---|---|---|
| Minutes after EHS | Control | Irradiated |
| 50 | 4.2 ± 0.3 | 1.5 ± 0.2 |
| 75 | 4.6 ± 0.4 | 2.1 ± 0.2 |
| 100 | 3.4 ± 0.4 | 1.6 ± 0.2 |
| 110 | 2.95 ± 0.3 | 1.4 ± 0.1 |
| 120 | 2.3 ± 0.2 | 1.2 ± 0.3 |
| 130 | 2.3 ± 0.2 | 1.3 ± 0.3 |

[a] Peak of cell division in controls occurs at EHS + 75 minutes. In cultures irradiated with 5000 R, the peak occurs at EHS + 110 minutes.

**Table II**   Concentrations of cGMP in Synchronized Cultures of *Tetrahymena pyriformis*[a]

| | $pM/10^6$ cells $\pm$ SE | |
|---|---|---|
| Minutes after EHS | Control | Irradiated |
| 50 | $3.2 \pm 0.5$ | $10.9 \pm 1.0$ |
| 75 | $3.9 \pm 0.3$ | $5.1 \pm 0.5$ |
| 100 | $324.8 \pm 16$ | $7.4 \pm 0.4$ |
| 110 | $197.6 \pm 2.0$ | $173.4 \pm 26$ |
| 120 | $225.7 \pm 39$ | $952.8 \pm 24$ |
| 130 | $184.2 \pm 2.3$ | $1932 \ \ \pm 166$ |

[a] Peak of cell division in control cultures occurs at EHS + 75 minutes. In cultures receiving 5000 R X-irradiation, the peak of division occurred at EHS + 110 minutes.

possess a $G_1$ phase, thus cAMP is accumulated during S and $G_2$ without affecting the subsequent mitotic event.

Studies involving synchronized *Tetrahymena* cultures have shown that cAMP increased until a point just before cell division and then decreased (Dickinson *et al.*, 1977) while during the same time cGMP levels increased just before cell division (Gray *et al.*, 1977). Using a different synchrony method with *Tetrahymena*, Charp and Whitson (1978) found similar results. However, after X-irradiating cultures with 5000 R, the intracellular levels of cAMP decreased at the same time as the controls but only the intracellular levels of cGMP increased as the treated cells divided, 30 minutes after the control cultures divided (Tables I and II). They state that cGMP is actually the signal that shows cell division is approaching and that cAMP is probably necessary for some other physiological process to occur. Earlier, Hadden *et al.* (1972) used mitogen-stimulated human lymphocytes and showed that cAMP levels increased very little after mitogen treatment while the intracellular levels of cGMP increased greatly.

Taylor-Papadimitriou (1974) showed that when cAMP levels are artificially elevated in Earle's L cells, that cell growth, as measured by mitotic index, is inhibited with accompanying change in cell morphology. Also there is a decrease in DNA synthesis after the second division. Once the cAMP levels are allowed to drop, an increase in DNA synthesis is seen followed by a subsequent increase in the division rate.

### D.   Summary

Intracellular levels of cyclic nucleotides vary considerably during the cell cycle in a variety of eukaryotic cell types. A summary of some of the

**Table III  Summary of Cyclic Nucleotide Levels during the Cell Cycle[a]**

| Cell line | Cyclic nucleotide | $G_1$ | S | $G_2$ | Division | References |
|---|---|---|---|---|---|---|
| *Tetrahymena* | cAMP | | | 27 pM/10⁶ cells | 15 p$M$/10⁶ cells | Dickinson *et al.* (1977) |
| | | | | | 4.6 p$M$/10⁶ cells | Charp and Whitson (1978) |
| | | | | | 2 p$M$/10⁶ cells | Gray *et al.* (1977) |
| | cGMP | | | 6 pM/10⁶ cells | 3.9 p$M$/10⁶ cells | Charp and Whitson (1978) |
| BALB/*c*3T3 | cAMP | 2.2 p$M$/10⁶ cells | 2.2 p$M$/10⁶ cells | | | Seifert and Rudland (1974b) |
| | cGMP | 0.8 p$M$/10⁶ cells | 0.5 p$M$/10⁶ cells | | | |
| | cAMP | 5 p$M$/10⁶ cells[b] | 15 p$M$/10⁶ cells[b] | | | Seifert and Rudland (1974a) |
| | cGMP | 2 p$M$/10⁶ cells[b] | 5 p$M$/10⁶ cells[b] | | | |
| | cGMP | 10[b] | | | | Yasuda *et al.* (1978) |
| CHO | cAMP | 44 (early) | 28 | | 16 | Sheppard and Prescott (1972) |
| | | 24 (late) | | | | |
| 3T3 | cAMP | 25 (confluent) | 22 | | 5 | Burger *et al.* (1972) |
| | cAMP | 59 (0.5% serum) | | | | |
| | cGMP | 0.1 (confluent) | | | | |
| | | 0.6 (0.5% serum) | | | | |
| SV3T3-101 | cAMP | 7 (confluent) | | | | Moens *et al.* (1975) |
| | | 20 (0.5% serum) | | | | |
| | cGMP | 0.1 (confluent) | | | | |
| | | 0.32 (0.5% serum) | | | | |
| V79 | cAMP | 3.5 p$M$/10⁶ cells | 5–6 p$M$/10⁶ cells | 2.5–3 p$M$/10⁶ cells | 3 p$M$/10⁶ cells | Moens *et al.* (1975)<br>Russell and Stambrook (1975) |
| Mouse lymphocytes | cAMP | 0.8 p$M$/10⁶ cells | 0.2 p$M$/10⁶ cells | | | Wang *et al.* (1978) |
| | cGMP | 1.8 p$M$/10⁶ cells[c] | 0.4 p$M$/10⁶ cells[c] | | | |

[a] Results are expressed as picomoles per milligram protein except where noted.

[b] Levels were measured upon addition of missing nutrients ($G_0 \rightarrow G_1$).

[c] Cultures were treated with concanavalin A.

recorded changes are shown in Table III. Note that there is also reported variability of cyclic nucleotide levels of specific cell cycle stages within the same cell types under varying growth conditions. One can only extract general trends in cyclic nucleotide levels from these existing data and, moreover, there are specific phases in the cell cycle where cyclic nucleotide levels remain to be quantified. The greatest fluctuations in cyclic nucleotide levels, however, appear to occur during the $G_1$ phase and also division phase of the cell cycle, particularly in mammalian cells. In general, high cAMP is associated with early $G_1$ and in cells stalled in $G_0$. Low cAMP is associated with cell division, whereas cGMP is highest in cells in $G_2$ or during cell division. In spite of all of these findings, the biochemical and physiological significance of these changes in cyclic nucleotide levels during the cell cycle remains a mystery.

## III. CALCIUM–CYCLIC NUCLEOTIDE INTERACTIONS

### A.  General Considerations

Recent interest in calcium has stemmed from Heilbrunn's (1956) statements that calcium is an important regulator in cellular metabolism, ranging from diverse enzymatic activities (Kretsinger, 1976; Dedman *et al.*, 1977; Beale *et al.*, 1977) to growth control (Bowen-Pope *et al.*, 1979; Rubin and Bowen-Pope, 1979) and possibly cell division (Wolfe, 1973; Steinhardt and Epel, 1974; Paul and Ristow, 1979; Sisken, this volume, Chapter 10). Presently, there are many reviews that cover the various biological roles of calcium (Rassmussen *et al.*, 1972; Berridge, 1975; Duncan, 1976; Kretsinger, 1976; Rebhun, 1977). Therefore, our intentions here are only to review relevant aspects of calcium in relation to cell division and cyclic nucleotide metabolism.

Calcium interaction with cyclic nucleotides presently focuses much attention on the modulation or regulation of enzymatic pathways affecting cyclic nucleotide metabolism, more specifically, the modes of action of CDR proteins. These subjects will be covered in Section IV, C.

### B.  Calcium Changes Associated with Cell Division

Since calcium has been shown to regulate a myriad of cell functions, the logical course of action is to study what effects, if any, calcium has on cell division and how these effects are mediated through some of the known calcium pathways. However, few reports have surfaced dealing with this important subject area.

Using sea urchin embryos, Clothier and Timourian (1972) reported that immediately following fertilization, there is a marked increase in calcium. This rapid increase that they described has also been substantiated by Nakamura and Yasumasu (1974) who also reported a large increase in free calcium after fertilization of sea urchin eggs. Steinhardt and Epel (1974) even showed that if unfertilized sea urchin eggs were treated with the calcium-transporting ionophore A23187, that division could be induced. It is possible, therefore, that a transient increase in calcium provides a "signal" or "trigger" calcium, which is necessary in order for cells to divide. One model proposing this hypothesis has been presented by Harris (1978) in which some control, such as cyclic nucleotides, free radicals, and sulfhydryls accumulate to a critical level thus triggering either calcium uptake or release of calcium from intracellular stores. Harris believes it is the increase in cytoplasmic calcium which could lead to cell division.

In the never-ending search for an exceptional model to study the regulatory events responsible for cell division, the heat-shock synchronized *Tetrahymena* system surfaces as an extremely well-studied model. The synchronizing treatment places the majority of cells (80–90%) in the process of visible cytokinesis 65–75 minutes after the synchronizing treatment, requiring about 25 minutes duration once division has been initiated (Scherbaum and Zeuthen, 1954). London and Whitson (1980) and Charp *et al.* (1980) have reported intracellular fluctuations of calcium in relation to cyclic nucleotides during the division cycle in this system. Using the photoprotein aequorin to determine free calcium and atomic absorption spectroscopy for total calcium levels, London and Whitson (1980) reported that intracellular calcium concentrations follow specific trends leading to cell division. First, there is an influx of free calcium just prior to cell division, possibly the "trigger" or "signal" calcium. After the onset and during the duration of cell division, the free calcium and total calcium decrease (Fig. 1) but bound calcium remains relatively constant. Similar results using calcium-45 have been reported by London *et al.* (1979). Flux studies show that there is an uptake of $^{45}$Ca during a 10–15 minute interval immediately prior to cell division. At the time when cell division is initiated, there is an efflux of $^{45}$Ca which continues during the division cycle and levels off at the end of cell division. During the same time periods, the cyclic nucleotides are showing definite changes in their intracellular levels as well (Fig. 2). As the intracellular free calcium levels are increasing, cAMP levels show an increase of 7% and cGMP levels remain relatively constant. As the calcium levels begin to decrease, cAMP decreases by 21% at the peak of cell division and the cGMP levels show a marked increase of over 100-fold.

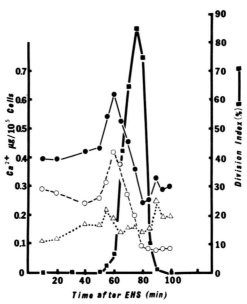

**Fig. 1.**  Intracellular calcium fluctuations in synchronized *Tetrahymena*. Total calcium (solid circles) was measured by atomic absorption spectroscopy. Free calcium (open circles) was determined with the photoprotein aequorin. The bound calcium (triangles) was calculated as total minus free. The division index (squares) was determined microscopically.

Coincidentally, the intracellular pH at the time of cell division in *Tetrahymena* as measured by Gillies and Deamer (1979) also increase during the time of our reported influx in calcium. They also reported a decrease in pH during the time we observed a calcium efflux. Thus, the closely associated pH shifts and calcium fluxes occurring during the division cycle may be indicative of a coupled calcium–hydrogen ion exchange mechanism.

Woods and Whitson (1980) have shown that in synchronized *Tetrahymena* cultures treated with the calcium transporting ionophore A23187 added at the end of the synchronizing protocol (EHS), there was a decrease in the percentage of cells dividing at the peak of division (75 minutes after EHS) in a dose-dependent manner. Furthermore, there appears to be a transition point located between EHS + 50 and EHS + 60 minutes after which the ionophore has little effect on the division process. Also, if EGTA is added to cultures at EHS at a concentration of 20 $\mu M$ in a calcium-free buffer, cell division proceeds at a reduced rate, whereas if higher concentrations are used, the percentage of dividing cells approaches control levels (P. A. Charp, unpublished observations). The hypothesis we put forth to explain these findings involves the possible activation and inactivation of AC and GC by changing concentrations of calcium (see Section V,A). In this model, AC and GC are stimulated by low calcium levels. As calcium is effluxing, the localized calcium concentration at the inner surface of the membrane changes. The calcium concentration would thus be high at this site, inhibiting AC, whereas in the cytoplasm, the concentration of calcium is low resulting in the autostimulation of GC. This hypothesis is based on the evidence presented in other experimental systems as well. In isolated renal cortical tubules treated with either parathyroid hormone or prostaglandin $E_1$, high cAMP levels were associated with an efflux of calcium (Wrenn and Biddulph, 1979). High levels of calcium have been shown to inhibit membrane-bound AC (Schultz *et al.*, 1973) and interfere with PDE and CDR activity (Jones *et al.*, 1979; Dedman *et al.*, 1977).

In *Tetrahymena,* there is evidence for the presence and synthesis of norepinephrine and serotonin (Janakidevi *et al.*, 1966a,b; Brizzi and Blum, 1970), and it has been suggested that there is release of catecholamine following calcium influx which could activate AC (Nandini-Kishore and Thompson, 1979). Still there is no evidence *in vivo* for the activation of AC by β-adrenergic stimulators in *Tetrahymena.* In detergent-dispersed AC of *Tetrahymena,* Kassis and Kindler (1975), however, showed that this preparation was not responsive to epineph-

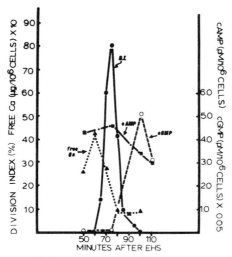

**Fig. 2.** Cyclic nucleotide fluctuations and calcium flux in synchronized *Tetrahymena.* Calcium-45 was used to measure the flux (triangles). Cyclic nucleotides were measured as described by Charp and Whitson (1978). The open circles represent cGMP while the squares display cAMP. The division index (solid circles) was determined microscopically.

rine but was active in the presence of fluoride ions and serotonin. Thus they state *Tetrahymena* AC "plays a role only in internal regulation and need not be accessible to external hormones." In support of these findings, Brizzi and Blum (1970) did indeed find high intracellular levels of serotonin in stationary phase *Tetrahymena* cultures which are also shown to have associated high levels of cAMP.

Thus, it appears from these data that calcium as well as the cyclic nucleotides play a very important role in the regulation of cell division; however, at the present time, the precise role is unknown.

## IV.  CONCEPTS OF CALCIUM MODULATION

### A.  General Considerations

Calcium, like cyclic nucleotides, has earned for itself a role as second messenger in mediating diverse cellular activities. The role of calcium in the regulation of cell division remains obscure, but the intracellular changes in calcium that occur during cell division are probably neither trivial nor coincidental. There is good evidence that calcium modulates various cellular activities by binding to cytosolic proteins (Kretsinger, 1976, 1979). In some instances, the binding of calcium to proteins, particularly enzymes, results in inhibition of function, while in others, binding of calcium results in activation. As will be pointed out in Sections IV and V, the fluctuations in intracellular levels of cyclic nucleotides are due in large part to both inhibition and activation of enzymes by calcium.

### B.  Effects of Free Calcium

Free calcium is believed to play a major role in the physiology of all living organisms from the most primitive prokaryote to the most highly evolved eukaryote (Kretsinger, 1979; Wolff and Brostrom, 1979). Both increases and decreases in free calcium are responsible for changes associated with cell motility functions (Dedman *et al.*, 1979a). Movement has been said, repeatedly by biologists in general, to be the "hallmark of living things" and it appears that where movement involves both contractile proteins and energy that the availability of free calcium as well as its spatial and temporal distribution in cells plays a major role. Cell division is the result of interactions of contractile elements and is a form of cell movement. As we already mentioned, marked changes in free intracellular calcium are associated with this event.

The determination of free intracellular calcium in cells is neither easy nor without inherent problems because they usually involve indirect means (Caswell, 1979). According to Kretsinger (1979) who has summarized the data obtained from a variety of cell types in several different laboratories, "unstimulated cells" contain free calcium in the range of $10^{-8}$ to $10^{-7}$ $M$. Total calcium in cells varies considerably and depends largely on the number of reservoirs and internal sequesters for calcium. Kretsinger (1979) has recently proposed an interesting hypothesis for why it is necessary for cells to maintain low levels of free calcium and why there is often a calcium efflux from stimulated cells. Bound calcium is usually in the form of $Ca_3(PO_4)_2$ which is insoluble and "cells initially extruded calcium to utilize phosphate as their basic energy currency." Kretsinger (1979), however, admits that this hypothesis may be oversimplified and that "evolutionary arguments are particularly irritating because they are difficult to disprove." The efflux of calcium associated with dividing cells most probably involves more than just the release of bound phosphate and may involve several interrelated biochemical changes altering enzyme functions involved in cyclic nucleotide metabolism, including changes in CDR protein (calmodulin).

## C. Calcium-Dependent Regulator Protein—Calmodulin

The properties and functions of CDR (calmodulin), both in relation to cyclic nucleotide metabolism (Wolff and Brostrom, 1979) and in the regulation of the assembly and disassembly of microfilaments and microtubules (Dedman et al., 1979a), is beginning to shed new light on the possible role of calcium in the control of cell division. There are a large variety of calcium-binding proteins, including troponin C which has been purified and characterized in cells (Kretsinger, 1976). The CDR (calmodulin) is one that is stirring up a great deal of current interest because it is a cytosolic protein with 17,000 MW first isolated from brain (Vanaman et al., 1977) and has many structural and physiological similarities to troponin C found in muscle.

Cheung (1971) was the first to observe that PDE, which attacks cAMP, was stimulated by a CDR activator. Later Cheung and co-workers (1978) suggested that this protein be called "calmodulin." It now appears that calmodulin in the presence of calcium is an activator of several cell activities and has widespread distribution (Wolff and Brostrom, 1979). Some recent ascribed functions for calmodulin are the stimulation of AC (Brostrom et al., 1975) and GC (Nagao et al., 1979), activation of myosin light chain kinase (Dabrowska et al., 1978), and the calcium-dependent ATPase of erythrocyte plasma membrane (Jarrett and Penniston, 1977;

Gopinath and Vincenzi, 1977; Larsen and Vincenzi, 1979). Activation of ATPase by calmodulin is probably responsible for the efflux of calcium during cell division, but this has not yet been proven.

## V. ROLE OF FREE CALCIUM AND CALMODULIN IN ALTERING CYCLIC NUCLEOTIDE LEVELS

### A. Effects on Cyclases

One way that cyclic nucleotide levels are altered in cells is through inhibition of AC activity by free calcium (Birnbaumer, 1973). There is some evidence, however, that low concentrations of calcium activate AC, whereas high concentrations inhibit the activity (Wolff and Brostrom, 1979). It has been estimated by Bär and Hechter (1969) that $10^{-9}$ to $10^{-7}$ $M$ free calcium stimulates AC in adrenal cells and that inhibition occurs at 1 m$M$ free calcium. Studies by Steer and Levitzki (1975) and Hanski *et al.* (1977) on the effects of calcium on AC in turkey erythrocytes indicate that when calcium is inhibitory, it functions as a negative allosteric effector without altering the affinity of the hormone receptor component of the enzyme or the catalytic site for ATP.

Brostrom *et al.* (1975) were the first to observe that AC from brain was stimulated by a protein activator (calmodulin) in the presence of low calcium. The recent separate findings by Valverde *et al.* (1979) and Sugden *et al.* (1979) that pancreatic islet cell AC is also stimulated by calmodulin may be indicative of a more universal feature of calmodulin. It has been found, however, that a calcium-binding protein in *Tetrahymena* with similar characteristics as calmodulin activate particulate GC in the presence of calcium but not AC (Suzuki *et al.*, 1979; Nagao *et al.*, 1979).

Information concerning the role of calcium in the modulation of GC is often paradoxical. Proponents for a positive role of calcium as the "second messenger" invoking cell regulatory processes (e.g., cell division) indicate that calcium is essential for activation of GC in intact cells (Schultz *et al.*, 1973; Ferrendelli *et al.*, 1973; Goldberg *et al.*, 1973; Hardman *et al.*, 1974). It is now generally accepted that GC is not a plasmalemmal enzyme, but it can be obtained in both soluble and particulate form from cytoplasmic fractions (White, 1975; Durham, 1976; Mittal and Murad, 1977; Takenawa and Sacktor, 1979). Divalent cations such as manganese, magnesium, and calcium are necessary at low concentration for the activation of GC. However, various combinations of these metal cofactors have yielded inconsistent results, sometimes indicating a stimulatory response or an inhibitory response when added to GC (Böhm, 1970; Marks, 1973; Nakazawa and Sano, 1974; Kimura and Murad, 1975). Apparently the concentrations of calcium varies for stimulation of detergent-soluble forms and particulate forms of this enzyme as well as for different tissue types (Takenawa and Sacktor, 1979). These investigators found that free calcium "(3 $\mu M$) significantly enhances the $Mg^{2+}$-dependent guanyl cyclase, but higher concentration of $Ca^{2+}$ (30 $\mu M$) were inhibitory." All of the above information indicates a need for further characterization of the role of calcium involved in cyclase function.

## B. Effects on Phosphodiesterase

A calcium-dependent PDE was first described by Kakiuchi and Yamazaki (1970) and then found also to be stimulated by a heat-stable protein activator (Cheung, 1970) now recognized as calmodulin. Activation of PDE by calmodulin results in the breakdown of cyclic nucleotides. The mechanism by which calmodulin in the presence of physiological levels of calcium results in the stimulation of hydrolysis of cyclic nucleotides by PDE is not completely understood (Wolff and Brostrom, 1979). The interaction of PDE with calmodulin is reversible by chelation of calcium with EGTA (Wolff and Brostrom, 1974). The most widely used assay for the identification of calmodulin is based on the ability of this protein activator in the presence of calcium to stimulate cyclic nucleotide PDE to hydrolyze cAMP to 5'-AMP (Dedman et al., 1979b). However, since calmodulin is now known to have other enzyme-stimulatory properties this may not be the best method for identification of this protein.

## C. Effects on Kinases

It is now generally accepted that the major role of cyclic nucleotides as second messengers is to activate cyclic nucleotide-dependent protein kinases, which in turn phosphorylate proteins. Greengard (1978) suggests that it is the phosphorylated proteins which are the ultimate effectors of changes in various biological responses. Although it is not known whether cyclic nucleotides or calcium involve the phosphorylation of specific proteins responsible for cell division, there is good evidence that calmodulin as well as cyclic nucleotides activate certain PK (Wolff and Brostrom, 1979). There is also evidence that free calcium alone inhibits some PK (Kretsinger, 1976). Contractility in both muscle (Ebashi and Endo, 1968) and nonmuscle cells (Pollard and Weihing,

1974) results from actin–myosin interactions and depends on the concentration of free calcium. Pires and Perry (1977) found a calcium-dependent myosin light chain kinase in striated muscle and Yagi *et al.* (1978) found that calmodulin was the binding regulatory subunit of this protein. The recent finding that platelet light chain kinase is calmodulin regulated (Hathaway *et al.,* 1978) may indicate that calmodulin is important wherever contractile proteins are found, perhaps even in microfilament assemblies involving cytokinesis. Studies by Marcum *et al.* (1978) and Dedman *et al.* (1979a) indicate that microfilament and microtubule assemblies are regulated by cAMP and calcium. Disassembly of microtubules is promoted both by high calcium and calmodulin, whereas assembly is promoted by low calcium and cAMP.

### D. Alteration in Membrane Transport

Calcium and calmodulin are also important regulators of membrane transport. The discovery that erythrocyte membrane ATPase was calcium dependent (Dunham and Glynn, 1961) has led to the discovery that calmodulin activates the calcium transport in a variety of cells. Where peptide hormones are involved, influxes of calcium and elevations in cAMP often follow. As proposed by Rasmussen *et al.* (1972), "it could be argued that this uptake of calcium is a secondary consequence of a rise in intracellular cAMP concentration." It is not known by what mechanism the efflux of calcium is triggered by "signal calcium," but there is little doubt that a membrane calcium-dependent ATPase must be active or stimulated when this occurs. It will be interesting to determine whether calmodulin is universally involved in the membrane transport of calcium.

### VI. CONCLUSIONS AND PERSPECTIVES

There are marked changes in intracellular levels of cyclic nucleotides associated with progress through the cell cycle in a variety of eukaryotic cells. As we have shown, these are well documented, but there are a few exceptions for there are some cells deficient in one or more of the cyclic nucleotides. The general trend indicates that, where cAMP is involved, high levels of this cyclic nucleotide are associated with cells stalled in $G_1$ or are in $G_0$. Yet it appears that the maintenance of moderate levels of cAMP is a necessary prerequisite for cells to enter DNA synthesis. It is known that cAMP activates kinases which phosphorylate histones, but cAMP-dependent kinases also phosphorylate many other different proteins so it can be questioned as to how specific such an event is in governing DNA synthesis.

Another general trend is that cAMP levels are lowest in dividing cells. There is no proof that a drop in cAMP is a necessary prerequisite for cell division, however, or whether lowered levels in dividing cells are due to: (1) inhibition of AC; (2) activation of PDE; (3) lack of available substrate for synthesis, or any specific combination of the above. In hormonal-stimulated cells and other cells stalled in $G_0$, such as nondividing lymphocytes treated with plant lectins, there seems to be a transient elevation in cAMP associated with reentry into the cell cycle, but as these cells approach the division phase, there often is observed a large increase in cGMP.

The authors and others have suggested that elevation of cGMP is the signal for cell division, but this may not be a specific trigger. There are innumerable substances that can catalyze the synthesis of cGMP by activation of GC (for review, see Murad *et al.*, 1979); some may act as natural endogenous stimulators and some obviously do not. Some cGMP-dependent PK's phosphorylate the same proteins as cAMP-dependent PK, but often not to the same extent. If there are any specific phosphorylated proteins directed by cGMP-dependent PK's which regulate cell division, they have not yet been identified. However, as Greengard (1978) states, phosphorylated proteins are the physiological effectors probably as final products formed by regulatory agents in cyclic nucleotide pathways.

Virtually nothing is known as to how phosphorylated substrates alter cell functions or whether there are specific sequential regulatory steps for their activation by specific kinases or, in the case of phosphorylated proteins, their deactivation by phosphoprotein phosphatases, which we have not discussed.

It is known, however, that certain external stimuli, which result in internal cyclic nucleotide fluctuations, also promote the entry of calcium into cells and may also stimulate the release of stored calcium from internal reservoirs. There is also good evidence that there is a signal calcium influx or elevation of free intracellular calcium in dividing cells (see also Sisken, Chapter 10, this volume). We have attempted to show in this chapter how calcium, both free $Ca^{2+}$ and calcium-dependent regulatory proteins, could account for the cyclic nucleotide fluctuations associated with cell division. Much concerning the role of calcium and calcium-dependent regulatory proteins (calmodulin) remains to be elucidated, particularly in relation to activities such as the regulation of reentry and progress through the cell cycle and the cytoskeletal changes associated with cell division.

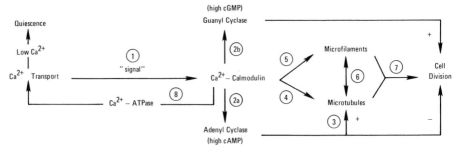

**Fig. 3.** (1) A calcium "signal," influx from outside or release from an internal reservoir, activates calmodulin. (2) Calmodulin in presence of low calcium stimulates cyclases [when present near the plasmalemma, activates AC (2a), raises cAMP; when present in the cytosol, activates GC (2b), raises cGMP]. (3) High cAMP promotes assembly of microtubules. (4) $Ca^{2+}$-calmodulin promotes disassembly of microtubules in the region of cell constriction. (5) $Ca^{2+}$-calmodulin at the same time promotes assembly of microfilaments (or change in structural alignment). (6) Microfilaments in the cytokinetic region interact with (perhaps cross-link) remaining microtubules at boundary. (7) All of the above interactions lead to cell division. (8) Efflux of high free $Ca^{2+}$ by $Ca^{2+}$-calmodulin stimulation of $Ca^{2+}-ATPase$ leads to quiesence and restoration of cells to interphase state and lowers activity of calmodulin.

New studies on the role of calcium and cyclic nucleotides in the control of nonmuscle cell motility are gaining considerable attention because of the possible cortical changes involving microtubule–microfilament interactions during cell division (for more detail, see Harris, 1978; Dedman *et al.*, 1979a). In this regard, certain evidence has led to the following general assumptions. Low levels of cAMP promote or stimulate microtubule assembly, at least *in vitro*. High levels of calcium promote the disassembly of microtubules, or low levels of calcium in the presence of calmodulin. Alterations between states of assembly and disassembly of microtubules are involved in many nonmuscle cell motility functions. Furthermore, interactions of microtubules with actin-like filaments provide the motive force required for mitosis. Ringlike arrays of microfilaments have been observed in large numbers in close association with the cell membrane in the region of constriction of many eukaryotic cells during cytokinesis. Although direct evidence for interactions between microtubule–microfilament complexes with membrane components is still lacking, there is some belief that such complexes alter cell surfaces and could play an important role in cell division.

The following question remains: Are the calcium–cyclic nucleotide fluctuations that occur at the time of cell division somehow responsible for activation of protein complexes involved in cell division? If so, how can one go about resolving this difficult problem? Where does one begin? We propose, because of paradoxical and often controversial results obtained concerning the cyclic nucleotide fluctuations associated with the effects of calcium on cyclase activity that this is a good place to begin. As we pointed out in Section IV,C, calcium and calmodulin play a major role in the regulation of cyclic nucleotide levels both by the activation of cyclases and PDE's. Means and co-workers (Dedman *et al.*, 1979b) have provided us with a working model where calmodulin and calcium are central to this problem. Our modification of this model, to suit the purpose of our perspectives is shown in Fig. 3.

This model is more than likely an oversimplification of the precise biochemical and physiological chain of events which occur involving calcium–cyclic nucleotide interactions during cell division, but it must suffice for now or until further insights are gained or new results prove it wrong.

## ACKNOWLEDGMENT

Paul A. Charp is a predoctoral fellow of the National Institutes of Health Cell and Molecular Biology Training Grant GM 7431-03 to the University of Tennessee School of Biomedical Sciences, Biology Division, Oak Ridge National Laboratory, Oak Ridge, Tennessee.

## REFERENCES

Abell, C. W., and Monahan, T. M. (1973). *J. Cell Biol.* **59,** 549.
Anderson, W. B., Johnson, G. S., and Pastan, I. (1973). *Proc. Natl. Acad. Sci. U.S.A.* **70,** 1055.
Bär, H. P., and Hechter, O. (1969). *Proc. Natl. Acad. Sci. U.S.A.* **63,** 350.
Beale, E. G., Dedman, J. R., and Means, A. R. (1977). *Endocrinology* **101,** 1621.
Berridge, M. J. (1975). *Adv. Cyclic Nucleotide Res.* **6,** 1–98.
Birnbaumer, L. (1973). *Biochim. Biophys. Acta* **300,** 129.
Böhm, E. (1970). *Eur. J. Biochem.* **14,** 422.
Bowen-Pope, D. F., Vadair, C., Sancii, H., and Rubin, A. H. (1979). *Proc. Natl. Acad. Sci. U.S.A.* **76,** 1308.
Brizzi, G., and Blum, J. J. (1970). *J. Protozool.* **17,** 553.
Brønstad, G. O., Elgjo, K., and Øye, I. (1971). *Nature (London) New Biol.* **233,** 78.
Brooker, G., Greengard, P., and Robison, G. A., eds. (1979). *Adv. Cyclic Nucleotide Res.* **10.**
Brostrom, C. O., Huang, Y. C., Brenkenridge, B. M., and Wolf, D. J. (1975). *Proc. Natl. Acad. Sci. U.S.A.* **72,** 64.
Burger, M. M., Bombik, B. M., Breckenridge, B. M., and Sheppard, J. R. (1972). *Nature (London) New Biol.* **239,** 161.
Caswell, A. H. (1979). *Int. Rev. Cytol.* **56,** 145.
Charp, P. A., and Whitson, G. L. (1978). *Radiat. Res.* **74,** 323.

Cheung, W. Y. (1970). *Biochem. Biophys. Res. Commun.* **38,** 533.

Cheung, W. Y. (1971). *J. Biol. Chem.* **246,** 2859.

Cheung, W. Y., Lynch, T. J., and Wallace, R. W. (1978). *Adv. Cyclic Nucleotide Res.* **9,** 233.

Clothier, G., and Timourian, H. (1972). *Exp. Cell Res.* **75,** 105.

Costa, M. (1977). *Biochem. Biophys. Res. Commun.* **78,** 1311.

Costa, M., Gerner, E. W., and Russell, D. H. (1976). *J. Biol. Chem.* **251,** 3313.

Dabrowska, R., Sherry, J. M., Amatorio, D. K., and Hartshorne, D. (1978). *Biochemistry* **17,** 253.

Dedman, J. R., Potter, J. D., Jackson, R. L., Johnson, J. D., and Means, A. R. (1977). *J. Biol. Chem.* **252,** 8415.

Dedman, J. R., Brinkley, B. R., and Means, A. R. (1979a). *Adv. Cyclic Nucleotide Res.* **11,** 131.

Dedman, J. R., Tash, J. S., and Means, A. R. (1979b). *In* "Laboratory Methods Manual for Hormonal Action and Molecular Endocrinology" (W. T. Schrader and B. O. O'Malley, eds.), 3rd ed., p. 1044. Endocrine Soc., Bethesda, Maryland.

Dickinson, J. R., Graves, M. G., and Swoboda, B. E. P. (1977). *Eur. J. Biochem.* **78,** 83.

DiPasquale, A., and McGuire, J. (1977). *J. Cell. Physiol.* **93,** 395.

Drummond, G. I., Greengard, P., and Butcher, G. A., eds. (1975). *Adv. Cyclic Nucleotide Res.* **5.**

Duncan, C. J., ed. (1976). "Calcium in Biological Systems" Symp. Soc. Exp. Biol., Cambridge Univ. Press, London and New York.

Dunham, E.T., and Glynn, I. M. (1961). *J. Physiol.* **156,** 274.

Durhan, J. P. (1976). *Eur. J. Biochem.* **61,** 535.

Ebashi, S., and Endo, M. (1968). *Prog. Biophys. Mol. Biol.* **18,** 123.

Eker, P. (1974). *J. Cell Sci.* **16,** 301.

Ferrendelli, J. A., Kinscherf, D. A., and Chang, M. M. (1973). *Mol. Pharmacol.* **9,** 445.

Froehlich, J. E., and Rachmeler, M. (1972). *J. Cell Biol.* **55,** 19.

Froehlich, J. E., and Rachmeler, M. (1974). *J. Cell Biol.* **60,** 249.

Gillies, R. J., and Deamer, D. W. (1979). *J. Cell. Physiol.* **100,** 23.

Goldberg, N. D., Haddox, M. K., Estensen, R., White, J. G., Lopez, C., and Hadden, J. W. (1973). *In* "Cyclic AMP in Immune Response and Tumor Growth" (L. Lichtenstein and C. Parker, eds.). pp. 247–262. Springer-Verlag, Berlin and New York.

Gopinath, R. M., and Vincenzi, F. F. (1977). *Biochem. Biophys. Res. Commun.* **77,** 1203.

Gray, N.C. C., Dickinson, J. R., and Swoboda, B. E. P. (1977). *FEBS Lett.* **81,** 311.

Greengard, P. (1978). *Science* **199,** 146.

Grimes, W. J., and Schroeder, J. L. (1973). *J. Cell Biol.* **56,** 487.

Hadden, J. W., Hadden, E. M., Haddox, M. K., and Goldberg, N. D. (1972). *Proc. Natl. Acad. Sci. U.S.A.* **69,** 3024.

Hanski, E., Sevilla, N., and Levitzki, A. (1977). *Eur. J. Biochem.* **76,** 513.

Hardman, J. G., Schultz, G., and Sutherlan, E. W. (1974). *In* "Cyclic AMP, Cell Growth and the Immune Response" (W. Braun, L. M. Lichenstein and C. W. Parker, eds.), pp. 223–226. Springer-Verlag, Berlin and New York.

Harris, P. (1978). *In* "Cell Cycle Regulation" (J. R. Jeter, Jr., I. L. Cameron, G. M. Padilla, and A. M. Zimmerman, eds.), pp. 75–104. Academic Press, New York.

Hathaway, D. R., Sobieszed, A., Eaton, C. E., and Adelstein, R. S. (1978). *Fed. Proc. Fed. Am. Soc. Exp. Biol.* **37,** 1328.

Heidrick, M. L., and Ryan, W. L. (1971). *Biochim. Biophys. Acta* **237,** 301.

Heilbrunn, L. V. (1956). "The Dynamics of Living Protoplasm." Academic Press, New York.

Hsie, A. W., O'Neill, J. P., Schoder, C. H., Kohtar, K., Borman, L. S., and Li, A. P. (1976). *In* "Control Mechanisms In Cancer" (W. E. Criss, T. Ono, and J. R. Sabine, eds.), pp. 183–203. Raven, New York.

Janakidevi, K., Dewey, V. C., and Kidder, G. W. (1966a). *J. Biol. Chem.* **241**, 2576.

Janakidevi, K., Dewey, V. C., and Kidder, G. W. (1966b). *Arch. Biochem. Biophys.* **113**, 758.

Jarrett, H. W., and Penniston, J. T. (1977). *Biochem. Biophys. Res. Commun.* **77**, 1210.

Jones, H. P., Matthews, J. C., and Cormier, M. J. (1979). *Biochem.* **18**, 55.

Kakiuchi, S., and Yamazaki, R. (1970). *Biochem. Biophys. Res. Commun.* **41**, 1104.

Kassis, S., and Kindler, S. H. (1975). *Biochim. Biophys. Acta* **391**, 513.

Kimura, H., and Murad, F. (1975). *J. Biol. Chem.* **250**, 4810.

Kram, R., Mamont, P., and Tomkins, G. M. (1973). *Proc. Natl. Acad. Sci. U.S.A.* **70**, 1432.

Kretsinger, R. H. (1976). *Int. Rev. Cytol.* **46**, 323.

Kretsinger, R. H. (1977). *In* "Calcium Binding Proteins and Calcium Function" (R. H. Wasserman, R. A. Corrinadino, E. Carafoli, R. H. Kretsinger, D. H. MacLennan, and F. L. Siegel, eds.), pp. 63–72. Elsevier, New York.

Kretsinger, R. H. (1979). *Adv. Cyclic Nucleotide Res.* **11**, 2.

Kuehn, G. D. (1971). *J. Biol. Chem.* **246**, 6366.

Kuehn, G. D. (1972). *Biochem. Biophys. Res. Commun.* **49**, 414.

Larsen, F. L., and Vincenzi, F. F. (1979). *Science* **204**, 306.

London, J. F., and Whitson, G. L. (1980). *Eur. J. Cell Biol.* (accepted).

London, J. F., Charp, P. A., and Whitson, G. L. (1979). *J. Cell Biol.* **83**, 9a.

Marcum, J. M., Dedman, J. R., Brinkley, B. R., and Means, A. R. (1978). *Proc. Natl. Acad. Sci. U.S.A.* **75**, 3771.

Marks, F. (1973). *Biochim. Biophys. Acta* **309**, 349.

Millis, A. J. T., Forrest, G., and Pious, D. A. (1972). *Biochem. Biophys. Res. Commun.* **49**, 1645.

Mittal, C. K., and Murad, F. (1977). *Proc. Natl. Acad. Sci. U.S.A.* **74**, 4360.

Moens, W., Vokaer, A., and Kram, R. (1975). *Proc. Natl. Acad. Sci. U.S.A.* **72**, 1063.

Murad, F., Arnold, W. P., Mittal, C. K., and Braughler, J. M. (1979). *Adv. Cyclic Nucleotide Res.* **11**, 175.

Murofushi, H. (1974). *Biochim. Biophys. Acta* **370**, 130.

Nagao, S., Suzuki, Y., Watanabe, Y., and Nozawa, Y. (1979). *Biochem. Biophys. Res. Commun.* **90**, 261.

Nakamura, M., and Yasumasu, I. (1974). *J. Gen. Physiol.* **63**, 374.

Nakazawa, K., and Sano, M. (1974). *J. Biol. Chem.* **249**, 4207.

Nandini-Kishore, S. G., and Thompson, G. A. (1979). *Proc. Natl. Acad. Sci. U.S.A.* **76**, 2708.

Nose, K., and Katsuta, H. (1975). *Biochem. Biophys. Res. Commun.* **64**, 983.

Oler, A., Iannaccone, P. M., and Gordon, G. B. (1973). *In Vitro* **9**, 35.

O'Neill, C. W., Schroder, C. H., Riddle, J. C., and Hsie, A. W. (1976). *Exp. Cell Res.* **97**, 213.

Otten, J., Boder, J., Johnson, G. S., and Pastan, I. (1972). *J. Biol. Chem.* **247**, 1632.

Paul, D., and Ristow, H. J. (1979). *J. Cell. Physiol.* **98**, 31.

Pires, E. M. V., and Perry, S. V. (1977). *Biochem. J.* **167**, 137.

Pollard, T. D., and Weihing, R. R. (1974). *Crit. Rev. Biochem.* **2**, 1.

Rassmussen, H. (1970). *Science* **170**, 404.

Rassmussen, H., Goodman, D. B. P., and Tenenhouse, A. (1972). *Crit. Rev. Biochem.* **1**, 95.

Rebhun, L. I. (1977). *Int. Rev. Cytol.* **49**, ].

Robinson, J. H., Smith, J. A., and King, R. J. B. (1974). *Cell* **3**, 361.

Rosenberg, H. (1966). *Exp. Cell Res.* **41**, 397.

Rubin, A. H., and Bowen-Pope, D. F. (1979). *J. Cell. Physiol.* **98**, 81.

Russell, D. H., and Stambrook, P. J. (1975). *Proc. Natl. Acad. Sci. U.S.A.* **72**, 1482.

Scherbaum, O., and Zeuthen, E. (1954). *Exp. Cell Res.* **6**, 221.

Schultz, G., Hardman, J. G., Schultz, K., Baird, C. E., and Sutherland, E. W. (1973). *Proc. Natl. Acad. Sci. U.S.A.* **70**, 3889.

Seifert, W. E., and Rudland, P. S. (1974a). *Proc. Natl. Acad. Sci. U.S.A.* **71**, 4920.

Seifert, W. E., and Rudland, P. S. (1974b). *Nature (London)* **248**, 138.

Sheppard, J. R., and Prescott, D. M. (1972). *Exp. Cell Res.* **75**, 293.

Steer, M. L., and Levitzki, A. (1975). *J. Biol. Chem.* **250**, 2080.

Steinhardt, R. A., and Epel, D. (1974). *Proc. Natl. Acad. Sci. U.S.A.* **71**, 1915.

Sugden, M. C., Christie, M. R., and Achcroft, S. J. H. (1979). *FEBS Lett.* **105**, 95.

Sutherland, E. W., and Rall, T. W. (1958). *J. Biol. Chem.* **232**, 1077.

Suzuki, Y., Hirabayashi, T., and Watanabe, Y. (1979). *Biochem. Biophys. Res. Commun.* **90**, 253.

Takenawa, T., and Sacktor, B. (1979). *Biochim. Biophys. Acta* **566**, 371.

Taylor-Papadimitriou, J. (1974). *Int. J. Cancer* **13**, 404.

Valverde, I., Vandemeers, A., Amjaneyulu, R., and Malaisse, W. J. (1979). *Science* **206**, 225.

Vanaman, T. C., Sharief, F., and Watterson, D. M. (1977). *In* "Calcium Binding Proteins and Calcium Functions" (R. H. Wasserman, R. A., Corrinadin, E. Carafoli, R. H. Kretsinger, D. H. MacLennan, and F. L. Siegel, eds.), pp. 107–117. Elsevier, New York.

Van Wijk, R., Wicks, W. D., Bevers, M. M., and Van Rijn, J. (1973). *Cancer Res.* **33**, 1331.

Wang, T., Sheppard, J. R., and Foker, J. E. (1978). *Science* **201**, 155.

White, A. A. (1975). *Adv. Cyclic Nucleotide Res.* **5**, 353.

Willingham, M. C., Johnson, G. S., and Pasten, I. (1972). *Biochem. Biophys. Res. Commun.* **48**, 743.

Willingham, M. C. et al. (1974).

Wolfe, J. (1973). *J. Cell. Physiol.* **82**, 39.

Wolff, D. J., and Brostrom, C. O. (1974). *Arch. Biochem. Biophys.* **163**, 349.

Wolff, D. J., and Brostrom, C. O. (1979). *Adv. Cyclic Nucleotide Res.* **11**, 27.

Woods, M. A., and Whitson, G. L. (1980). In preparation.

Wrenn, R. W., and Biddulph, D. M. (1979). *J. Cyclic Nucleotide Res.* **5**, 239.

Yagi, K., Yazawa, M., Kakiuchi, S., Ohshima, M., and Uenishi, K. (1978). *J. Biol. Chem.* **253**, 1338.

Yasuda, H., Hanai, N., Kurata, M., and Yamada, M. (1978). *Exp. Cell Res.* **114**, 111.

Yasumasu, I., Fujiwara, A., and Ishida, K. (1973). *Biochem. Biophys. Res. Commun.* **54**, 628.

Zeilig, C. E., Johnson, R. A., Sutherland, E. W., and Friedman, D. L. (1974). *Fed. Proc., Fed. Am. Soc. Exp. Biol.* **33**, 1391.

*Note added in proof:*

Since the original writing of this manuscipt, we have been made aware of several possible pitfalls regarding internal measurements of free calcium by the aequorin technique dealing with cell homogenates. First of all, disruption of cells and the length of time required for homogenization could release calcium from internal stores such as mitochondria and other storage organelles, namely volutin granules in *Tetrahymena* (Rosenberg, 1966), which could mask any rapid changes in flux of free calcium as well as the true levels. Secondly, volume measurements were not performed in our initial experiments indicating that what we considered to be free calcium was of the magnitude of 100 times that found in any other eukaryotic cells.

We have corrected our original findings and observe by the aequorin assay that calcium in the cell homogenates is about $10^{-4}$ $M$. This level is still considerably higher than that reported for free calcium by Kretsinger (1979) as reviewed in this paper in Section IV,B. Even though our measurements for calcium by the aequorin method are high, we have observed by flux studies with calcium-45 that the same trend of an influx in calcium occurs prior to cell division. In preliminary observations we also find that verapamil, a known inhibitor of calcium transport by acting as a blocker of a slow calcium channel, delays synchronized cell division in *Tetrahymena* ($200\mu M$) and inhibits it altogether at $500\mu M$ concentration. We are planning to investigate whether this drug action inhibits the transport, both influx and/or efflux, of calcium by using calcium-45 during the synchronized division cycle.

We have also found that crude extracts of adenyl cyclase from *Tetrahymena* are affected by free calcium *in vitro* and that low levels ($10^{-9}$ $M$) are nonstimulatory, $5 \times 10^{-7}$ $M$ maximally stimulates it and $10^{-5}$ $M$ inhibits its activity. We still conclude therefore, that calcium may play an important role as an internal regulator involving cyclic nucleotide metabolism and that all of these moieties in some way interact in the control of cell division.

# Index

# CELL BIOLOGY: A Series of Monographs

## EDITORS

**D. E. BUETOW**

*Department of Physiology*
*and Biophysics*
*University of Illinois*
*Urbana, Illinois*

**I. L. CAMERON**

*Department of Anatomy*
*University of Texas*
*Health Science Center at San Antonio*
*San Antonio, Texas*

**G. M. PADILLA**

*Department of Physiology*
*Duke University Medical Center*
*Durham, North Carolina*

**A. M. ZIMMERMAN**

*Department of Zoology*
*University of Toronto*
*Toronto, Ontario, Canada*

**G. M. Padilla, G. L. Whitson, and I. L. Cameron** (editors). THE CELL CYCLE: Gene-Enzyme Interactions, 1969

**A. M. Zimmerman** (editor). HIGH PRESSURE EFFECTS ON CELLULAR PROCESSES, 1970

**I. L. Cameron and J. D. Thrasher** (editors). CELLULAR AND MOLECULAR RENEWAL IN THE MAMMALIAN BODY, 1971

**I. L. Cameron, G. M. Padilla, and A. M. Zimmerman** (editors). DEVELOPMENTAL ASPECTS OF THE CELL CYCLE, 1971

**P. F. Smith.** The BIOLOGY OF MYCOPLASMAS, 1971

**Gary L. Whitson** (editor). CONCEPTS IN RADIATION CELL BIOLOGY, 1972

**Donald L. Hill.** THE BIOCHEMISTRY AND PHYSIOLOGY OF *TETRA-HYMENA*, 1972

**Kwang W. Jeon** (editor). THE BIOLOGY OF AMOEBA, 1973

**Dean F. Martin and George M. Padilla** (editors). MARINE PHARMACOGNOSY: Action of Marine Biotoxins at the Cellular Level, 1973

**Joseph A. Erwin** (editor). LIPIDS AND BIOMEMBRANES OF EUKARYOTIC MICROORGANISMS, 1973

**A. M. Zimmerman, G. M. Padilla, and I. L. Cameron** (editors). DRUGS AND THE CELL CYCLE, 1973

Stuart Coward (editor). DEVELOPMENTAL REGULATION: Aspects of Cell Differentiation, 1973

I. L. Cameron and J. R. Jeter, Jr. (editors). ACIDIC PROTEINS OF THE NDCLEUS, 1974

Govindjee (editor). BIOENERGETICS OF PHOTOSYNTHESIS, 1975

James R. Jeter, Jr., Ivan L. Cameron, George M. Padilla, and Arthur M. Zimmerman (editors). CELL CYCLE REGULATION, 1978

Gary L. Whitson (editor). NUCLEAR-CYTOPLASMIC INTERACTIONS IN THE CELL CYCLE, 1980

*In preparation*

Danton H. O'Day and Paul A. Horgen (editors). SEXUAL INTERACTIONS IN EUKARYOTIC MICROBES, 1981

Ivan L. Cameron and Thomas B. Pool (editors). THE TRANSFORMED CELL, 1981